T0292557

Studies in Computational Intelligence

Volume 626

Series editor

Janusz Kacprzyk, Polish Academy of Sciences, Warsaw, Poland
e-mail: kacprzyk@ibspan.waw.pl

About this Series

The series "Studies in Computational Intelligence" (SCI) publishes new developments and advances in the various areas of computational intelligence—quickly and with a high quality. The intent is to cover the theory, applications, and design methods of computational intelligence, as embedded in the fields of engineering, computer science, physics and life sciences, as well as the methodologies behind them. The series contains monographs, lecture notes and edited volumes in computational intelligence spanning the areas of neural networks, connectionist systems, genetic algorithms, evolutionary computation, artificial intelligence, cellular automata, self-organizing systems, soft computing, fuzzy systems, and hybrid intelligent systems. Of particular value to both the contributors and the readership are the short publication timeframe and the worldwide distribution, which enable both wide and rapid dissemination of research output.

More information about this series at http://www.springer.com/series/7092

Donald Davendra · Ivan Zelinka
Editors

Self-Organizing Migrating Algorithm

Methodology and Implementation

Editors
Donald Davendra
Department of Computer Science
Central Washington University
Ellensburg, WA
USA

Ivan Zelinka
Faculty of Electrical Engineering and Computer Science, Department of Computer Science
VŠB—Technical University of Ostrav
Ostrava-Poruba
Czech Republic

ISSN 1860-949X ISSN 1860-9503 (electronic)
Studies in Computational Intelligence
ISBN 978-3-319-28159-9 ISBN 978-3-319-28161-2 (eBook)
DOI 10.1007/978-3-319-28161-2

Library of Congress Control Number: 2015958861

Printed on acid-free paper

This Springer imprint is published by SpringerNature
The registered company is Springer International Publishing AG Switzerland

Donald Davendra would like to dedicate this book to his father Michael Davendra.

Foreword

Since the beginning of our civilization, the human race in its engineering challenges has had to confront numerous technological problems such as finding optimal solutions for various problems in civil engineering, scheduling, control technologies, and in many other fields. These examples encompass both ancient and modern technologies such as automatic theater controlled by special programs in ancient Greece (Heron of Alexandria), the first electrical energy distribution network in the USA, mechanical, electronic as well as computational controllers, and control and scheduling of the space exploration. Technology development of these and related areas has had and continues to have a profound impact on our civilization and our everyday lifestyle.

A special class of algorithms that plays an important role in the solution process of the above-mentioned problems is the so-called nature-inspired algorithms. The oldest in this class are evolutionary algorithms that are based on Darwinian evolution theory and Mendel's theory of propagation of genetic information. These algorithms are simple, flexible, mathematically unrestrictive, and very powerful. This book discusses one of such algorithms that was proposed in 1999 and subsequently further developed and published as conference articles, journal articles, and book chapters. It is SOMA: Self-Organizing Migrating Algorithm that mimics competitive–cooperative behavior of a pack of intelligent agents. SOMA can be regarded as a member of the family of swarm intelligence algorithms and is based on effective combination of exploration and exploitation. The SOMA has been used during its existence by numerous researchers from different countries for solving diverse tasks such as controller design, chaos control, synthesis and identification, electronic circuit synthesis, synthesis of control program for an artificial ant (Santa Fe trail), aircraft wing design, mathematical model synthesis for astrophysical data, artificial neural network synthesis, and learning among many others.

The book you are holding in your hands consists of a detailed description of SOMA principles, its history with all relevant references and selected new as well as summarized application of this algorithm. Authors of the chapters are well-experienced practitioners and researchers in their respective fields.

The topics discussed in this book cover the above-mentioned areas and they are cohesively joined into a comprehensive text, which while discussing the specific selected topics gives a deeper insight into the interdisciplinary fusion of those modern and promising areas of emerging technologies in computer science.

Therefore, this book titled Self-Organizing Migrating Algorithm: Methodology and Implementation, edited by Donald Davendra and Ivan Zelinka, is a timely volume to be welcomed by the community focused on innovative algorithms of optimization, computational intelligence, and beyond. This book is devoted to the studies of common and related subjects in intensive research fields of nature-inspired algorithms. For these reasons, I enthusiastically recommend this book to our students, scientists, and engineers working in the aforementioned fields of research and applications.

Singapore
October 2015

Ponnuthurai Nagaratnam Suganthan

Preface

Swarm-based algorithms have become one of the foremost researched and applied heuristics in the field of evolutionary computation within the past decade. One of the new and novel approaches is that of the self-organizing migrating algorithm (SOMA). Initially developed and published in 2001 by Prof. Ivan Zelinka, SOMA has been actively researched by a select group of researchers over the past decade and a half.

SOMA is conceptualized on a predator/prey relationship, where the sampling of the search space is conducted on a multidimensional facet, with the dimension selection conducted pre-sampling, using a randomly generated PRT vector. Two unique aspects of SOMA, which differentiate it from other swarm-based algorithms, are the creation and application of the PRT vector, and the path length, which specifies the distance and sampling required within a particular dimension.

Over the past few years, SOMA has been modified to solve combinatorial optimization problems. This discrete variant so-called discrete self-organizing migrating algorithm (DSOMA) has been proven to be robust and efficient.

With its ever-expanding applications and utilization, it was thought beneficial and timely to produce a collated work of all the active applications of SOMA, which shows its current state of the art. To this effect, we have reached out and have obtained original research topics in SOMA and its application from a very diverse group of academics and researchers. This provides a rich source of material and ideas for both students and researchers.

Chapter authors' background: Chapter authors are to the best of our knowledge the originators or closely related to the originators of the different variants and applications of SOMA.

Organization of the Chapters

The book is divided into two parts. The first part methodology is divided into two chapters. The first chapter "SOMA—Self-organising Migrating Algorithm" written by the originator of SOMA, Ivan Zelinka, introduces SOMA to the broad audience. The second chapter "DSOMA—Discrete Self-Organising Migrating Algorithm" by Davendra, Zelinka, Pluhacek, and Senkerik describes the discrete variant of SOMA.

The second part of the book describes the different implementations of SOMA. The chapters in this section are given in the following order. Chapter "SOMA and Strange Dynamics" by Zelinka introduces the concepts of chaos and complex networks in SOMA.

Chapter "Multi-objective Self-organizing Migrating Algorithm" by Kadlec and Raida introduces multi-objective SOMA (MOSOMA), whereas chapter "Multi-objective Design of EM Components" describes its application to EM component design.

Chapter by Běhálek, Gajdŏs, and Davendra shows the "Utilization of Parallel Computing for Discrete Self-organizing Migration Algorithm" using OpenMP and CUDA.

Chapter "C-SOMAQI: Self-organizing Migrating Algorithm with Quadratic Interpolation Crossover Operator for Constrained Global Optimization" by Singh, Agarway, and Deep introduces another variant of SOMA, C-SOMAQI, to solve constrained optimization problems. Another hybrid variant C-SOMGA also used to solve constrained optimization problems is given in chapter "Optimization of Directional Overcurrent Relay Times Using C-SOMGA" by Deep and Singh. SOMAGA is further expanded in chapter "SOMGA for Large Scale Function Optimization and its Application" to solve large-scale and real-life problems.

Chapter "Solving the Routing Problems with Time Windows" by Čičková, Brezina, and Pekár describes the application of SOMA to the vehicle routing problem. The same authors apply SOMA to financial modeling in chapter "SOMA in Financial Modeling."

The final two chapters deal with SOMA parameters and influences. Chapter "Setting of Control Parameters of SOMA on the Base of Statistics" by Čičková and Lukáčik looks at different statistical bases for SOMA parameter settings. The final chapter "Inspired in SOMA: Perturbation Vector Embedded into the Chaotic PSO Algorithm Driven by Lozi Chaotic Map" by Pluhacek, Zelinka, Senkerik, and Davendra looks at the influences of the PRT vector in the PSO algorithm.

Audience: The book will be an instructional material for senior undergraduate and entry-point graduate students in computer science, applied mathematics, statistics, management and decision sciences, and engineering, who are working in the area of modern optimization. The book will also serve as a resource handbook and material for practitioners who want to apply SOMA to solve real-life problems and challenging applications.

USA and Czech Republic
October 2015

Donald Davendra
Ivan Zelinka

Acknowledgments

Donald Davendra would like to acknowledge and thank his parents, Michal Davendra and Manjula Devi, his sister Annjelyn Shalvina, and his wife Magdalena Bialic-Davendra. He would also like to thank his friends, especially in the Czech Republic for all their support over the years. He would like to dedicate this work to his father.

Ivan Zelinka would like to acknowledge the following grants for the financial support provided for this research: Grant Agency of the Czech Republic—GACR P103/15/06700S and by the research grant of VŠB—TU Ostrava No. SP2015/142.

Contents

Contributors

Seema Agrawal Department of Mathematics, S.S.V.P.G. College, Hapur, India

Ivan Brezina Department of Operations Research and Econometrics, University of Economics in Bratislava, Bratislava, Slovakia

Marek Běhálek FEECS, Department of Computer Science, VŠB—Technical University of Ostrava, Ostrava, Czech Republic; IT4Innovations National Supercomputing Center, VŠB—Technical University of Ostrava, Ostrava, Czech Republic

Zuzana Čičková Department of Operations Research and Econometrics, University of Economics in Bratislava, Bratislava, Slovakia

Donald Davendra Department of Computer Science, Central Washington University, Ellensburg, WA, USA; Department of Computer Science, Faculty of Electrical Engineering and Computer Science, VŠB—Technical University of Ostrava, Ostrava-Poruba, Czech Republic

Kusum Deep Indian Institute of Technology Roorkee, Roorkee, India

Petr Gajdoš FEECS, Department of Computer Science, VŠB—Technical University of Ostrava, Ostrava, Czech Republic; IT4Innovations National Supercomputing Center, VŠB—Technical University of Ostrava, Ostrava, Czech Republic

Petr Kadlec Brno University of Technology, Brno, Czech Republic

Martin Lukáčik Department of Operations Research and Econometrics, University of Economics in Bratislava, Bratislava, Slovakia

Juraj Pekár Department of Operations Research and Econometrics, University of Economics in Bratislava, Bratislava, Slovakia

Michal Pluhacek Department of Informatics and Artificial Intelligence, Faculty of Applied Informatics, Tomas Bata University in Zlin, Zlin, Czech Republic

Zbyněk Raida Brno University of Technology, Brno, Czech Republic

Roman Senkerik Department of Informatics and Artificial Intelligence, Faculty of Applied Informatics, Tomas Bata University in Zlin, Zlin, Czech Republic

Dipti Singh Department of Applied Sciences, Gautam Buddha University, Greater Noida, India

Ivan Zelinka Department of Computer Science, Faculty of Electrical Engineering and Computer Science, VŠB—Technical University of Ostrava, Ostrava-Poruba, Czech Republic

Part I
Methodology

SOMA—Self-organizing Migrating Algorithm

Ivan Zelinka

Abstract This chapter discuss basic principles of Self-Organizing Migrating Algorithm (SOMA) that has been firstly proposed in 1999 and published consequently in various journals, book chapters and conferences. Algorithm itself is, from today classification point of view, between memetic and swarm algorithms and is based on competetive-cooperative strategies, that generate new solutions. During its existence it has been tested on various problems, including real-time + black box ones, it has been parallelized and used with such algorithms like genetic programming, grammatical evolution or/and analytic programming in order to synthesize complex structures—solutions of different problems. In this chapter are discussed basics of algorithm, its use and selected applications. All mentioned SOMA use is completely referenced for detailed reading and further research.

1 Introduction

In recent years, a broad class of algorithms has been developed for stochastic optimization, i.e. for optimizing systems where the functional relationship between the independent input variables x and output (objective function) y of a system S is not known. Using stochastic optimization algorithms such as Genetic Algorithms (GA), Simulated Annealing (SA) and Differential Evolution (DE), a system is confronted with a random input vector and its response is measured. This response is then used by the algorithm to tune the input vector in such a way that the system produces the desired output or target value in an iterative process. Most engineering problems can be defined as optimization problems, e.g. the finding of an optimal trajectory for a robot arm, the optimal thickness of steel in pressure vessels, the optimal set of parameters for controllers, optimal relations or fuzzy sets in fuzzy

I. Zelinka (✉)
Department of Computer Science, Faculty of Electrical Engineering and Computer Science, VSB-Technical University of Ostrava, 17. Listopadu 15, 708 33 Ostrava-Poruba, Czech Republic
e-mail: ivan.zelinka@vsb.cz

D. Davendra and I. Zelinka (eds.), *Self-Organizing Migrating Algorithm*,
Studies in Computational Intelligence 626, DOI 10.1007/978-3-319-28161-2_1

models, etc. Solutions to such problems are usually difficult to find their parameters usually include variables of different types, such as floating point or integer variables. Evolutionary algorithms (EAs), such as the Genetic Algorithms and Differential Evolutionary Algorithms, have been successfully used in the past for these engineering problems, because they can offer solutions to almost any problem in a simplified manner: they are able to handle optimizing tasks with mixed variables, including the appropriate constraints, and they do not rely on the existence of derivatives or auxiliary information about the system, e.g. its transfer function. Evolutionary algorithms work on populations of candidate solutions that are evolved in generations in which only the best-suited—or fittest—individuals are likely to survive. This article introduces SOMA ('Self-Organizing Migrating Algorithm'), a new class of stochastic optimization algorithms. It explains the principles behind SOMA and demonstrates how this algorithm can assist in solving of various optimization problems. Functions on which SOMA have been tested can be found in this chapter. SOMA, which can also works on a population of individuals, is based on the self-organizing behavior of groups of individuals in a "social environment". It can also be classified as an evolutionary algorithm, despite the fact that no new generations of individuals are created during the search (based on philosophy of this algorithm). Only the positions of the individuals in the search space are changed during a generation, called a 'migration loop'. Individuals are generated by random according to what is called the 'specimen of the individual' principle. The specimen is in a vector, which comprises an exact definition of all those parameters that together lead to the creation of such individuals, including the appropriate constraints of the given parameters. SOMA is not based on the philosophy of evolution (two parents create one new individual—the offspring), but on the behavior of a social group of individuals, e.g. a herd of animals looking for food. One can classify SOMA as an evolutionary algorithm, because the final result, after one migration loop, is equivalent to the result from one generation derived by the classic EA algorithms—individuals hold new positions on the N dimensional hyper-plane. When the group of individuals is created, then the rule mentioned above governs the behavior of all individuals so that they demonstrate 'self-organization' behavior. Because no new individuals are created, and only existing ones are moving over the N dimensional hyper-plane, this algorithm has been termed the Self-Organizing Migrating Algorithm, or SOMA for short. In the following text the principle of the SOMA algorithm including its constraint handling and testing will be explained. The description is divided into short sections to increase the understandability of principles of the SOMA algorithm.

2 Historical Background and Algorithm Classification

Evolutionary algorithms are based on principles of evolution which have been observed in nature long time before they were applied to and transformed into algorithms to be executed on computers. When next reviewing some historical facts that led to evolutionary computation as we know it now, we will mainly focus on the basic ideas, but will also allow to glimpse at the people who did the pioneering work and established the field. Maybe the two most significant persons whose research on evolution and genetics had the biggest impact on modern understanding of evolution and its use for computational purposes are Gregor Johann Mendel and Charles Darwin.

Gregor Johann Mendel, July 20, 1822–January 6, 1884) was an Augustinian priest and scientist, and is often called the father of genetics for his study of the inheritance of certain traits in pea plants. He was born in the family of farmers in Hyncice (Heinzendorf bei Odrau) in Bohemia (that time part of Austrian—Hungary empire). The most significant contribution of Mendel for science was his discovery of genetic laws which showed that the inheritance of these traits follows particular laws (published in [1]), which were later named after him. All his discoveries were done in Abbey of St. Thomas in Brno (Bohemia). Mendel published his research at two meetings of the Natural History Society of Brünn in Moravia (east part of Bohemia) in 1865 [1]. When Mendel's paper was published in 1866 in Proceedings of the Natural History Society of Brünn, it had little impact and was cited only about three times over the next thirty-five years. His paper was criticized at the time, but is now considered a seminal work. The significance of Mendel's work was not recognized until the turn of the 20th century. Its rediscovery (thanks to Hugo de Vries, Carl Correns and Erich von Tschermak) prompted the foundation of the discipline of genetics. Very peculiar historical fact about Mendel's research is also that his letters about his discovery, sent to many of scientific societies, had been found after many years in their libraries unopened. Mendel died on January 6, 1884, at age 61, soon after his death the succeeding abbot burned all papers in Mendel's collection, to mark an end to the disputes over taxation [2].

The other important (and much more well-known and therefore here only briefly introduced) researcher whose discoveries founded the theory of evolution was the British scientists Charles Darwin. Darwin published in his work [3] the main ideas of the evolutionary theory. The full and original title was "*On the Origin of Species by Means of Natural Selection, or the Preservation of Favoured Races in the Struggle for Life*". Word "*races*" refers here to biological varieties. The title has been changed to [3] for the 6th edition of 1872. In Darwin's book On the Origin of Species (1859) established evolutionary descent with modification as the dominant scientific explanation of diversification in the nature.

The above mentioned ideas of genetics and evolution have been formulated long before the first computer experiments with evolutionary principles had been done. The beginning of the ECT is officially dated to the 70s of the 20th century, when famous genetic algorithms were introduced by Holland [4, 5] or to the late 60s with evolutionary strategies, introduced by Schwefel [6] and Rechenberg [7] and evolutionary programming by Fogel [8]. However, when certain historical facts are taken into consideration, then one can see that the main principles and ideas of ECT as well as its computer simulations had been done earlier than mentioned above. Conceptionally, ECT can be traced back to the famous Turing [9], first numerical experiments to the (far less famous) N.A. Barricelli and others. For more see [10].

At the present time, there is a broad spectrum of publications dealing with optimization algorithms, for example, [11]. The purpose of this chapter is to outline only the principles of some selected algorithms for better information for the reader. The discussed algorithms are in the next section.

2.1 SOMA in the Context of Selected Evolutionary Algorithms

Since the first introduction of evolutionary algorithms in 50s by Barricelli [12–14] and in 70s genetic algorithms by Holland [4, 5] has been developed a rich class of so called genetic algorithms, later on memetic algorithms and swarm algorithms that copied less more bio/natural processes on different scales, i.e. micro/mezo/macro level (genetic operations, interactions amongst intelligent agents,…). In order to better understand SOMA position amongst today existing algorithms and its classes, it is better to briefly mention a few main representative of them.

Genetic algorithm (GA) This algorithm is one of the first successful applied ECT methods [4, 15]. In GAs the main principles of ECT are applied in their purest form. The individuals are encoded as binary strings (mostly over the alphabet [0, 1]), which can be understood as a model of the biological counterpart, the genome,[1] and represent possible solutions to the optimization problem under study. After initially a population of binary strings is created randomly, the circle as given in Fig. 1 is carried out with the steps fitness evaluation, selection, offspring generation (crossover) and mutation until the algorithm terminates. The application area of these algorithms are wide and it seem particularly sensible to use them if the problem description allows a straightforward coding of the objects to optimize as binary string over a finite alphabet, for instance in combinatorial optimization problem timetabling and scheduling.

Evolutionary strategy (ES) This algorithm also belongs to the first successful stochastic algorithms in history. It was proposed at the beginning of the sixties by Rechenberg [7] and Schwefel [6]. It is based on the principles of natural selection similarly as the genetic algorithms. Contrary to genetic algorithms, the evolutionary strategy works directly with individuals described by vectors of real values. Its core is to use candidate solutions in the form of vectors of real numbers, which are recombined and then mutated with the help of a vector of random numbers. The problem of accepting a new solution is strictly deterministic. Another distinctive feature is that ES use self-adaptation, that is the mutation strength for each individual is variable over the generational run and subject to an own evolutionary adaption and optimization process.

[1]The genome is coded over the alphabet $[A, C, G, T]$, which stand for the amino acids adenine A, cytosine C, guanine G, thymine T.

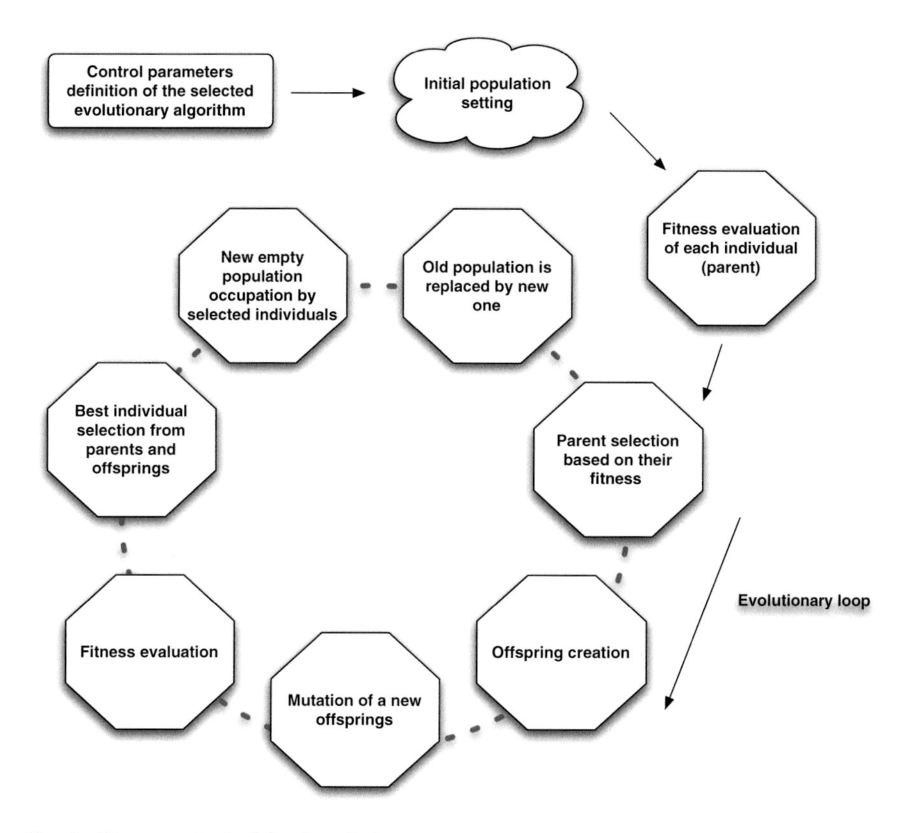

Fig. 1 The general principle of evolution

Evolutionary programming (EP) EP algorithms [8] have much similarity to ES in using vectors of real numbers as representation. The main operator in the generational circle is mutation, in most (particularly early) implementations no recombination is carried out. In recent years, by adopting elements of their algorithmic structure EP more and more tends to become similar to ES.

Learning classifier systems (LCS) LCS [16] are machine learning algorithms which are based on GAs and reinforcement learning techniques. Interestingly, LCS were introduced by Holland[2] [4] and for a certain time regarded as a generalization of GAs. LCS optimize over a set of rules that are intended to best-fit inputs to outputs. The rules are coded binary and undergo an adaption using GA-like optimization that modifies and selects the best rules. The fitting of the rules is determined by reinforcement learning methods.

Population-based incremental learning (PBIL) PBIL was proposed by Baluja [17] and combines ideas from evolutionary computation with methods from

[2]Holland is also known as the father of GAs.

statistical learning [18]. It uses a real valued representation that is usually restricted to the interval [0, 1] and can be interpreted as the probability to have a "1"-bit at a certain place in a binary string. From these probabilities, a collection of binary strings is created. These strings are subjected to a standard evolutionary circle with fitness evaluation, selection and discarding of inferior samples. In addition, based on the evaluation of the fitness, a deterministic statistical-learning-like updating of the probability vector takes place, which afterwards is also altered by random mutation.

Ant Colony Optimization (ACO), [19] This is an algorithm whose action simulates the behaviour of ants in a colony. It is based on the following principle. Let there be a source of ants (colony) and the goal of their activity (food). When they are released, all the ants move after some time along the shorter (optimum) route between the source and goal. The effect of finding the optimum route is given by the fact that the ants mark the route with pheromones. If an ant arrives to the crossroads of two routes that lead to the same goal, his decision along which route to go is random. Those ants that found food start marking the route and when returning, their decision is influenced thanks to these marks in favor of this route. When returning, they mark it for the second time, which increases the probability of the decision of further ants in its favor. These principles are used in the ACO algorithm. Pheromone is here represented by the weight that is assigned to a given route leading to the goal. This weight is additive, which makes it possible to add further "pheromones" from other ants. The evaporation of pheromones is also taken into account in the ACO algorithm in such a way that the weights fade away with time at individual joints. This increases the robust character of the algorithm from the point of view of finding the global extreme. ACO was successfully used to solve optimization problems such as the traveling salesman problem or the design of telecommunication networks, see [20].

Immunology System Method (ISM) This algorithm is unusual by its algorithm based on the principles of functioning of the immunology system in living organisms. As indicated in [20], there are several principles based on this model. In this work, the immunology system is considered as a multivalent system, where individual agents have their specific tasks. These agents have various competencies and ability to communicate with other agents. On the basis of this communication and a certain "freedom" in making decisions of individual agents, a hierarchic structure is formed able to solve complicated problems. As an example of using this method, antivirus protection can be mentioned in large and extensive computer systems [21, 22].

Memetic Algorithms (MA) This term represents a broad class of metaheuristic algorithms [20, 23–25]. The key characteristics of these algorithms are the use of various approximation algorithms, local search techniques, special recombination operators, etc. These metaheuristic algorithms can be basically characterized as competitive-cooperative strategies featuring attributes of synergy. As an example of memetic algorithms, hybrid combinations of genetic algorithms and simulated annealing or a parallel local search can be indicated.

Memetic algorithms were successfully used for solving such problems as the traveling salesman problem, learning of a neural multilayer network, maintenance planning, nonlinear integer number programming and others (see [20]).

Scatter Search (SS) This optimization algorithm differs by its nature from the standard evolutionary diagrams. It is a vector oriented algorithm that generates new vectors (solutions) on the basis of auxiliary heuristic techniques. It starts from the solutions obtained by means of a suitable heuristic technique. New solutions are then generated on the basis of a subset of the best solutions obtained from the start. A set of the best solutions is then selected from these newly found solutions and the entire process is repeated. This algorithm was used for the solution of traffic problems, such as traffic control, learning neural network, optimization without limits and many other problems [20, 26].

Particle Swarm (PSO) The "particle swarm" algorithm is based on work with the population of individuals, whose position in the space of possible solutions is changed by means of the so-called velocity vector. According to the description in [20, 27, 28, 29], there is no mutual interaction between individuals in the basic version. This is removed in the version with the so-called neighborhood. In the framework of this neighborhood, mutual interaction occurs in such a manner that individuals belonging to one neighborhood migrate to the deepest extreme that was found in this neighborhood.

Differential Evolution (DE) Differential Evolution [30, 31] is a population-based optimization method that works on real-number coded individuals. For each individual $\vec{x}_{i,G}$ in the current generation G, DE generates a new trial individual $\vec{x'}_{i,G}$ by adding the weighted difference between two randomly selected individuals $\vec{x}_{r1,G}$ and $\vec{x}_{r2,G}$ to a third randomly selected individual $\vec{x}_{r3,G}$. The resulting individual $\vec{x'}_{i,G}$ is crossed-over with the original individual $\vec{x}_{i,G}$. The fitness of the resulting individual, referred to as perturbated vector $\vec{u}_{i,G+1}$, is then compared with the fitness of $\vec{x}_{i,G}$. If the fitness of $\vec{u}_{i,G+1}$ is greater than the fitness of $\vec{x}_{i,G}$, $\vec{x}_{i,G}$ is replaced with $\vec{u}_{i,G+1}$, otherwise $\vec{x}_{i,G}$ remains in the population as $\vec{x}_{i,G+1}$. Differential Evolution is robust, fast, and effective with global optimization ability. It does not require that the objective function is differentiable, and it works with noisy, epistatic and time-dependent objective functions.

Fire Fly algorithm (FF) is a metaheuristic algorithm, inspired by the flashing behaviour of fireflies. The primary purpose for a firefly's flash is to act as a signal system to attract other fireflies, as defined by Yang in [32].

Bat algorithm (BA) Bat-inspired algorithm is a metaheuristic optimization algorithm developed by Yang in [31]. This bat algorithm is based on the echolocation behaviour of microbats with varying pulse rates of emission and loudness.

Cuckoo search (CS) is an optimization algorithm developed by Yang and Deb in [30]. As inventors reported, it was inspired by the obligate brood parasitism of some cuckoo species by laying their eggs in the nests of other host birds. Some host birds can engage direct conflict with the intruding cuckoos.

Chaos based and hybridized algorithms (CHA) are already mentioned (and another) algorithms that are hybridized with deterministic chaos used like generators of randomness and are used instead of pseudorandom number generators, see for example [34–39].

In this context it can be stated that SOMA is a stochastic optimization algorithm that is modeled on the social behavior of cooperating individuals [33], as swarm algorithms are, e.g. PSO or later on BA, FF amongst the others. SOMA works on a population of candidate solutions in loops called *migration loops*. The population is initialized randomly distributed over the search space at the beginning of the search. In each loop, the population is evaluated and the solution with the highest fitness becomes the *Leader L*. Apart from the leader, in one migration loop, all individuals will traverse the input space in the direction of the leader. Mutation, the random perturbation of individuals, is an important operation for evolutionary algorithms. It ensures the diversity amongst the individuals and it also provides the means to restore lost information in a population. Mutation is different in SOMA compared with other EAs. SOMA uses a parameter called *PRT* to achieve perturbation. This parameter has the same effect for SOMA as mutation has for GA. The novelty of this approach is that the *PRT* Vector is in canonical version created before an individual starts its journey over the search space. The *PRT* Vector defines the final movement of an active individual in search space. The randomly generated binary perturbation vector controls the allowed dimensions for an individual. If an element of the perturbation vector is set to zero, then the individual is not allowed to change its position in the corresponding dimension. An individual will travel a certain distance (called the *path length*) towards the *Leader* in n steps of defined length. If the path length is chosen to be greater than one, then the individual will overshoot the leader. This path is perturbed randomly.

The evolutionary algorithms can be essentially used for the solution of very heterogeneous problems. Of course, for the solution of the optimization problems, there are many more algorithms than were indicated here. Because their description would exceed the framework of this text, we can only refer to the corresponding literature, where the algorithms indicated above are described in more details.

3 SOMA Applicability

The set of functions on which a given algorithm shows good performance (i.e. its algorithm domain) should be clearly defined. This definition is, however, very general and hence not satisfactory for these purposes here. Based on experiences from artificial and real world test functions can be classified like for example:

1. None-fractal type: theoretically, the geometrical complexity of the cost function shall be finite, i.e. under repeated zoom no more complex structures should be discovered. If zoom is going to infinity and complex structures are still visible, then we can say that the function is a fractal, which are complex structures

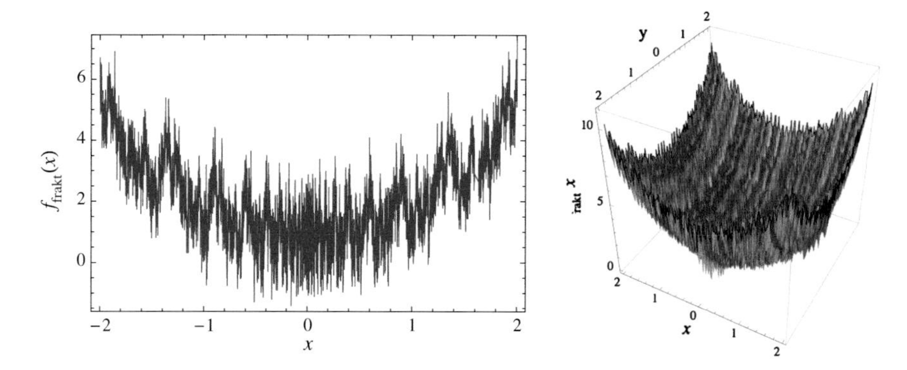

Fig. 2 Pseudo-fractal function: fractal "noise" added to the 1st DeJong function, see Eq. (1) and Fig. 4

repeated in itself, see for example [40]. Due to the physical reality, where infinity does not exists (everything has its own limits) we can model fractal functions as defined on Fig. 2.

2. Defined at real, integer or discrete argument spaces: function can be based on the mixture of the various arguments of different nature (i.e. real numbers, integer numbers etc.).
3. Constrained, multiobjective, nonlinear: functions can be constrained by different constrains, can represent multiobjective problem where solutions lie on the so called Pereto front. Also nonlinearity in combinations with previous attributes is allowed.
4. Needle-in-haystack problems: this problem usually can be understood like search for a global extreme which is represented like Dirac (or better pseudo-Dirac) peak in the wide and flat area, see Fig. 3. Finding of such a extreme is matter of randomness rather than of sophisticated search procedure.
5. NP problems: i.e. problems where exists a huge number of possible solutions and only one is the right one. Typical example is the traveling salesman.
6. Permutative problems: problem of the permutative nature like flow shop scheduling, no wait problems, etc.

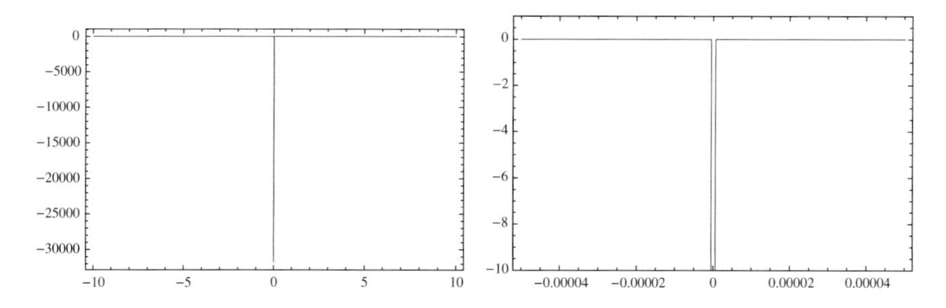

Fig. 3 Possible vizualization of the needle in haystack problem: graphical approximation. On the *right side* is a zoom. Only one solution is the right one

SOMA has been successfully tested on functions of all types reported here. Generally, the SOMA algorithm should be able to work on any system which provides an objective function, i.e. one that returns cost value. No auxiliary information, such as gradients etc. are needed.

Beside real-world problems and artificial permutative test problems, SOMA was also tested on classical test functions that were selected from the test bed of 17 test functions. In total 16 test function were selected as a representative subset of functions which shows geometrical simplicity and low complexity as well as functions from the "opposite side of spectra". Selected functions (see Figs. 4, 5, 6 and 7) were: 1st DeJong (1), Schwefel function (6), Rastrigin function (5), Ackley function (10) amongst the others [see (1)–(16)]. Dimension is in the formulas (1)–(16) represented by variable D, so as one can see, it is easy to calculate selected functions for an arbitrary dimension. Functions (1)–(16) has been selected due to their various complexity and mainly for the fact that this functions are widely used by researchers working with evolutionary algorithms. Another reason was, that speed of convergence and thus evolutionary dynamics itself, is different for simple functions like (1) or more complex example (13).

$$f(\mathbf{x}) = \sum_{i=1}^{D} x_i^2 \tag{1}$$

$$f(\mathbf{x}) = \sum_{i=1}^{D-1} 100\left(x_i^2 - x_{i+1}\right)^2 + (1 - x_i)^2 \tag{2}$$

$$f(\mathbf{x}) = \sum_{i=1}^{D} |x_i| \tag{3}$$

$$f(\mathbf{x}) = \sum_{i=1}^{D} i x_i^4 \tag{4}$$

$$f(\mathbf{x}) = 2D\sum_{i=1}^{D} x_i^2 - 10\cos(2\pi x_i) \tag{5}$$

$$f(\mathbf{x}) = \sum_{i=1}^{D} -x_i \sin(\sqrt{|x_i|}) \tag{6}$$

$$f(\mathbf{x}) = 1 + \sum_{i=1}^{D} \frac{x_i^2}{4000} - \prod_{i=1}^{D} \cos\left(\frac{x_i}{\sqrt{i}}\right) \tag{7}$$

$$f(\mathbf{x}) = -\sum_{i=1}^{D-1} \left(0.5 + \frac{\sin(x_i^2 + x_{i+1}^2 - 0.5)^2}{(1 + 0.001(x_i^2 + x_{i+1}^2))^2}\right) \tag{8}$$

$$f(\mathbf{x}) = \sum_{i=1}^{D-1} \left(\sqrt{4}(x_i^2 + x_{i+1}^2) \sin(50 \sqrt[10]{(x_i^2 + x_{i+1}^2)})^2 + 1 \right) \tag{9}$$

$$f(\mathbf{x}) = \sum_{i=1}^{D-1} \left(\frac{1}{e^5} \sqrt{(x_i^2 + x_{i+1}^2)} + 3(\cos(2x_i) + \sin(2x_{i+1})) \right) \tag{10}$$

$$f(\mathbf{x}) = \sum_{i=1}^{D-1} \left(20 + e - \frac{20}{e^{0.2\sqrt{\frac{(x_i^2 + x_{i+1}^2)}{2}}}} - e^{0.5(\cos(2\pi x_i) + \cos(2\pi x_{i+1}))} \right) \tag{11}$$

$$f(\mathbf{x}) = \sum_{i=1}^{D-1} \left(\begin{array}{l} -x_i \sin(\sqrt{|x_i - x_{i+1} - 47|}) \\ -(x_{i+1} + 47) \sin(\sqrt{\left|x_{i+1} + 47 + \frac{x_i}{2}\right|}) \end{array} \right) \tag{12}$$

$$f(\mathbf{x}) = \sum_{i=1}^{D-1} \left(\begin{array}{l} x_i \sin(\sqrt{|x_{i+1} + 1 - x_i|}) \cos(\sqrt{|x_{i+1} + 1 + x_i|}) \\ + (x_{i+1} + 1) \cos(\sqrt{|x_{i+1} + 1 - x_i|}) \sin(\sqrt{|x_{i+1} + 1 + x_i|}) \end{array} \right) \tag{13}$$

$$f(\mathbf{x}) = \sum_{i=1}^{D-1} \left(0.5 + \frac{\sin(\sqrt{100x_i^2 - x_{i+1}^2})^2 - 0.5}{(1 + 0.001(x_i^2 - 2x_i x_{i+1} + x_{i+1}^2)^2)} \right) \tag{14}$$

$$f(\mathbf{x}) = \sum_{i=1}^{D-1} \left(-1 \left(\sin(x_i) \sin(\frac{x_i^2}{\pi})^{20} + \sin(x_{i+1}) \sin(\frac{2x_i^2}{\pi})^{20} \right) \right) \tag{15}$$

$$f(\mathbf{x}) = \sum_{i=1}^{D-1} \left(e^{\frac{-(x_i^2 + x_{i+1}^2 + 0.5x_i x_{i+1})}{8}} \cos(4\sqrt{x_i^2 + x_{i+1}^2 + 0.5x_i x_{i+1}}) \right) \tag{16}$$

Fig. 4 Selected test functions: 1st DeJong, (1)

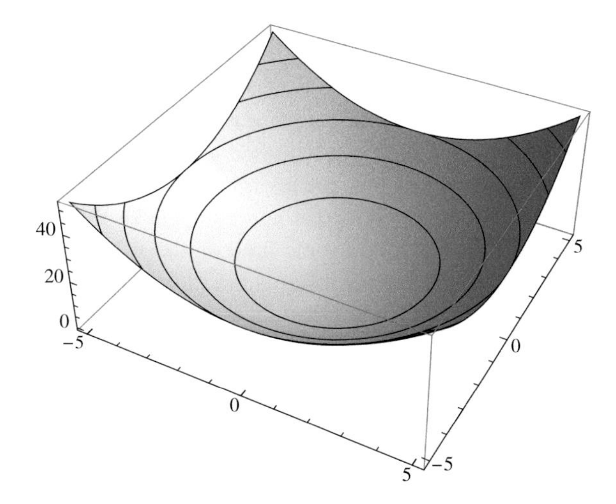

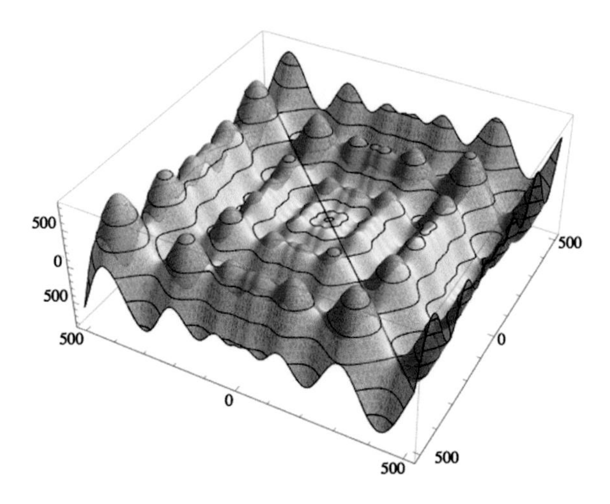

Fig. 5 Schwefel function (6)

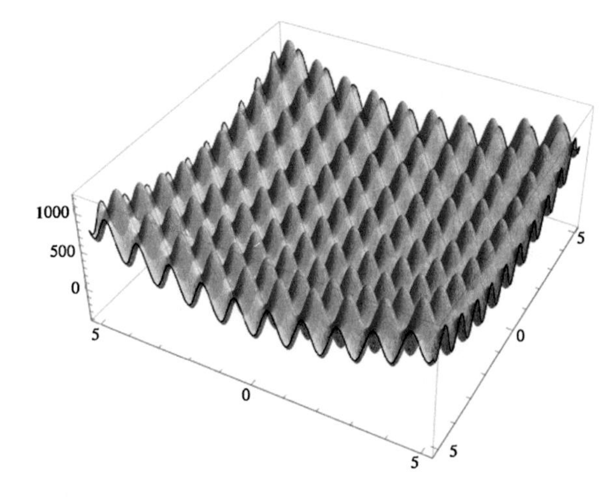

Fig. 6 Selected test functions: Rastrigin function, (5)

4 SOMA Principles and Control Parameters

In the previous sections it was mentioned that SOMA was inspired by the competitive-cooperative behavior of intelligent creatures solving a common problem. Such a behavior can be observed anywhere in the world. A group of animals such as wolves or other predators may be a good example. If they are looking for food, they usually cooperate and compete so that if one member of the group is successful (it has found some food or shelter) then the other animals of the group change their trajectories towards the most successful member. If a member of this group is more successful than the previous best one (is has found more food, etc.)

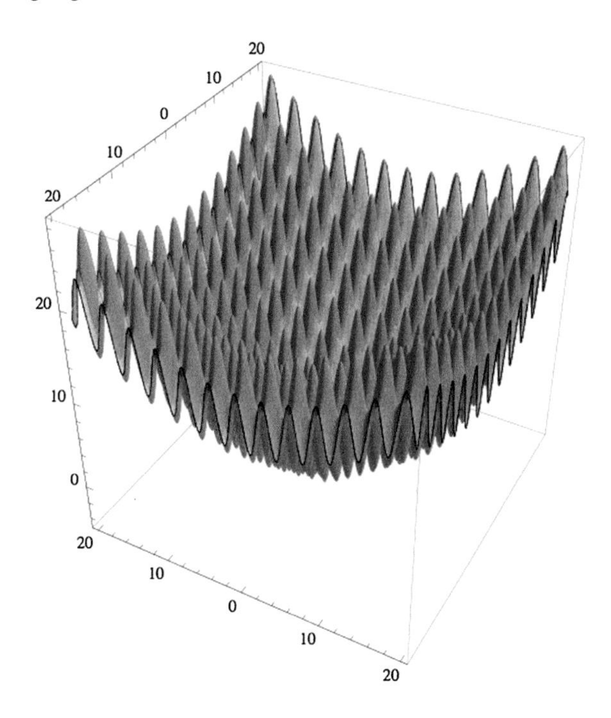

Fig. 7 Ackley function (11)

then again all members change their trajectories towards the new successful member. It is repeated until all members meet around one food source. This principle from the real world is of course strongly simplified. Yet even so, it can be said it is that competitive-cooperative behavior of intelligent agents that allows SOMA to carry out very successful searches. For the implementation of this approach, the following analogies are used:

1. Members of herd/pack $\Leftrightarrow$ individuals of population, *PopSize* parameter of SOMA.
2. Member with the best source of food $\Leftrightarrow$ Leader, the best individual in population for actual migration loop.
3. Food $\Leftrightarrow$ fitness, local or global extreme on N dimensional hyper-plane.
4. Landscape where pack is living $\Leftrightarrow$ N dimensional hyper-plane given by cost function.
5. Migrating of pack members over the landscape $\Leftrightarrow$ migrations in SOMA.

The following section explains in a series of detailed steps how SOMA actually works. SOMA works in loops—so called *Migration* loops. These play the same role as *Generations* in classic EAs. The difference between SOMA's *Migrationloops* and EA's *Generations* come from the fact that during a *Generations* in classic EA's offspring is created by means of at least two or more parents (two in GA, four in DE for example). In the case of SOMA, there is no newly created offspring based on parents crossing. Instead, new positions are calculated for the individuals traveling towards the current Leader. The term *Migrations* refers to their movement over the landscape-hyper-plane. It can be demonstrated that SOMA can be viewed as an algorithm based on offspring creation. The *Leader* plays the role of roe-buck (male), while other individuals play the role of roe (female); note that this has the characteristics of pack reproduction with one dominant male. Hence, GA, DE, etc. may be seen as a special case of SOMA and vice versa (see later SOMA strategy AllToAll). Because the original idea of SOMA is derived from competitive-cooperative behavior of intelligent beings, we suppose that this background is the most suitable one for its explanation. The basic version of SOMA consists of the following steps:

1. Parameter definition. Before starting the algorithm, SOMA's parameters, e.g. *Specimen*, *Step*, *PathLength*, *PopSize*, *PRT*, *MinDiv*, *Migrations* and the cost function needs to be defined. Cost function is simply the function which returns a scalar that can directly serve as a measure of fitness. The cost function is then defined as a model of real world problems, (e.g. behavior of controller, quality of pressure vessel, behavior of reactor, etc.).
2. Creation of Population. A population of individuals is randomly generated. Each parameter for each individual has to be chosen randomly from the given range [Lo, Hi] by using Eq. (17). The population (Fig. 8) then consists of columns—individuals which conform with the specimen.
3. Migrating loop. Each individual is evaluated by cost function and the *Leader* (individual with the highest fitness) is chosen for the current migration loop. Then all other individuals begin to jump, (according to the *Step* definition) towards the *Leader*. Each individual is evaluated after each jump using the cost function. The jumping (Eq. 18) continues, until a new position defined by the *PathLength* has been reached. The new position after each jump is calculated by Eq. (18). This is shown graphically in Fig. 13. The individual returns then to that position where it found the best fitness on its trajectory. Before an individual begins jumping towards the *Leader*, a random number is generated (for each individual's component), and then compared with *PRT*. If the generated random number is larger than *PRT*, then the associated component of the individual is set to 0 by means of the *PRTVector* (see Eq. (19) otherwise set to 1. Hence, the individual moves in the *N*-k dimensional subspace, which is perpendicular to the original space. This fact establishes a higher robustness of the algorithm. Earlier experiments have demonstrated that, without the use of *PRT*, SOMA tends to determine a local optimum rather than the global one. *Migrations* can be also viewed as a competitive-cooperative phase. During the competitive phase each individual tries to find the best position on its way and also the best from all individuals. Thus during migration, all individuals compete among themselves. When all individuals are in new positions, they release information as to their cost value. This can be regarded as a cooperative phase. All individuals cooperate

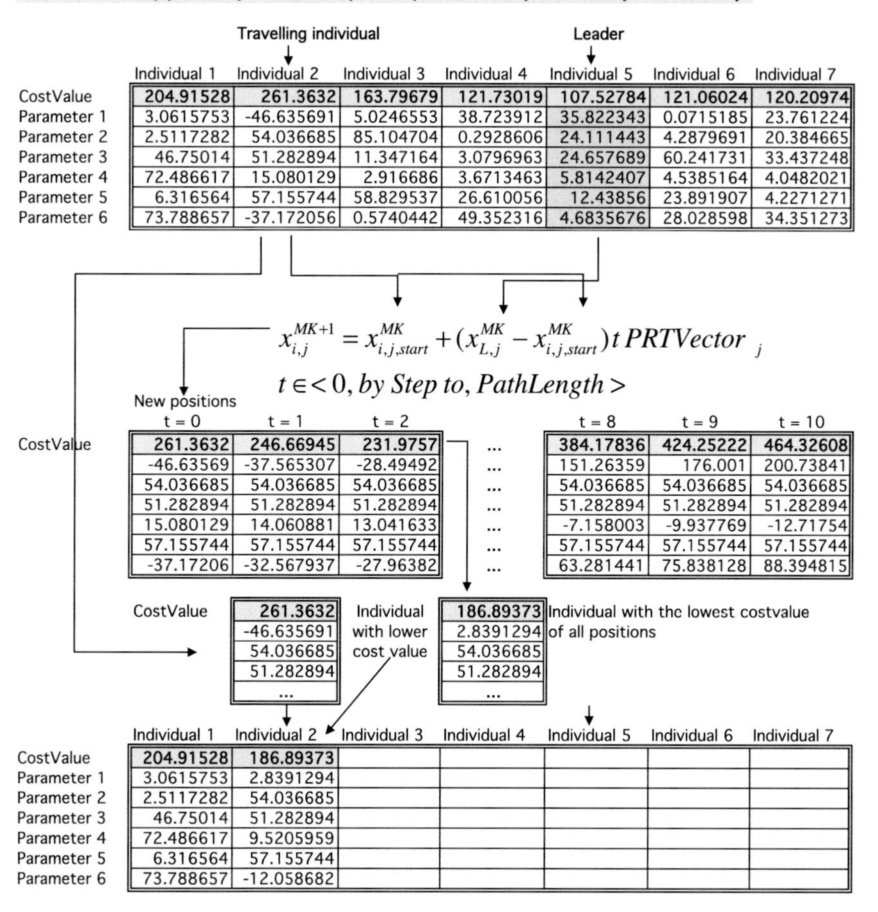

Fig. 8 SOMA Principle

so that the best individual (*Leader*) is chosen. Competitive-cooperative behavior is one of the other important attributes typical for memetic algorithms.

4. Test for stopping condition. If the difference between *Leader* and the worst individual is not lower than the *MinDiv* and the maximum number of *Migrations* has not been reached, return to step 3 otherwise go to step 5.
5. Stop. Recall the best solution(s) found during the search.

$$InitilPopulation = x_j^{(lo)} + rand_j[0, 1] \times (x_j^{(hi)} - x_j^{(lo)}) \tag{17}$$

$$x_{i,j}^{ML+1} = x_{i,j,start}^{ML} + (x_{L,j}^{ML} - x_{i,j,start}^{ML})\ t\ PRTVector_j \tag{18}$$

$$if \quad rnd_j < PRT \quad then \quad PRTVector_j = 1\ else\ 0, \quad j = 1, \ldots, N \tag{19}$$

Steps 1–5 are graphically depicted in Fig. 8 or in pseudocode in Eq. (20). The pseudocode of SOMA can be written like this:

SOMA AllToOne input parameters :
x : *the initial randomly generated population*
Controlling and stopping parameters – see Tab. 1.1
$\mathbf{f_{cost}}$: *cost function (fitness function)*
Specimen : *an individual structure (parameters range, its "nature" i.e. real, integer, discrete, ...)*

```
for i ≤ Migration do
    begin
    Selection of the best individual – Leader
      for j ≤ PopSize do
        selection of j_th individual
        calculate f_cost of the new positions see Eq.1.18
        save the best solution of the j_th individual on its trajectory in a new population
      end
    if MinDiv < |best_individual − worst_individual|
      then begin
        Stop SOMA and return the best solution (or last calculated population)
      end
    end
```

or more technically see Eq. (20). SOMA principle can be graphically visualized as it is done at the Fig. 8.

Input: N, $Migrations(ML)$, $PopSize \geq 2$, $PRT \in [0, 1]$, $Step \in (0, 1]$, MinDiv $\in (-\infty, \infty)$, PathLength $\in (1, 5]$, $Specimen$ with upper and lower bound $x_j^{(hi)}, x_j^{(lo)}$

$$\text{Inicialization:} \begin{cases} \forall i \leq PopSize \wedge \forall j \leq N : x_{i,j}^{ML_0} = x_j^{(lo)} + rand_j[0, 1]\left(x_j^{(hi)} - x_j^{(lo)}\right) \\ i = \{1, 2, \ldots, Migrations\}, \quad j = \{1, 2, \ldots, N\} \end{cases}$$

$$\begin{cases} \text{While} \quad i < Migrations \\ \forall i \leq PopSize \begin{cases} \text{While} \quad t \leq PathLength \\ if \quad rnd_j < PRT\ then\ PRTVector_j = 1 \quad else\ 0, \ j = 1, \ldots, N \\ x_{i,j}^{ML+1} = x_{i,j,start}^{ML} + (x_{L,j}^{ML} - x_{i,j,start}^{ML})t\ PRTVector_j \\ f\left(x_{i,j}^{ML+1}\right) = \text{if} \quad f\left(x_{i,j}^{ML}\right) \leq f\left(x_{i,j,start}^{ML}\right) \quad \text{else}\ f\left(x_{i,j,start}^{ML}\right) \\ t = t + Step \end{cases} \\ i = i + 1 \end{cases} \tag{20}$$

Based on above described principles, SOMA can be also regarded as a member of swarm intelligence class algorithms. In the same class is the algorithm particle swarm, which is also based on population of particles, which are mutually influenced amongst themselves. Some similarities as well as differences are between SOMA and particle swarm, for details see [33, 41].

5 SOMA Strategies

Currently, a few variations—strategies of the SOMA algorithm exist. All versions are almost fully comparable with each other in the sense of finding of global optimum. These versions are:

1. 'AllToOne': This is the basic strategy, that was previously described. Strategy AllToOne means that all individuals move towards the Leader, except the Leader. The *Leader* remains at its position during a *Migration* loop. The principle of this strategy is shown in Fig. 9.
2. 'AllToAll': In this strategy, there is no Leader. All individuals move towards the other individuals. This strategy is computationally more demanding. Interestingly, this strategy often needs less cost function evaluations to reach the global optimum than the AllToOne strategy. This is caused by the fact that each individual visits a larger number of parts on the N dimensional hyper-plane during one *Migration* loop than the AllToOne strategy does. Figure 10 shows the AllToAll strategy with $PRT = 1$.
3. 'AllToAll Adaptive': The difference between this and the previous version is, that individuals do not begin a new migration from the same old position (as in AllToAll), but from the last best position found during the last traveling to the previous individual.

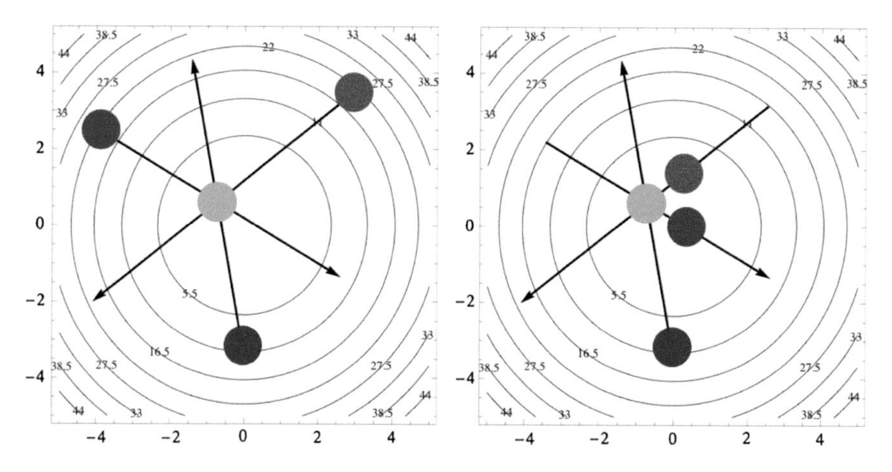

Fig. 9 SOMA AllToOne, the principle of migrating (*left*) and new individual position (*right*) after one migration

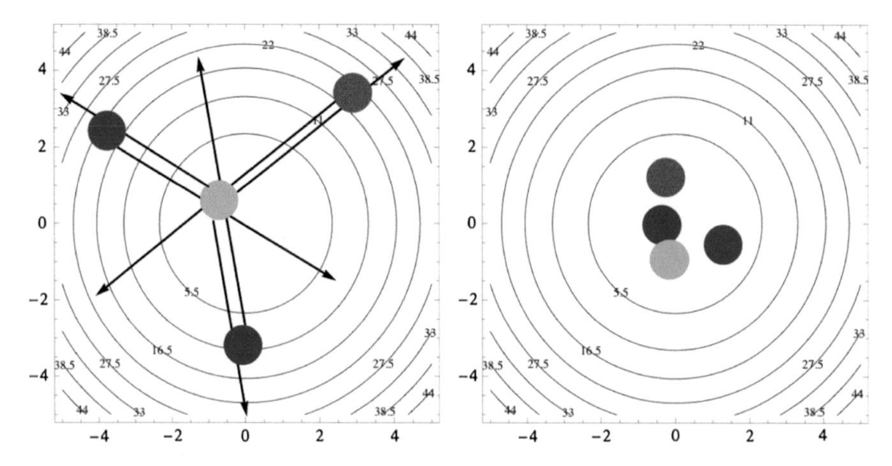

Fig. 10 SOMA AllToAll, the principle of migrating (*left*) and new individual position (*right*) after one migration

4. 'AllToRand': This is a strategy, where all individuals move towards a randomly selected individual during the migration loop, no matter what cost value this individual has. It is up to the user to decide how many randomly selected individuals there should be. Here are two sub-strategies:

   - The number of randomly selected individuals is constant during the whole SOMA process.
   - For each migration loop, (in intervals of [1, *PopSize*]) the actual number of individuals is determined randomly. Thus, the number of randomly chosen individuals in the second sub-strategy is different in each migration loop.

5. Clusters: This version of SOMA with Clusters can be used in any of the above strategies. The word 'Cluster' refers to calculated clusters. Each individual from the population is tested for the cluster to which it belongs, according to Eq. (21) expressed below, where IND_i is the ith parameter of the individual; CC_i is the ith parameter of the *Leader* (Cluster Center); HB_i and LB_i are the allowed bounds for the given parameter (see Specimen); and *CD* is the Cluster Distance given by the user. The result is that after a cluster calculation, clusters with their leaders are derived, and each individual belongs to one cluster. In the case that all individuals create their own cluster (1 individual = 1 cluster), then each individual will jump toward all others, (this is identical with the 'AllToAll' strategy). Some clusters may be created or annihilated during migration loops.

$$CD > \sqrt{\sum_{i=1}^{D} \left(\frac{IND_i - CC_i}{HB_i - LB_i} \right)^2} \tag{21}$$

By using SOMA with clusters, the user must define a so-called 'Cluster Distance'—the parameter, which says how large (how many of individuals) the cluster should be, and the domain of attraction of the local cluster Leader. Using this basic parameter, SOMA breaks itself up into more local SOMAs, each focusing on the contained *Leader*. Therefore, independent groups of individuals are carrying out the search. These local SOMAs create clusters, which can join together or split into new clusters. This strategy has not been studied in detail yet, because of its increased complexity of computation compared with the low quality improvement of the optimization process. Other possible strategies or variations of SOMA are, for example, that individuals need not move along a straight line-vectors, but they can travel on curves, etc.

5.1 SOMA Parameters

SOMA, as other EAs, is controlled by a special set of parameters. Some of these parameters are used to stop the search process when one of two criteria are fulfilled; the others are responsible for the quality of the results of the optimization process. The parameters are shown in Table 1.

A sensitivity of SOMA, as well as of other EAs, is that it has a slight dependence on the control parameter setting. During various tests it was found that SOMA is sensitive on the parameter setting as well as others algorithms. On the other side there was found setting that is almost universal, i.e. this setting was used almost in all simulations and experiments with very good performance of SOMA. The control parameters are described below and recommended values for the parameters, derived empirically from a great number of experiments, are given:

- *PathLength* $\in$ [1.1, 5]. This parameter defines how far an individual stops behind the *Leader* (*PathLength* = 1: stop at the leader's position, *PathLength* = 2: stop behind the leader's position on the opposite side but at the same distance as at the starting point). If it is smaller than 1, then the Leader's position is not overshotted, which carries the risk of premature convergence. In that case SOMA may get trapped in a local optimum rather than finding the global optimum. Recommended value is **3–5**.

Table 1 SOMA parameters

Parameter name	Recommended range	Remark
ParthLength	[1.1, 5]	Controlling parameter
Step	[0.11, *ParthLength*]	Controlling parameter
PRT	[0, 1]	Controlling parameter
Dim	Given by problem	Number of arguments in cost function
PopSize	[10, up to user]	Controlling parameter
Migrations	[10, up to user]	Stopping parameter
MinDiv	[arbitrary negative, up to user]	Stopping parameter

- *Step* ∈ [0.11, *PathLength*]. The step size defines the granularity with what the search space is sampled. In case of simple objective functions (convex, one or a few local extremes, etc.), it is possible to use a large *Step* size in order to speed up the search process. If prior information about the objective function is not known, then the recommended value should be used. For greater diversity of the population, it is better if the distance between the start position of an individual and the *Leader* is not an integer multiple of the *Step* parameter. That means that a *Step* size of 0.11 is better than a *Step* size of 0.1 (that lead jumping directly on the *Leader* position), because the active individual will not reach exactly the position of the *Leader*. Recommended value is **0.11**.
- *PRT* ∈ [0, 1]. *PRT* stands for perturbation. This parameter determines whether an individual will travel directly towards the *Leader*, or not. It is one of the most sensitive control parameters. The optimal value is near 0.1. When the value for *PRT* is increased, the convergence speed of SOMA increases as well. In the case of low dimensional functions and a great number of individuals, it is possible to set *PRT* to 0.7–1.0. If *PRT* equals 1 then the stochastic component of SOMA **disappears** and it performs only deterministic behavior suitable for local search.
- *Dim*—the dimensionality (number of optimized arguments of cost function) is given by the optimization problem. Its exact value is determined by the cost function and usually cannot be changed unless the user can reformulate the optimization problem. Recommended value is **0.1–0.2**.
- *PopSize* ∈ [10, up to the user]. This is the number of individuals in the population. It may be chosen to be 0.5–0.7 times of the dimensionality (*Dim*) of the given problem. For example, if the optimization function has 100 arguments, then the population should contain approximately 30–50 individuals. In the case of simple functions, a small number of individuals may be sufficient; otherwise larger values for *PopSize* should be chosen. It is recommended to use at least 10 individuals (two are theoretical minimum), because if the population size is smaller than that, SOMA will strongly degrade its performance to the level of simple and classic optimization methods. Recommended value is **10**>.
- *Migrations* ∈ [10, up to user]. This parameter represents the maximum number of iterations. It is basically the same as generations for GA or DE. Here, it is called *Migrations* to refer to the nature of SOMA—individual creatures move over the landscape and search for an optimum solution. *Migrations* is a stopping criterion, i.e. it tells the optimizing process when to stop. Recommended value is up to user experience, generally **10**>.
- *MinDiv* ∈ [arbitrary negative (switch off this criterion), up to the user]. The *MinDiv* defines the largest allowed difference between the best and the worst individual from actual population. If the difference is smaller then defined *MinDiv*, the optimizing process is will stop (see Fig. 11). It is recommended to use small values. It is safe to use small values for the *MinDiv*, e.g. *MinDiv* = 1. In the worst case, the search will stop when the maximum number of migrations is reached. Negative values **are also possible** for the *MinDiv*. In this case, the stop condition for *MinDiv* will not be satisfied and thus SOMA will pass through all *Migrations*.

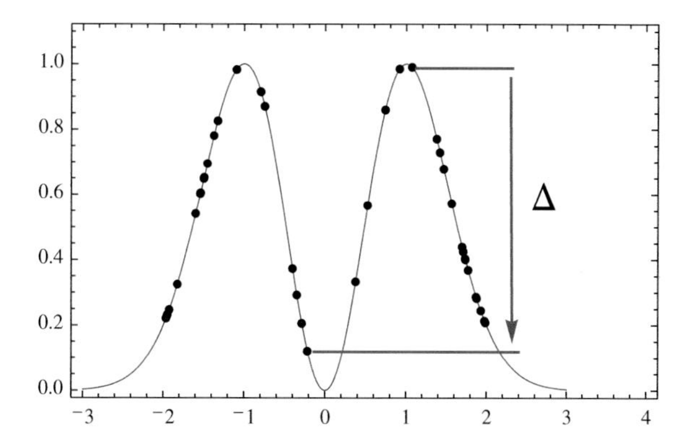

Fig. 11 *MinDiv* principle

When recommended values are taken into consideration like acceptable, then they can be included into algorithm or permanently set to be constant and number of control parameters will decrease from 6 to 1 (*Migrations*). The problem of deterministic finding suitable SOMA parameter settings for a given optimization problem is not absolutely solved (as for another algorithms) and can be regarded as one of the future research activities.

5.2 Standard Evolutionary Operations in SOMA

5.2.1 Population

SOMA, as well as the other algorithms mentioned above, is working on a population of individuals. A population can be viewed as a matrix of size $N \times M$ (Table 2) where the columns represent individuals. Each individual in turn represents one candidate solution—or input vector—for the given problem or system, i.e. a set of arguments for the cost function. Associated with each individual is also a so-called cost value, i.e. the system response to the input vector. The cost value represents the fitness of the evaluated individual. It does not take part in the evolutionary process itself—it only guides the search process.

Table 2 Population (of the *N*x*M* size), I_x is the *x*-th individual, P_y is the *y*-th individual Fitness—individual quality measured by means of the objective function

	I_1	I_2	I_3	I_4	…	…	…	I_M
Fitness	**55.2**	**68.3**	**5.36**	**9.5**	…	…	…	**0.89**
P_1	2.55	549.3	−55.36	896.5	…	…	…	1.89
P_2	0.25	66.2	2	−10	…	…	…	−2.2
P_3	−66.3	56	4	15.001	…	…	…	−83.66
…	…	…	…	…	…	…	…	…
P_N	259.3	−10	22.22	536.22	…	…	…	−42.22

A population is usually randomly initialized at the beginning of the evolutionary process. Before that, a so-called Specimen (Eq. 22) has to be defined on which the generating of the population is based.

$$\text{Specimen} = \{\{\text{Real},\{\text{Lo},\text{Hi}\}\},\ \{\text{Integer},\{\text{Lo},\text{Hi}\}\},\ldots\} \tag{22}$$

The Specimen defines for each parameter the type (e.g. integer, real, discrete, etc.) and its borders. For example, Integer, *Lo*, *Hi* defines an integer parameter with an upper border *Hi* and a lower border *Lo*. In other words, the borders define the allowed range of values for that particular cost function parameter. The careful selection of these borders is crucial for engineering applications, because without well-defined borders one can get solutions which are not applicable to the real physical system. For example one could get a negative thickness of the wall of a pressure vessel as an optimal result. The borders are also important for the evolution process itself. Without them, the evolutionary process could go to infinity (author's experience with Schwefel's function—extremes are further and further away from the original). When a Specimen is properly defied then the population (Table 2) is generated as follows

$$P^{(0)} = x_{i,j}^{(0)} = rnd_{i,j}(x_j^{(Hi)} - x_j^{(Lo)}) + x_j^{(Lo)} \quad i = 1,\ldots,n_{PopSize},\ j = 1,\ldots,m_{param} \tag{23}$$

Meaning of parameters is following—$P^{(0)}$ is the initial population and x is jth parameter of individual which consist of n parameters. Population then consist of $n_{PopSize}$ individuals. Equation (12) ensures that the parameters of all individuals are randomly generated within the allowed borders, i.e. that the initial candidate solutions are chosen from that area within the search space that contains a feasible solution to the optimization problem, see Fig. 12.

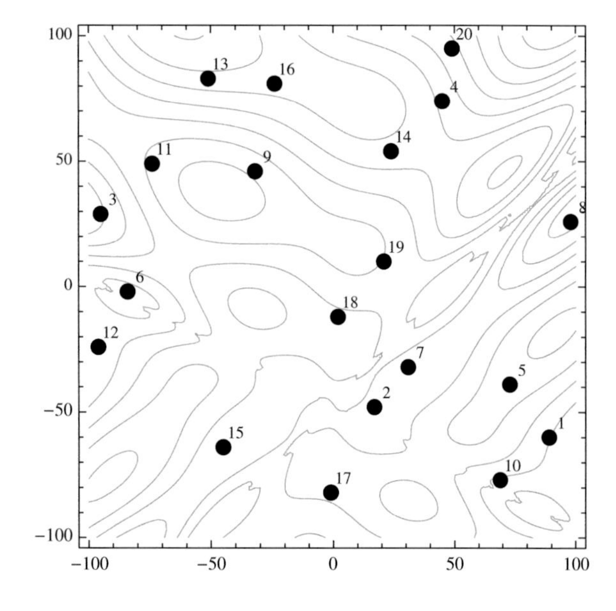

Fig. 12 Randomly generated population inside searched space

5.2.2 Mutations

Mutation, the random perturbation of individuals, is an important operation for EA strategies. It ensures the diversity amongst the individuals and it also provides the means to restore lost information in a population. Mutation in SOMA is different compared to other EA strategies. SOMA uses a *PRT* parameter to achieve perturbation. This parameter has the same effect for SOMA as mutation has for GA. It is defined in the range [0, 1] and is used to create a perturbation vector (PRTVector, see Table 3) as follows:

$$\textit{if} \quad rnd_j < PRT \quad \textit{then}\ PRTVector_j = 1 \quad \textit{else}\ 0, \quad j = 1, \ldots, n_{param} \tag{24}$$

The "novelty" of this approach was that the PRTVector is created before an individual starts its journey over the search space (in standard EA terminology "before crossover"). The PRTVector defines the final movement of an active individual in $N - k$ dimensional subspace (see next section. Later on, PRTVector creation has been changed so that it was generated after each jump of each individual, as reported later. This improved SOMA performance significantly and individual trajectory was no longer straight line but stepwise one.

5.2.3 Crossover

In standard EAs the Crossover operator usually creates new individuals based on information from the previous generation. Geometrically speaking, new positions are selected from an N dimensional hyper-plane. In SOMA, which is based on the simulation of cooperative behavior of intelligent beings, sequences of new positions in the N dimensional hyper-plane are generated. They can be thought of as a series of new individuals obtained by the special crossover operation. This crossover operation determines the behavior of SOMA. The movement of an individual is thus given as follows:

$$\begin{aligned} &\vec{r} = \vec{r_0} + \vec{m}\, t\, \overrightarrow{PRT}\, Vector \\ &\textit{where} \quad t \in \langle 0,\ \textit{by Step to},\ \textit{PathLength} \rangle \end{aligned} \tag{25}$$

Table 3 An example of perturbation vector for four parameter individual with $PRT = 0.3$

rnd_j	PRTVector_j
0.231	1
0.456	0
0.671	0
0.119	1

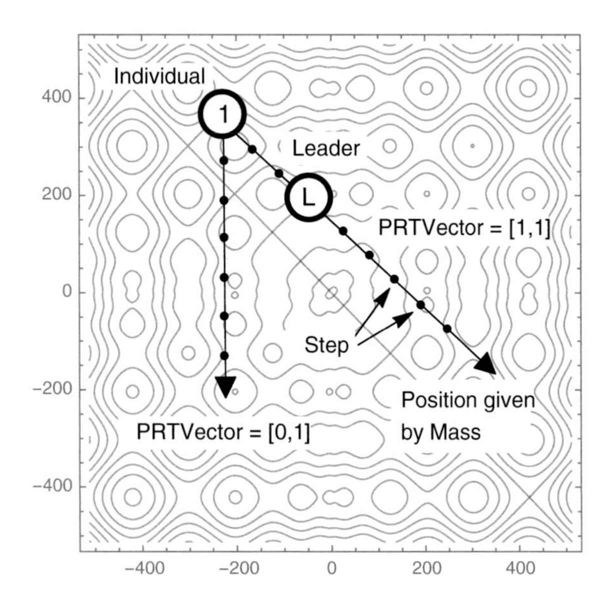

Fig. 13 PRTVector and its action on individual movement (an artificial example) when calculated before jumping sequence and kept without changes during jumping of one individual

or, more precisely:

$$x_{i,j}^{MLnew} = x_{i,j,start}^{ML} + (x_{L,j}^{ML} - x_{i,j,start}^{ML})t\ PRTVector_j \\ where \quad t \in \langle 0,\ by\ Step\ to,\ PathLength \rangle \\ and \quad ML\ is\ actual\ migration\ loop \tag{26}$$

It can be observed from Fig. 13 that the PRTVector causes an individual to move toward the leading individual (the one with the best fitness) in *N-k* dimensional space. If all *N* elements of the PRTVector are set to 1, then the search process is carried out in an *N* dimensional hyper-plane (i.e. on a *N* + 1 fitness landscape). If some elements of the PRTVector are set to 0 (see Eq. 24) then the second terms on the right hand side of equation equal 0. This means those parameters of an individual that are related to 0 in the PRTVector are 'frozen', i.e. not changed during the search. The number of frozen parameters *k* is simply the number of dimensions which are not taking part in the actual search process. Therefore, the search process takes place in a *N-k* dimensional subspace.

There is one important issue about SOMA sampling by *Step* jumping. This jumping is scale-free, i.e. for all trajectories there is for example 30 jumps per trajectory of one individual. On the beginning, when path of jumping can be over the entire space as well as at the end, when all individuals are in small limited part of space of possible solutions. That means that SOMA, when approaching (global if possible) extreme then sampling is more dense and searched space of **more intensively explored**.

Another, newer and better strategy for perturbation, was set according to the idea that perturbation vector is generated after each *Step*. This caused dynamics as

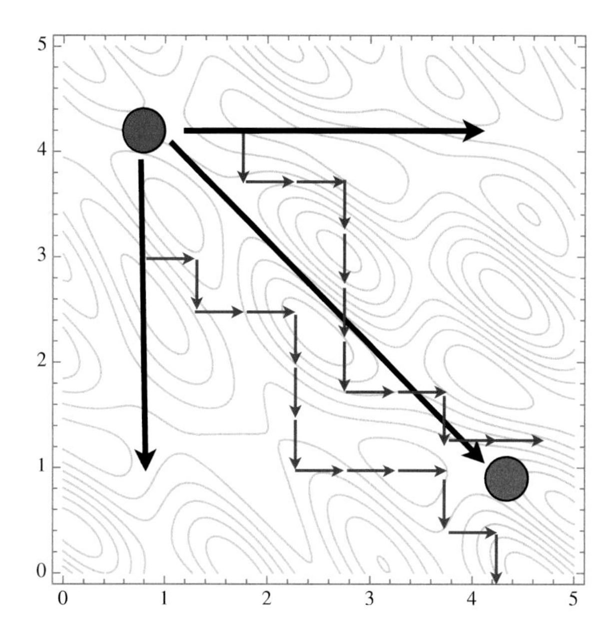

Fig. 14 PRTVector and its action on individual movement when recalculated after each *Step*. *Solid thick black arrows* shows vector of direction and its components (selected according to the *PRT*) and *red* and *blue stepwise lines* shows one on many possible trajectories when PRTVector is recalculated after each *Step*. There are possible two kind of trajectories (depend on style recalculation), the first one as in figure and second, where the individual jumps on vector components

Table 4 An example of perturbation vector for four parameter individual with $PRT = 0.5$

rnd_j	$PRTVector_j$
0.231	1
0.456	1
0.812	−1
0.671	1
0.119	−1

depicted on Fig. 14. Also negative steps are allowed, that lead to more random-like behavior, based on probability of occurrence of negative perturbations. see Table 4. Another possibility is to include 0 between −1 and 1 etc.

$$if \quad rnd_j < PRT \quad then\ PRTVector_j = 1 \quad else\ -1, \quad j = 1, \ldots, n_{param} \tag{27}$$

5.2.4 Constraint Handling

SOMA, as well as other such evolutionary algorithms, can be used to solve optimization problems, sometimes called mixed integer-discrete-continuous, non-linear programming problems, etc. These can (see also [42]) be expressed as follows:

$$\begin{array}{l} \text{Find} \\ \mathrm{X} = \{\mathrm{x}_1, \mathrm{x}_2, \mathrm{x}_3, \ldots, \mathrm{x}_\mathrm{n}\} = \left[X^{(i)}, X^{(d)}, X^{(c)}\right]^T \\ \text{to minimize} \\ f(X) \\ \text{subject to constraints} \\ g_j(X) \leq 0 \quad j = 1, \ldots, m \\ \text{and subject to boundary constraints} \\ x_i^{(Lo)} \leq x_i \leq x_i^{(Hi)} \quad i = 1, \ldots, n \\ \text{where} \\ X^{(i)} \in R^i, \quad X^{(d)} \in R^d, \quad X^{(c)} \in R^c \end{array} \tag{28}$$

$X(i)$, $X(d)$ and $X(c)$ denote feasible subsets of integer, discrete and continuous variables respectively. The above formulation is general and basically the same for all types of variables. Only the structure of the design domain distinguishes one problem from another. However, it is worth noticing here the principal differences between integer and discrete variables. While both integer and discrete variables have a discrete nature, only discrete variables can assume floating-point values. For example, Discrete = −1, 2.5, 20, −3, −5.68 …. In practice, the discrete values of the feasible set are often unevenly spaced. These are the main reasons why integer and discrete variables require different handling. SOMA can be categorized as belonging to the class of floating-point encoded, 'memetic' optimization algorithms. Generally, the function to be optimized, *f*, is of the form:

$$f(X) : R^n \to R \tag{29}$$

The optimization target is to minimize the value of this objective function f(X),

$$\min(f(X)) \tag{30}$$

by optimizing the values of its parameters:

$$X = \left(x_1, \ldots, x_{n_{param}}\right) \quad x \in R \tag{31}$$

where X denotes a vector composed of n_{param} objective function parameters. Usually, the parameters of the objective function are also subject to lower and upper boundary constraints, $x^{(Lo)}$ and $x^{(Hi)}$, respectively:

$$x_j^{(Lo)} \le x_j \le x_j^{(Hi)} \quad j = 1, \ldots, n_{param} \tag{32}$$

5.2.5 Boundary Constraints

With boundary-constrained problems, it is essential to ensure that the parameter values lie within their allowed ranges after recalculation. A simple way to guarantee this, is to replace the parameter values that violate boundary constraints with random values generated within the feasible range:

$$x_{i,j}^{\prime(\mathrm{ML}+1)} \begin{cases} r_{i,j}(x_j^{(Hi)} - x_j^{(Lo)}) + x_j^{(Lo)} & \text{if} \quad x_{i,j}^{\prime(\mathrm{ML}+1)} < \mathrm{x}_j^{(\mathrm{Lo})} \vee \mathrm{x}_{i,j}^{\prime(\mathrm{ML}+1)} > \mathrm{x}_j^{(\mathrm{Hi})} \\ x_{i,j}^{\prime(\mathrm{ML}+1)} & \text{otherwise} \end{cases}$$
$$\text{where,} \quad i = 1, \ldots, PopSize, \quad j = 1, \ldots, n_{param}. \tag{33}$$

5.2.6 Constraint Functions

A soft-constraint (penalty) approach was applied for the handling of the constraint functions. The constraint function introduces a distance measure from the feasible region, but is not used to reject unfeasible solutions, as is the case with hard-constraints. One possible soft-constraint approach is to formulate the cost-function as follows:

$$\begin{aligned} & f_{cost}(X) = (f(X) + a) \cdot \prod_{i=1}^{m} c_i^{b_i} \\ & \text{where} \\ & c_i = \begin{cases} 1.0 + s_i \cdot g_i(X) & \mathit{if} \quad g_i(X) > 0 \\ 1 & \text{otherwise} \end{cases} \\ & s_i \ge 1 \\ & b_i \ge 1 \\ & \min(f(X)) + a > 0 \end{aligned} \tag{34}$$

The constant, a, is used to ensure that only non-negative values will be assigned to fcost. When the value of a is set high enough, it does not otherwise affect the search process. The constant, s, is used for appropriate scaling of the constraint function value. The exponent, b, modifies the shape of the optimization hyper-plane. Generally, higher values of s and b are used when the range of the constraint function, $g(X)$, is expected to be low. Often setting $s = 1$ and $b = 1$ works satisfactorily and only if one of the constraint functions, $g_i(X)$, remains violated after the optimization run, will it be necessary to use higher values for s_i and/or b_i. In many real-world engineering optimization problems, the number of constraint

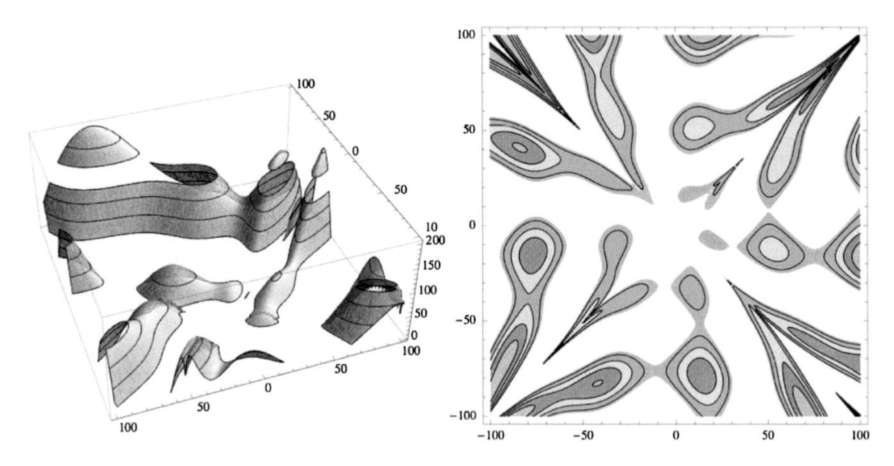

Fig. 15 Separated islands around the search space

functions is relatively high and the constraints are often non-trivial. It is possible that the feasible solutions are only a small subset of the search space. Feasible solutions may also be divided into separated islands around the search space, Fig. 15. Furthermore, the user may easily define totally conflicting constraints so that no feasible solutions exist at all. For example, if two or more constraints conflict, so that no feasible solution exists, EAs are still able to find the nearest feasible solution. In the case of non-trivial constraints, the user is often able to judge which of the constraints are conflicting on the basis of the nearest feasible solution. It is then possible to reformulate the cost-function or reconsider the problem setting itself to resolve the conflict. A further benefit of the soft-constraint approach is that the search space remains continuous. Multiple hard constraints often split the search space into many separated islands of feasible solutions. This discontinuity introduces stalling points for some genetic searches and also raises the possibility of new, locally optimal areas near the island borders. For these reasons, a soft-constraint approach is considered essential. It should be mentioned that many traditional optimization methods are only able to handle hard-constraints. For evolutionary optimization, the soft-constraint approach was found to be a natural approach.

5.2.7 Handling of Integer and Discrete Variables

In its canonical form, SOMA (as well as DE) is only capable of handling continuous variables. However extending it for optimization of integer variables is rather easy. Only a couple of simple modifications are required. First, for evaluation of cost-function, integer values should be used. Despite this, the SOMA algorithm itself may still work internally with continuous floating-point values. Thus,

$$\begin{aligned}&f_{\text{cost}}(y_i) \quad i = 1, \ldots, n_{param} \\ &\text{where} \\ &y_i = \begin{cases} x_i & \text{for continuous variables} \\ INT(x_i) & \text{for integer variables} \end{cases} \\ &x_i \in X\end{aligned} \tag{35}$$

INT() is a function for converting a real value to an integer value by truncation. Truncation is performed here only for purposes of cost function value evaluation. Truncated values are not elsewhere assigned. Thus, EA works with a population of continuous variables regardless of the corresponding object variable type. This is essential for maintaining the diversity of the population and the robustness of the algorithm. Secondly, in case of integer variables, the population should be initialized as follows:

$$\begin{aligned}P^{(0)} = x_{i,j}^{(0)} = r_{i,j}(x_j^{(Hi)} - x_j^{(Lo)} + 1) + x_j^{(Lo)} \\ i = 1, \ldots, PopSize, \quad j = 1, \ldots, n_{param}\end{aligned} \tag{36}$$

Additionally, instead of Eq. (35), the boundary constraint handling for integer variables should be performed as follows:

$$x_{i,j}^{\prime(\mathrm{ML}+1)} = \begin{cases} r_{i,j}(x_j^{(Hi)} - x_j^{(Lo)} + 1) + x_j^{(Lo)} \\ \text{if} \quad INT\left(x_{i,j}^{\prime(\mathrm{ML}+1)}\right) < x_j^{(Lo)} \vee INT\left(x_{i,j}^{\prime(\mathrm{ML}+1)}\right) > x_j^{(Hi)} \\ x_{i,j}^{\prime(\mathrm{ML}+1)} \quad \text{otherwise} \end{cases} \tag{37}$$
$$\text{where,} \quad i = 1, \ldots, n_{pop}, \quad j = 1, \ldots, n_{param}$$

Discrete values can also be handled in a straightforward manner. Suppose that the subset of discrete variables, $X(d)$, contains l elements that can be assigned to variable x:

$$X^{(d)} = x_i^{(d)} \quad i = 1, \ldots, l \quad \text{where} \quad x_i^{(d)} < x_{i+1}^{(d)} \tag{38}$$

6 Parameter Dependence

As already mentioned above, and reported in [43], the control parameters for SOMA are: *PRT*, *ParthLength*, *Step*, *MinDiv* and *PopSize*. The quality of the results for the optimization partially dependent on the selection of these parameters. To demonstrate, how this influences the algorithm, some simulations were performed to show SOMA's dependence on them (Figs. 16, 17, 18 and 19).

These simulations demonstrate the dependency of the quality of optimization on the parameters *PRT*, *ParthLength*, *Step*, *MinDiv* and *PopSize*. A total of 100

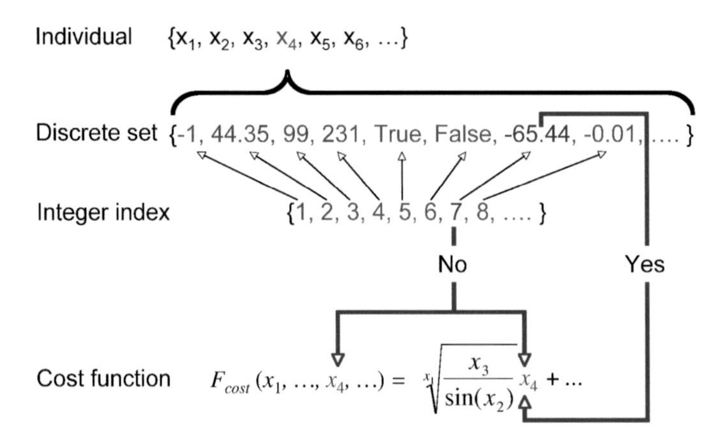

$$F_{cost}(x_1, \ldots, x_4, \ldots) = \sqrt[x_1]{\frac{x_3}{\sin(x_2)}}\, x_4 + \ldots$$

Fig. 16 Discrete parameter handling

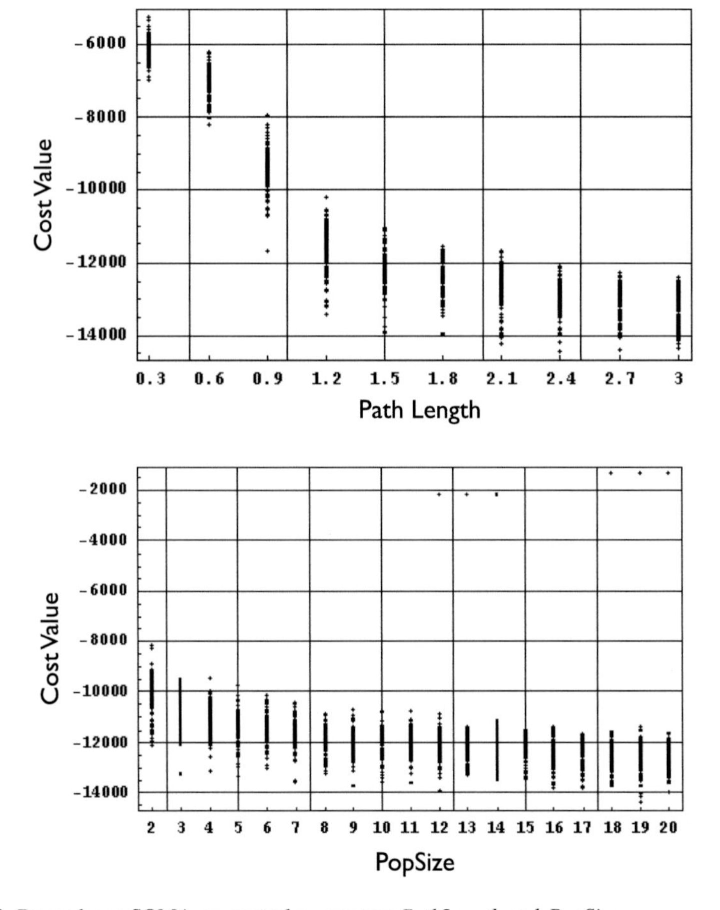

Fig. 17 Dependence SOMA on control parameters *PathLength* and *PopSizer*

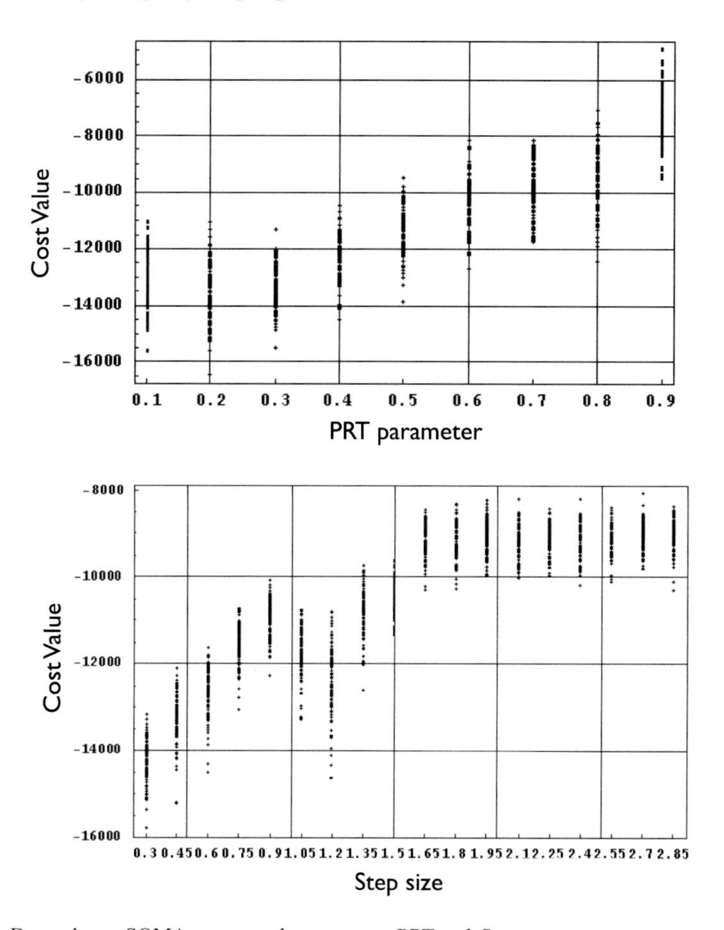

Fig. 18 Dependence SOMA on control parameters *PRT* and *Step*

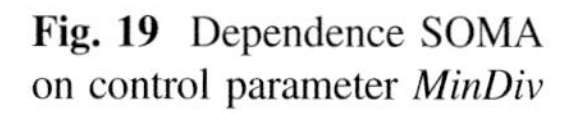

Fig. 19 Dependence SOMA on control parameter *MinDiv*

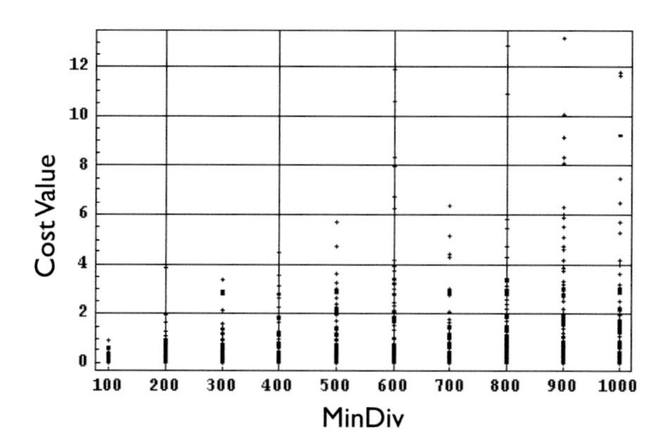

simulations were performed for each change of each parameter. The results are depicted as columns of points (point—the best value founded during actual simulation). On Fig. 17, left, shows the dependence on the *ParthLength* parameter. It can be observed that an increases of the *ParthLength* parameter resulted in deeper (better) extremes. This is logical since the search process passes through a lager search space. On Fig. 17 right shows the dependence on the *PopSize* parameter. In addition, there is a small improvement. On Fig. 18, left, shows the dependence on *PRT* from which it is visible, that in the case of low *PRT*, SOMA's efficiency is very good. On Fig. 17, right, shows the dependence on the *Step* parameter. From this picture, it is clear that in the case of big steps, results are poor due to individuals performing large jumps and thus the searched space is poorly searched. The last part Fig. 19 shows the dependence on the *MinDiv* parameter. From the picture it is visible that the *MinDiv* plays a role of something like a lens which determines the dispersion of final solutions in the case of repeated simulations (or/and solutions in the last population). *MinDiv* is the difference between the worst and the best individual and if it is quite large, then the process may stop before the optimal solution is found.

7 SOMA and Cost Function Evaluations

Number of cost function evaluation for SOMA and DE can be calculated quite easily. If principles of SOMA are taken into account, then number of cost function evaluations done during one individual run is given by

$$N_{eval\_1\_individual} = \frac{PathLength}{Step} \tag{39}$$

Because there is one *Leader* and thus (*PopSize* − 1) individuals will run in one migration loop then for one migration is done number of cost function evaluations by

$$N_{eval\_1\_migration} = \frac{(PopSize - 1) * PathLength}{Step} \tag{40}$$

Finally, during all migrations is total number of cost function evaluations given by

$$N_{eval} = \frac{(PopSize - 1) * PathLength * Migrations}{Step} \tag{41}$$

For strategy AllToAll is situation similar. Because there is no *Leader* and all individuals run toward themselves, then nominator is multiplied by *PopSize*. Term (*PopSize* − 1) means here the fact, that no one individual run toward itself. The number of cost function evaluations is for AllToAll strategy given by

$$N_{eval} = \frac{PopSize * (PopSize - 1) * PathLength * Migrations}{Step} \tag{42}$$

It is visible from Eq. (42), when compared with Eq. (41), that AllToAll is *PopSize* times harder (in cost function evaluations) than AllToOne. That is why (in previous experimentation, reported in various journals and conferences) it is possible to set *Step* = 0.4 and more because AllToAll search process is very dense and thus bigger step does not influence it so strongly as in the case of the AllToOne strategy.

Under some assumptions can be also calculated probability that global extreme will be found. Main assumption is that searched space is discrete, i.e. can be covered by grid, fine or rough. This discretization can be done if there are no individuals, who are exactly at one point, i.e. there are no individuals with exactly the same coordinates. This is true every time, because probability that two or more individuals will share the same position is almost 0 (there is an infinite number of real numbers). Discretization can be done a priori by estimation or a posteriori so that after evolution is grid size based on minimal distance of individuals in the population. Thus each axe of searched space consist of L discrete elements and for cost function with n arguments (i.e. n axes) is probability done by

$$P_{GE} = \frac{N_{eval}}{L^n} \tag{43}$$

If grid size is constant, then probability of global extreme retrieval is bigger if N_{eval} increase and vice versa. Values represented by N_{eval} can increase only if *PopSize*, *PathLength* and *Migrations* are bigger or/and *Step* is lover.

8 Selected SOMA Applications

Since 1999 when the first SOMA versions has been released, there was done a lot of various research experiments with this algorithm as well as a few applied research projects. In this section are briefly discussed selected experiments. The most interesting applications of SOMA were:

- Chemical reactor design and control. The SOMA algorithm was used for static optimization of a given chemical reactor with 5 inputs and 5 outputs. SOMA was used on this reactor for static optimization because the reactor, which was set by an expert, shows poor performance behavior. Participation consists of simulation results, which shows how expertly set reactor behaves. Also set of static optimization simulations of given reactor is presented here including results and conclusions.
- Plasma reactor control. In this research, the performance of SOMA, has been compared with simulated annealing (SA) and differential evolution (DE) for an

engineering application. This application is the automated deduction of fourteen Fourier terms in a radio-frequency (RF) waveform to tune a Langmuir probe. Langmuir probes are diagnostic tools used to determine the ion density and the electron energy distribution in plasma processes. RF plasmas are inherently nonlinear, and many harmonics of the driving fundamental can be generated in the plasma. RF components across the ion sheath formed around the probe distort the measurements made. To improve the quality of the measurements, these RF components can be removed by an active-compensation method. In this research, this was achieved by applying an RF signal to the probe tip that matches both the phase and amplitude of the RF signal generated from the plasma. Here, seven harmonics are used to generate the waveform applied to the probe tip. Therefore, fourteen mutually interacting parameters (seven phases and seven amplitudes) had to be tuned on-line. In previous work SA and DE were applied successfully to this problem, and hence were chosen to be compared with the performance of SOMA. In this application domain, SOMA was found to outperform SA and DE (Figs. 20 and 21).

- Aircraft wing design. The paper deals with a promising approach of modeling the real life systems, characterized with sets of measured/discrete data, by replacing them with analytical functions framework. The article is focused on neural network approximation of functional expressions. As an analyzed system a dynamic flight model has been chosen due to the necessity of considering several classes of large sets of aerodynamic lift, drag, speed, force, balance and

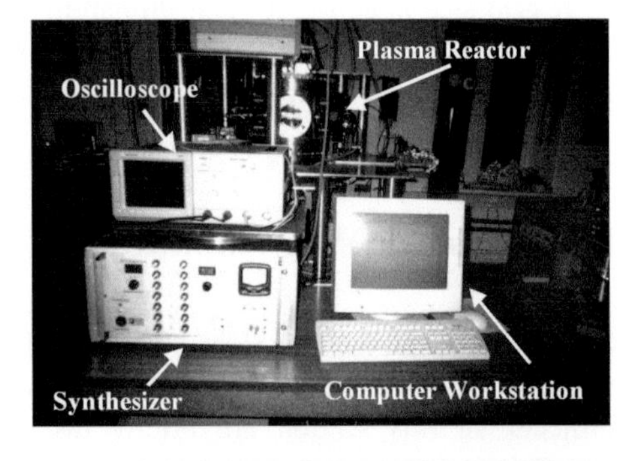

Fig. 20 Plasma reactor equipment

Fig. 21 Plasma reactor chamber

Fig. 22 Aircraft Cobra partially optimized by SOMA

Fig. 23 Another version of the Cobra aircraft

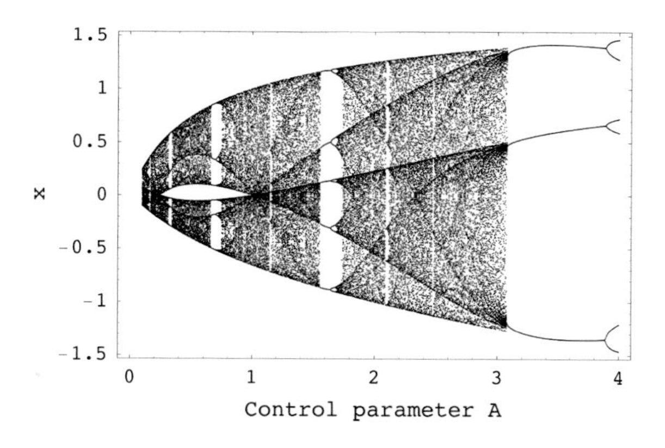

Fig. 24 Bifurcation diagram of artificially synthesized chaotic system by SOMA

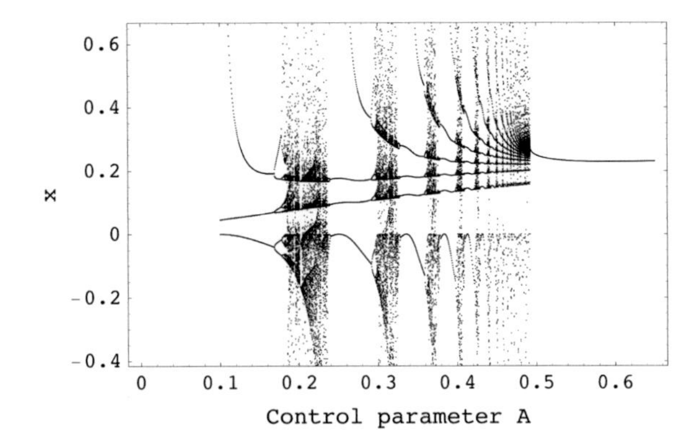

Fig. 25 Bifurcation diagram of another artificially synthesized chaotic system by SOMA

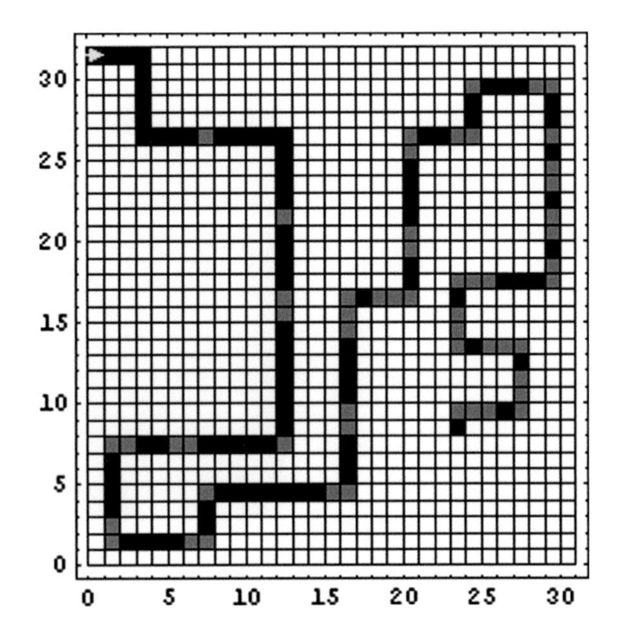

Fig. 26 The Santa Fe trail test problem

mass data to get a comparable mock-up response. Handling such type of model is naturally a huge computation time demanding process. Being able to substitute it with analytical functions system presenting a coincident behavior could dramatically improve computation time at all aspects of utilization (UAV/UAS, autopilot systems, flight simulators, real time control and stability response determination, etc.). Therefore first steps how to obtain analytical function are shown here. In this paper, sample case parameters were used to produce data that were then fitted with an exact function obtained from feedforward neural network (Figs. 22, 23, 24, 25 and 26).

- Chaos control. In this research was SOMA used (with another algorithms also) for the optimization of the control of chaotic system. The main aim of this paper is to show a new approach of solving this problem and constructing new cost

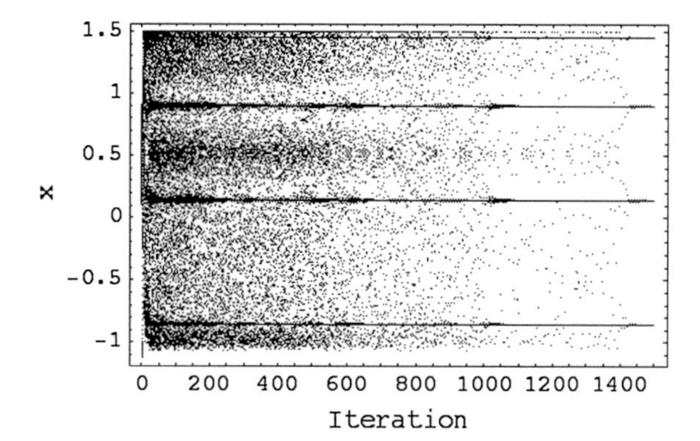

Fig. 27 Chaos stabilized in all 100 repeated experiments at the 4 period orbit

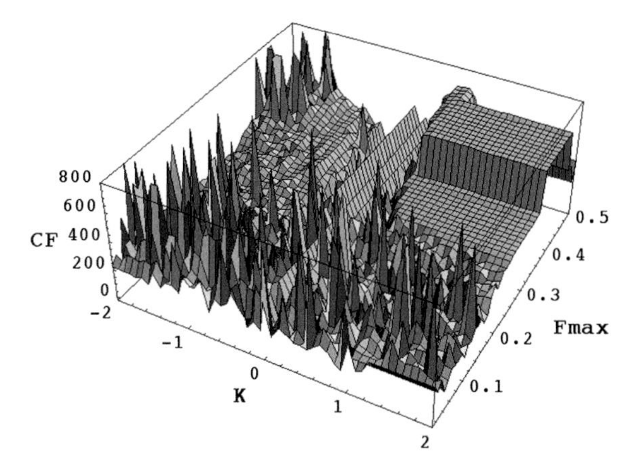

Fig. 28 Cost function surface of chaos control problem

functions operating in *blackbox* mode without previous exact mathematical analysis of the system, thus without knowledge of stabilizing target state. Three proposals of *blackbox* mode cost functions are tested in this paper. As a model of deterministic chaotic system, the two dimensional Henon map was used. The optimizations were realized in several ways, each one for another desired state of system. The evolutionary algorithms Self-Organizing Migrating Algorithm (SOMA) and Differential Evolution (DE) were used. For each version, repeated simulations were conducted to outline the effectiveness and robustness of used method and cost function (Figs. 27 and 28).

- Flow-shop scheduling problem. This paper introduces a novel Discrete Self-Organizing Migrating Algorithm for the task of flow-shop scheduling with no-wait makespan. The heuristic used in this research is the novel Discrete Self-Organizing Migrating Algorithm (SOMA). SOMA is a class of swarm heuristic which has been used to solve real domain problems. SOMA provided a new inputs to the swarm class of metaheuristics. In total four unique versions exist, namely the All to One, All to Best, All to All and All to All Adaptive.

Initial research has been conducted in the field on real domain problems; specifically in the field of Chaos Control. Initial work on SOMA in the discrete problem domain was conducted by Lampinen and Zelinka [42]. The new algorithm is tested with the small and medium Taillard benchmark problems and the obtained results are competitive with the best performing heuristics in the literature.

- Synthesis of robot control program. The paper deals with a novelty tool for symbolic regression–Analytic Programming (AP) which is able to solve various problems from the symbolic regression domain. One of tasks for it can be setting an optimal trajectory for artificial ant on Santa Fe trail which is the main application of Analytic Programming in this paper. In this contribution main principles of AP are described and explained. In second part of the article how AP was used for setting an optimal trajectory for artificial ant according the user requirements is in detail described. AP is a superstructure of evolutionary algorithms which are necessary to run AP. In this contribution 3 evolutionary algorithms were used–Self Organizing Migrating Algorithm, Differential Evolution and Simulated Annealing. The results show that the first two used algorithms were more successful than not so robust Simulated Annealing.
- Controller synthesis and setting of classical PID control. A novel tool for symbolic regression, Analytical Programming and its application for the synthesis of a new robust feedback control law are presented in this paper. This synthesized robust chaotic controller secures the fully stabilization of several selected sets containing one-dimensional, two-dimensional and evolutionary synthesized discrete chaotic systems. The paper consists of the descriptions of analytic programming as well as selected chaotic systems, used heuristic and cost function design. For experimentation, Self-Organizing Migrating Algorithm and Differential evolution were used.
- SOMA powered by deterministic chaos and periodic generators. Inherent part of evolutionary algorithms that are based on Darwin theory of evolution and Mendel theory of genetic heritage, are random processes. In participation [34–38, 44] we discuss whether random processes really are needed in evolutionary algorithms. We use n periodic deterministic processes instead of random number generators and compare performance of evolutionary algorithms powered by those processes and by pseudo-random number generators. Deterministic processes used in this participation are based on deterministic chaos and are used to generate periodical series with different length. Results presented here are numerical demonstration rather than mathematical proofs. We propose that certain class of deterministic processes can be used instead of random number generators without lowering the performance of evolutionary algorithms.
- Nonlinear dynamic system synthesis. This SOMA application introduces the notion of chaos synthesis by means of evolutionary algorithms and develops a new method for chaotic systems synthesis. This method is similar to genetic programming and grammatical evolution and is being applied along with three evolutionary algorithms: differential evolution, self-organizing migration and genetic algorithm. The aim of this investigation is to synthesize new and

"simple" chaotic systems based on some elements contained in a pre-chosen existing chaotic system and a properly defined cost function. The investigation consists of eleven case studies: the aforementioned three evolutionary algorithms in eleven versions. For all algorithms, 100 simulations of chaos synthesis were repeated and then averaged to guarantee the reliability and robustness of the proposed method. The most significant results were carefully selected, visualized and commented in this report.
- Evolutionary algorithm synthesis. This research was focused on evolutionary synthesis of another evolutionary algorithms. Different technique apart from genetic programming and grammatical evolution, called "analytic programming" [45] was used here. The main attention was given to possibility whether it is possible to synthesize another evolutionary algorithm by means of methods of symbolic regression, in this case by analytic programming, as was already mentioned. The results presented in [46, 47] clearly shows that it is possible, however powerful hardware and coding in low-level programmable language is needed.

Of course number of its use was more wider, let for example mention solution of partial differential equation of civil engineering, describing the beam in the wall under statical load, synthesis of electrical circuits (train control, house heating and traffic light control) [45], and more.

9 SOMA in Computer Games

SOMA has not been used only in various applications or interdisciplinary implementations, it has been used also in computer game Star Craft.[3] In this game was SOMA used in realtime regime so that trajectories of an individuals were one-to-one trajectories of game bot warriors.

In [48] SOMA application focused on techniques of artificial intelligence (AI) applications and practical utilization. The goal of the [48] is to implement computer player replacing human in real time strategy StarCraft: Brood War. The implementation uses conventional techniques from scope of artificial intelligence, as it at same time endeavors use of unconventional techniques, such as evolutionary computation. The computer player is provided by implementation of decision-making tree together with evolutionary algorithm called SOMA. Everything was written in programming language Java. I created system, which ensures behavior of computer player in an easy way in implementation of artificial intelligence. My particular implementation of SOMA algorithm provides an opportunity for efficient, coordinated movement of combat units over the map. The work has shown great benefit of evolutionary techniques in the field of real time strategy games.

[3]http://eu.blizzard.com/en-gb/games/hots/landing/.

The peculiarities of this algorithm implementation in Star Craft game is a variable size population (parameter PopSize). Combat units were treated as members of the population at the moment of creation to the population and deleted out of population when they are killed. This is not the only specific modification of the population in this application. The entire population is composed of two, say subspecies. One subspecies is a classic of the population of units, moving around hypersurface—battle field. This population is dynamic. Beside it contains implementations still static subspecies of the population, which is unique in that individuals of this subspecies does not move. These individuals are placed at all points on the battle field where there are natural sources of materials, and also to all points on the battle field that are starting points of the players. This subpopulation avoid its journey and are never destroyed. A leader can be selected from this subpopulation for most of the time. It is due to the fact that under the objective function on all of these positions represent the static local extremes, while dynamic local extremes (enemy combat units moving around the battle field) are found in motion of a dynamic subspecies (combat units). This implementation populations solves all the disadvantages of using the classic version of the algorithm SOMA. It consist of individuals that are dynamic warrior/scout subpopulation pseudorandomly set on battle field (static subpopulation is set according to the deterministic knowledge and is distributed according to the positions of natural materials and starting positions of players).

After the start of the algorithm is chosen as the leader of a static subject, corresponding to the second starting position on a map. There is the enemy's main base. After the warrior unity was generated (dynamic population algorithm SOMA) and begins its journey directly to the leader (the main enemy base). It is possible to implemented here, that the condition that his journey will start only with a certain dynamic population size to achieve a more concentrated combat force. On the way

Fig. 29 Combat screenshot—battle remoted by SOMA

to the leader (in this moment enemy base) can warriors encounter enemy units, which is a dynamic component of the objective function, and try to destroy them. In this moment can be applies classical combat algorithms as well as algorithms based on AI which will decide whether it is the strength of my troops found sufficient to destroy the enemy army, or put prefer to retreat and preferentially produce more units. Another role of leader, in the position of home base, play the alarm function representing a threat of the enemy, attacking one of our bases. This is a beneficial of that static population, because it contains individuals presenting a position of its own base. Since the weight of enemy units (in defined cost function) are set at a high value, although dynamic population currently attacking the enemy base and the enemy in a greater number attacks our base, a leader is selected from static population, representing the position base, which is currently under heavy attack, thus it is currently attracting dynamic population to the source of hostile attacks.

Race, which was chosen as a test has been Zerg race. Screenshot in Fig. 29 demonstrate one of many game phases. As reported in [48], SOMA driven combats has exhibit high statistical success rather than another, more classical, approaches.

10 SOMA and Interdisciplinary Research

The SOMA has been used not only in classical optimization tasks but also in interdisciplinary research as for example in evolutionary dynamics and its relations with complex networks structures. The latest research, inspired SOMA and its

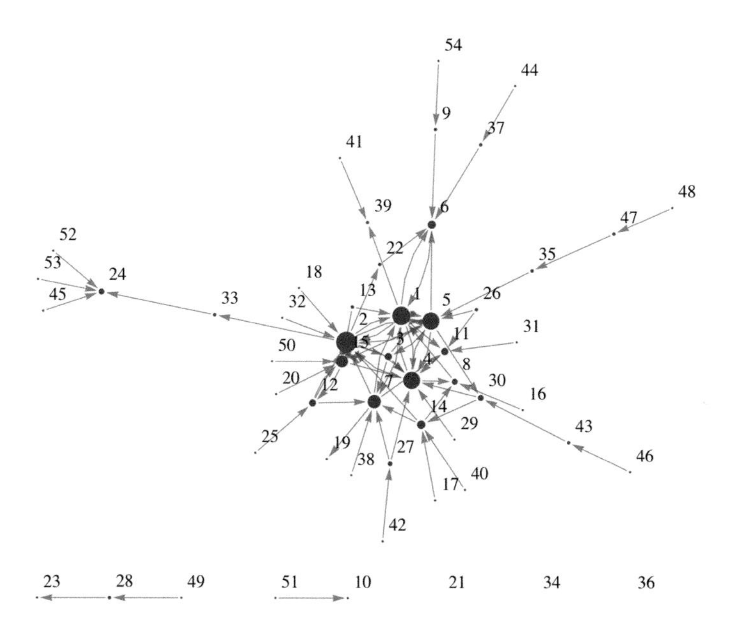

Fig. 30 The complex network based on GA dynamics

dynamics, is focused on clear analogy between individual interactions in population and interactions between people on social networks. As already reported in [49–51] then it can be stated that, in fact, dynamics of an arbitrary bio-inspired algorithm based on population philosophy, can be viewed as a complex network in which nodes are individuals and edges are related to its mutual interactions. During evolutionary process is then generated complex network, see Fig. 30 that show attributes of well known social networks and that can be used in EA performance

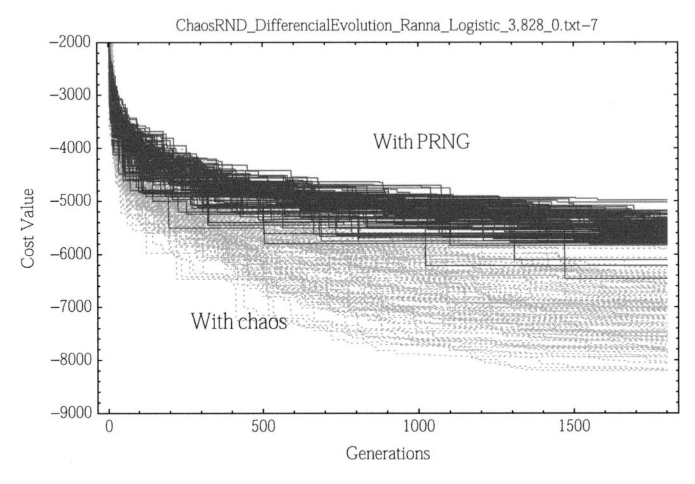

Fig. 31 Mutual comparison of the dependance of DE performance on different level of chaos in logistic equation (A = 3.828, numerical precision = 7). The *red lines* is DE with MT PRNG (100× repeated) and *blue* DE with chaotic generator instead of PRNG (100× repeated)

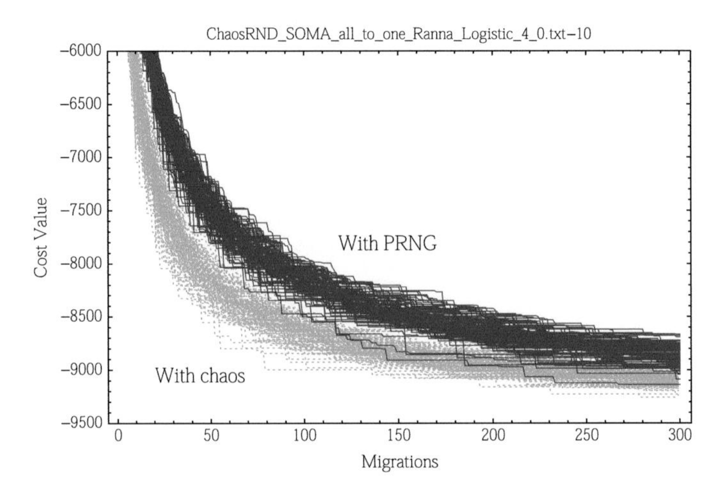

Fig. 32 Mutual comparison of the dependance of SOMA performance on different level of chaos in logistic equation (A = 4, numerical precision = 10). The *red lines* is SOMA with MT PRNG (100× repeated) and *blue* DE with chaotic generator instead of PRNG (100× repeated)

improvement. This approach has been successfully tested on DE, SOMA, GA and ABC (Artificial Bee Colony). All important results were described in [52–56].

Another future research, that is already in process, see [57], is focused on what are relations between chaos level (measured by Lyapunov exponent) and its impact on EAs performance, as preliminary reported in [57]. In this paper was control parameter A of so called logistic equation [19] varied and logistic equation then used instead of pseudorandom number generator under this setting. It was found that for some settings EAs gave much more better results than with Mersenne twister generator, see Figs. 31 and 32.

11 Conclusion

This chapter has been discussed SOMA, a new swarm class search algorithm for global optimization. In this chapter were introduced:

- Basic principles of SOMA algorithm
- Various strategies (versions) of the algorithm
- Testing for robustness
- Handling of various constraints
- Selected applications

The methods described for handling constraints are relatively simple, easy to implement and easy to use. They were introduced in [42] and used here because of their universality and easy implementation. A soft-constraint (penalty) approach is applied for the handling of constraint functions. Some optimization methods require a feasible initial solution as a starting point for a search. Preferably, this solution should be rather close to a global optimum to ensure convergence to it instead of to a local optimum. If nontrivial constraints are imposed, it may be difficult or impossible to provide a feasible initial solution. The efficiency, effectiveness and robustness of many methods are often highly dependent on the quality of the starting point. The combination of SOMA algorithm and the soft-constraint approach does not require an initial solution, but yet it can take advantage of a high quality initial solution if one is available. For example, this initial solution can be used for initialization of the population in order to establish an initial population that is biased towards a feasible region of the search space. If there are no feasible solutions in the search space, as is the case for totally conflicting constraints, SOMA algorithms with the soft-constraint approach are still able to find the nearest feasible solution. This is often important in practical engineering optimization applications, because many nontrivial constraints are involved. The test functions used for basic SOMA testing (amongst the others as combinatorial ones) had all negative attributes. Well known tested functions Eqs. (1)–(16) have been used. Each test was designed so that the global extreme was searched in 101 dimensions, i.e. the cost functions had 100 arguments. For each cost function, the success for SOMA and DE for 100 simulations was given. These tests have demonstrated that SOMA is capable of solving hard optimization problems. However it is important to remember some important facts:

1. All versions of SOMA has been tested against another algorithms like (DE, GA, PSO, SA…) and all algorithms were compared on heavy test problems as reported above and related papers. Problems on which SOMA has been tested during 16 years of its existence were of different nature, from real-encoded to permutative, from artificial to realtime problems of black-box systems.
2. Test functions (as well as the others problems) are sensitive to coordinates of the global extreme. This means that small differences in coordinates of the global extreme can cause a large change in the final cost values despite the fact that the evaluated position is not far from the position of the global extreme. This is especially important for high dimensional extremes. From this point of view, a difference of 10 % or 39.56 (say from 0) does not means that the optimizing process is far from the global extreme (in sense of the cost function—coordinates).
3. For some functions it is difficult to find the global optimum. Typical examples are Schwefel function and Rosenbrock saddle. Schwefel function being particularly difficult: The first problem is that the minimum is not in the origin as it usually is for other functions, but at the edge of the search space. The second problem is more interesting: in the case of symmetrical unfolding of the search space owing to the origin, the position of the global extreme cyclically changes its position. A second tricky function is Rosenbrock saddle. There are almost two identical extremes but only one is the global one. Hence, this is a quite an unpleasant test function for any optimization algorithm. Both functions (and the others as well as) provide rather difficult test conditions for EAs as well as problems where global extreme change its position (that was problem of real example on plasma reactor, [58].
4. The conditions for the optimization were set as highly difficult as possible i.e. a limited number of individuals was used, etc. Usually, 20–60 individuals were used for searching in 100 dimensional hyper-plane. It is for this reason that the global extreme was not exactly found in all cases. Different numbers of individuals, different parameter settings or different versions of the algorithms could improve this dramatically. Based on results can be declared that the SOMA performance was almost the same like DE. SOMA showed a very good, and sometimes better, ability to find extremes as DE or another algorithms.
5. Convergence speed. In the case of many test functions different EAs were less more of the same speed like SOMA. It can be stated that SOMA gives better performance on problems which shows higher dimensionality and/or irregularity like functions Eqs. (1)–(16)
6. Despite comparing only the basic version of the SOMA algorithm with usually of the best versions of selected EAs, it is visible that SOMA performance on the tested functions was very good. The SOMA algorithm works with minimum

assumptions with respect to the objective function. The algorithm requires only the cost value returned from the objective function for guidance of its seeking for the optimum. No derivatives or other auxiliary information are needed. Including the algorithm's extensions discussed in this article, the SOMA algorithm can be applied to a wide range of optimization problems, which practitioners in the field of modern optimization would like to solve.

Acknowledgments The following grants are acknowledged for the financial support provided for this research: Grant Agency of the Czech Republic—GACR P103/15/06700S, by the SP2015/142.

References

1. Mendel, J.G.: Versuche über Plflanzenhybriden Verhandlungen des naturforschenden Vereines in Brünn, Bd. IV für das Jahr, 1865 Abhandlungen: 3–47 (1866). For the English translation, see: Druery, C.T, Bateson, W.: Experiments in plant hybridization. J. Royal Hortic. Soc. **26**, 1–32 (1901). http://www.esp.org/foundations/genetics/classical/gm-65.pdf
2. Carlson, E.A.: Doubts about mendel's integrity are exaggerated. Mendel's legacy. Cold Spring Harbor, Cold Spring Harbor Laboratory Press, NY. pp. 48–49. ISBN 978-087969675-7
3. Darwin, C.R.: On the origin of species by means of natural selection, or the preservation of favoured races in the struggle for life. John Murray, London. 1st ed (1859)
4. Holland, J.: Adaptation in natural and artificial systems. University of Michigan Press, Ann Arbor (1975)
5. Holland, J.: Genetic algorithms. Scientific American, July 44–50 (1992)
6. Schwefel, H.: Numerische Optimierung von Computer-Modellen (PhD thesis). Reprinted by Birkhäuser, 1977 (1974)
7. Rechenberg, I. Evolutionsstrategie - Optimierung technischer Systeme nach Prinzipien der biologischen evolution (Ph.D. thesis), Printed in Fromman-Holzboog, 1973 (1971)
8. Fogel, L.J., Owens, A.J., Walsh, M.J.: Artificial intelligence through simulated evolution, Wiley, New York (1966)
9. Turing, A.M.: Intelligent machinery, unpublished report for National Physical Laboratory; published (ed. D. Michie) in Machine Intelligence 7 (1969), and in Volume 3 of The Collected Works of A. M. Turing (ed) Ince D, Amsterdam: North-Holland (1992)
10. Zelinka, I., Celikovsky, S., Richter, H., Chen, G.: (2010) Evolutionary algorithms and chaotic systems, (Eds), Springer, Germany, 550s (2010)
11. Back, T., Fogel, B., Michalewicz, Z.: Handbook of evolutionary computation. Institute of Physics, London (1997)
12. Barricelli, N.A.: Esempi Numerici di processi di evoluzione, Methodos, 45–68 (1954)
13. Barricelli, N.A.: Symbiogenetic evolution processes realized by artificial methods. Methodos **9** (35–36), 143–182 (1957)
14. Barricelli, N.A.: Numerical testing of evolution theories: Part I: theoretical introduction and basic tests. Acta. Biotheor. **16**(1–2), 69–98 (1962)
15. Goldberg, D.: Genetic algorithms in search, optimization, and machine learning. Addison-Wesley Publishing Company Inc. ISBN 0201157675. Optimization: methods and case studies. Springer, Verlag. ISBN 3-540-23022 BEN, Praha. ISBN 80-7300-069-5 (1989)
16. Bull, L., Kovacs, T.: Foundations of learning classifier systems. Springer, Berlin (2005)
17. Baluja, S.: Population-based incremental learning: a method for integrating genetic search based function optimization and competitive learning. Technical report CMU-CS-94–163, Carnegie Mellon University, USA (1994)

18. Larrañaga, P., Lozano, J.A.: Estimation of distribution algorithms: a new tool for evolutionary computation. Kluwer Academic Publishers, Berlin (2002)
19. Dorigo, M., Sutzle, T. Ant colony optimization. MIT Press, Cambridge. ISBN: 978-0262042192 (2004)
20. Onwubolu, G., Babu, B. New optimization techniques in engineering. Springer, Verlag. pp. 167–218. ISBN 3-540-20167X
21. Dasgupta, D.: Artificial immune systems and their applications. Springer, Verlag (1999). ISBN 3-540-64390-7
22. Castro L, Timmis J (2002) Artificial Immune Systems: A New Computational Intelligence Approach, Springer-Verlag, ISBN 978-1-85233-594-6
23. Hart, W., Krasnogor, N., Smith, J.: Recent advances in memetic algorithms. Springer, Verlag. ISBN 978-3-540-22904-9 (2005)
24. Goh, C., Ong, Y., Tan, K.: Multi-objective memetic algorithms. Springer, Verlag. ISBN 978-3-540-88050-9 (2009)
25. Schönberger, J.: Operational freight carrier planning, basic concepts, optimization models and advanced memetic algorithms. Springer, Verlag (2005). ISBN 978-3-540-25318-1
26. Laguna, M., Martí, R.: Scatter search—methodology and implementations in C. Springer, Verlag. ISBN 978-1-4020-7376-2 (2003)
27. Clerc, M.: Particle swarm optimization. ISTE Publishing Company, ISBN (2009). 1905209045
28. Li, X.: Particle swarm optimization—an introduction and its recent developments [online] 4.10.2006 [cit. 20. 2. 2007]. Available from www.nical.ustc.edu.cn/seal06/doc/tutorial_pso.pdf (2006)
29. Eberhart, R.C., Kennedy, J.: A new optimizer using particle swarm theory, pp. 39–43. Proceedings of the Sixth International Symposium on Micromachine and Human Science, Nagoya (1995)
30. Yang, X.-S.: Deb, S.: Cuckoo search via LŽvy flights. World Congress on Nature and Biologically Inspired Computing (NaBIC 2009). IEEE Publications. pp. 210Ð214 (December 2009) arXiv:1003.1594v1
31. Price, K.: An introduction to differential evolution. In: Dorigo, M., Glover, F. (eds.) Corne D, pp. 79–108. New ideas in optimisation, McGraw Hill, International (UK) (1999)
32. Yang, X.S.: Nature-inspired metaheuristic algorithms. Luniver Press, Frome (2008). ISBN 1-905986-10-6
33. Yang, X.S.: A new metaheuristic bat-inspired algorithm. In: Gonzalez et al., J.R. (eds.) Nature inspired cooperative strategies for optimization (NISCO 2010), Studies in computational intelligence vol. 284, pp. 65–74 Springer, Berlin (2010). http://arxiv.org/abs/1004.4170
34. Skanderova, L., Zelinka, I., Saloun, P.: Chaos powered selected evolutionary algorithms. In: Proceedings of Nostradamus 2013: international conference prediction, modeling and analysis of complex systems, Springer Series: advances in intelligent systems and computing, vol. 210, pp. 111–124 (2013)
35. Zelinka, I.: Petr saloun roman senkerik, chaos powered grammatical evolution, 13th international conference on computer information systems and industrial management applications—CISIM 2014. Springer, Ho Chi Minh City (2014)
36. Zelinka I., Senkerik R., Pluhacek M.: Do evolutionary algorithms indeed require randomness? In: IEEE congress on evolutionary computation. Cancun, Mexico (2013)
37. Caponetto, R., Fortuna, L., Fazzino, S., Xibilia, M.: Chaotic sequences to improve the performance of evolutionary algorithms. IEEE Trans. Evol. Comput. **7**(3), 289–304 (2003)
38. Lozi, R.: Emergence of randomness from chaos. Int. J. Bifurcation Chaos **22**(2), 1250021 (2012). doi:10.1142/S0218127412500216
39. Schuster, H.G.: Handbook of chaos control. Wiley-VCH, New York (1999)
40. Barnsley, M.F.: Fractals everywhere. Academic Press Professional. ISBN 0-12-079061-0 (1993)
41. Clerc M.: Particle swarm optimization. ISTE Publishing Company (2006). ISBN 1-905209-04-5

42. Lampinen, J., Zelinka, I.: Mechanical engineering design optimization by differential evolution. In: Corne, D., Dorigo, M., Glover, F. (eds.) New ideas in optimization, pp. 127–146. McGraw-Hill, London. ISBN 007-709506-5 (1999)
43. Zelinka, I.: SOMA—self organizing migrating algorithm. In: Onwubolu, B.B. (eds) New optimization techniques in engineering. Springer, New York. ISBN 3-540-20167X, pp. 167–218 (2004)
44. Zelinka I., Chadli M., Davendra D., Senkerik R., Pluhacek M., LampinenJ.: Do evolutionary algorithms indeed require random numbers? Extended study. In: Proceedings of Nostradamus 2013: international conference prediction, modeling and analysis of complex systems, Springer Series: advances in intelligent systems and computing, vol. 210, pp. 61–75 (2013)
45. Zelinka, I., Davendra, D., Jasek, R., Senkerik, R., Oplatkova, Z.: Analytical programming—a novel approach for evolution—ary synthesis of symbolic structures. In: Kita, E. (ed.) Evolutionary algorithms. ISBN: 978– 953-307-171-8, InTech, doi:10.5772/16166. Available from: http://www.intechopen.com/books/evolutionary-algorithms/analytical-programming-a-novel-approach-for-evolutionary-synthesis-of-symbolic-structures
46. Oplatkova, Z., Zelinka, I., Senkerik, R.: Santa fe trail for artificial ant by means of analytic programming and evolutionary computation. Int. J. Simul. Syst. Sci. Technol., vol. 9, no. 3, pp. 20Ð33 (2008)
47. Oplatkova, Z., Zelinka, I.: Investigation on artificial ant using analytic programming. In: Proceedings of genetic and evolutionary computation conference. Seattle, WA, p. 949Ð950 (2006)
48. Sikora L.: Intelligent bot for the game starcraft: brood war. Diploma thesis. VSB-TU Ostrava. Czech Republic (2015)
49. Zelinka I., Davendra D., Chadli M., Senkerik R., Dao T.T., Skanderova L.: Evolutionary dynamics and complex networks. In: Zelinka, I., Snasel, V., Ajith, A., (eds) Handbook of optimization. Springer, Germany, p 1100 s (2012)
50. Zelinka I., Davendra D., Senkerik R., Jasek R.: Do evolutionary algorithm dynamics create complex network structures? Complex Syst. **20**(2), 127–140, 0891–2513 (2011)
51. Zelinka I.: Mutual relations of evolutionary dynamics, deterministic chaos and complexity, tutorial at IEEE congress on evolutionary computation, Mexico (2013)
52. Zelinka, I., Davendra D., Lampinen J., Senkerik R., Pluhacek M., Evolutionary algorithms dynamics and its hidden complex network structures, congress on evolutionary computation (CEC) IEEE congress, pp. 3246– 3251, 6–11 July 2014, doi:10.1109/CEC.2014.6900441
53. Magdalena M., Davendra, D.: Chaos-driven discrete artificial bee colony. IEEE congress on evolutionary computation pp. 2947–2954
54. Davendra, D., Zelinka, I., Metlicka, M., Senkerik, R., Pluhacek, M.: Complex network analysis of differential evolution algorithm applied to flowshop with no-wait problem. IEEE Symposium on differential evolution, Orlando, USA, 9–12 December, pp. 65–72 (2014)
55. Davendra D., Metlicka M.: Ensemble centralities based adaptive artificial bee algorithm, IEEE congress on evolutionary computation (2015)
56. Zelinka I.: Evolutionary algorithms as a complex dynamical systems, tutorial at IEEE congress on evolutionary computation. Sendai (2015)
57. Skanderova, L., Zelinka, I., Saloun, P.: Chaos powered selected evolutionary algorithms. In: Proceedings of Nostradamus 2013: international conference prediction, modeling and analysis of complex systems, Springer Series: ÒAdvances in intelligent systems and computing, vol. 210, pp 111–124 (2013)
58. Zelinka, I., Nolle, L.: Plasma reactor optimizing using differential evolution. In Price, K., Lampinen, J., Storn, R. (eds.) Differential evolution: a practical approach to global optimization. Springer, New York, p. 499Ð512 (2005)

DSOMA—Discrete Self Organising Migrating Algorithm

Donald Davendra, Ivan Zelinka, Michal Pluhacek and Roman Senkerik

Abstract A discrete Self Organising Migrating Algorithm (DSOAM) is described in this chapter. This variant is specifically designed for the permutative based combinatorial optimisation problem, where the problem domain in generally NP-Hard. Specific sampling between individuals in the search space is introduced as a means of constructing new feasible individuals. These feasible solutions are improved using 2-Opt routines. DSOMA has proven successful in solving manufacturing scheduling and assignment problems.

1 Introduction

Complex engineering problems can be loosely defined into three main domains. Unimodal and multimodal real domain problems generally deal with floating point values. Combinatorial optimisation problems on the other hand deal with integer based values. In *strict sense* combinatorial optimisation problems, the values are

D. Davendra (✉)
Department of Computer Science, Central Washington University, 400 E. University Way, Ellensburg, WA 98926-7520, USA
e-mail: DonaldD@cwu.edu; donald.davendra@vsb.cz

D. Davendra · I. Zelinka
Department of Computer Science, Faculty of Electrical Engineering and Computer Science, VSB-Technical University of Ostrava, 17. Listopadu 15, 708 33 Ostrava-Poruba, Czech Republic
e-mail: ivan.zelinka@vsb.cz

M. Pluhacek · R. Senkerik
Department of Informatics and Artificial Intelligence, Faculty of Applied Informatics, Tomas Bata University in Zlin, Nad Stranemi 4511, 76005 Zlin, Czech Republic
e-mail: pluhacek@fai.utb.cz

R. Senkerik
e-mail: senkerik@fai.utb.cz

D. Davendra and I. Zelinka (eds.), *Self-Organizing Migrating Algorithm*, Studies in Computational Intelligence 626, DOI 10.1007/978-3-319-28161-2_2

inherently permutative. *Wide sense* combinatorial optimisation problems deal with integer values within specific ranges. The third domain in simply a mixture of the first two, and is generally called mixed integer-real optimisation.

Combinatorial optimisation problems are generally considered *NP-Hard*, with the associated decision problems formulated as *NP-Complete*. A range of different problems in engineering have been classified as combinatorial optimisation problems. The most common ones are scheduling problems, routing and assignment problems amongst others.

Scheduling problems such as flowshop, openshop and jobshop are readily identifiable as the most common manufacturing problems whereas routing problems such as vehicle routing and traveling salesman problems are classic mathematical problems, and forms the general basis for the *P* versus *NP Problem* Millennium problem [7].

Since these problems are *NP-Hard*, specific deterministic and stochastic algorithms have been developed to solve these problems in reasonable time with given resources. One of the powerful stochastic algorithms are evolutionary algorithms (EA). However, general canonical EA's for real-domain problems cannot be used to solve combinatorial optimisation problems. Therefore, a separate class of *discrete* EA's have been developed to deal with this domain of problem.

All established metaheuristics have developed discrete variants such as: Genetic algorithms [4, 8, 13], Ant Colony Optimisation [5], Tabu Search [14], Particle Swarm [11, 15], Differential Evolution [9, 10], Artificial Bee Colony [12, 16] and Harmony Search [1, 6].

A discrete variant of the Self-Organising Migrating Algorithm (SOMA) has been developed to solve combinatorial optimisation problems [2, 3]. This chapter details the discrete Self-Organising Migrating Algorithm (DSOMA), with population initialisation, creating jump sequences, constructing trial individual, repairment and selection.

2 Discrete Self-organising Migrating Algorithm

Discrete Self-Organising Migrating Algorithm (DSOMA) [3] is the discrete version of SOMA, developed to solve permutation based combinatorial optimisation problem. The same ideology of the sampling of the space between two individuals is retained. Assume that there are two individuals in a search space as given in Fig. 1. The objective for DSOMA is to transverse from one individual to another, while mapping each discrete space between these two individuals. Figures 2 and 3 are 3D representations, where the DSOMA mapping is shown as the surface joining these two.

The major input of this algorithm is the sampling of the jump sequence between the individuals in the populations, and the procedure of constructing new trial individuals from these sampled jump sequence elements.

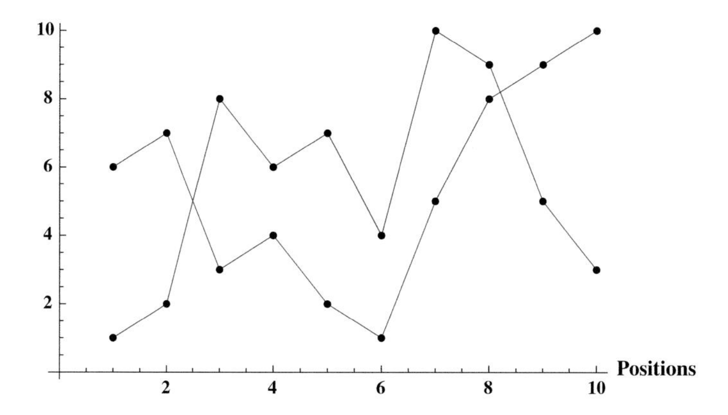

Fig. 1 Two individuals in search space

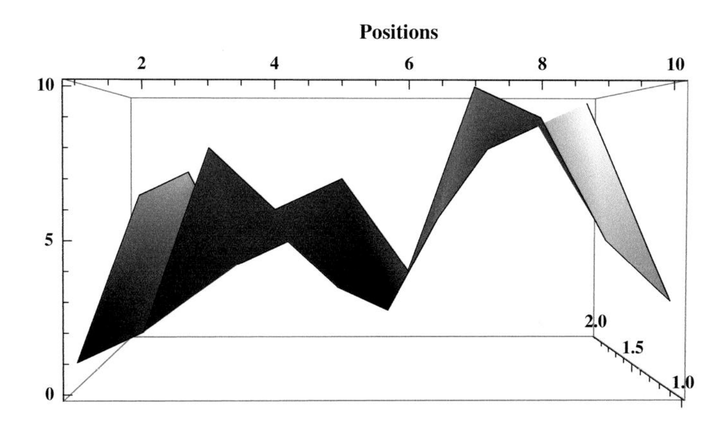

Fig. 2 End view of the two individuals in 3D search space

The overall outline for DSOMA can be given as:

1. **Initial Phase**

   a. *Population Generation*: An initial number of permutative trial individuals is generated for the initial population.
   b. *Fitness Evaluation*: Each individual is evaluated for its fitness.

2. **DSOMA**

   a. *Creating Jump Sequences*: Taking two individuals, a number of possible jump positions is calculated between each corresponding element.
   b. *Constructing Trial Individuals*: Using the jump positions; a number of trial individuals is generated. Each element is selected from a jump element between the two individuals.
   c. *Repairment*: The trial individuals are checked for feasibility and those, which contain an incomplete schedule, are repaired.

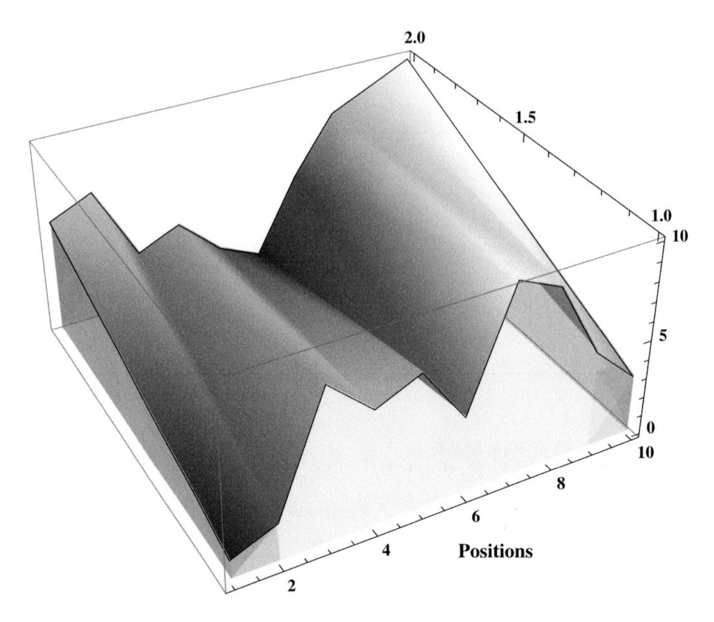

Fig. 3 Isometric view of the two individuals in 3D search space

3. **Selection**

   a. *New Individual Selection*: The new individuals are evaluated for their fitness and the best new fitness based individual replaces the old individual, if it improves upon its fitness.

4. **Generations**

   a. *Iteration*: Iterate the population till a specified migration.

DSOMA requires a number of parameters as given in Table 1. The major addition is the parameter J_{min}, which gives the minimum number of jumps (sampling) between two individuals. The SOMA variables PathLength, StepSize and PRT Vector are not initialised as they are dynamically calculated by DSOMA using the adjacent elements between the individuals.

Table 1 DSOMA parameters

Name	Range	Type	
J_{min}	(1+)	Control	Minimum number of jumps
Population	10+	Control	Number of individuals
Migrations	10+	Termination	Total number of iterations

3 Initialisation

The population is initialised as a permutative schedule representative of the size of the problem at hand (1). As this is the initial population, the superscript of x its set to 0. The *rand()* function obtains a value between 0 and 1, and the *INT()* function rounds down the real number to the nearest integer. The *if* condition checks to ensure that each element within the individual is unique.

$$x_{i,j}^{0} = \begin{cases} 1 + INT\left(rand() \cdot (N-1)\right) \\ \text{if } x_{i,j}^{0} \notin \left\{x_{i,1}^{0}, \ldots, x_{i,j-1}^{0}\right\} \end{cases} \quad (1)$$
$$i = 1, \ldots, \beta; j = 1, \ldots, N$$

Each individual is vetted for its fitness (2), and the best individual, whose index in the population can be assigned as L (leader) and it is designated the leader as X_L^0 with its best fitness given as C_L^0.

$$C_i^0 = \Im\left(X_i^0\right), \quad i = 1, \ldots, \beta \quad (2)$$

The pseudocode for generating a population is given in Fig. 4.

After the generation of the initial population, the migration counter t is set to 1 where $t = 1, \ldots, M$ and the individual index i is initialised to 1, where $i = 1, \ldots, \beta$. Using these values, the following sections (Sects. 4–7) are recursively applied, with the counters i and t being updated in Sects. 8 and 9 respectively.

Pseudocode for Generating Initial Population

Assume a population P of size β, a problem of size N, and a schedule given as $X = \{x_1, \ldots, x_N\}$. The fitness can be given as C, while the best individual is represented as X_L with its associated fitness C_L and its index b.

1. For $i = 1, 2, \ldots, \beta$ do the following:
   a. For $j = 1, 2, \ldots, N$ do the following:
      i. Randomly generate a value $x_j = $ rnd int $[1, N]$
      ii. **WHILE** $x_j \notin X_i$
         Randomly generate a value $x_j = $ rnd int $[1, N]$
      iii. Insert $x_j \rightarrow X_i$
   b. Insert $X_i \rightarrow P$
   c. Calculate the fitness of X_i as $C_i = \Im(X_i)$
2. Set $C_L = \min(C)$
3. Set best index $b = $ index of $\min(C)$
4. Set $X_L = X_b$

Fig. 4 Pseudocode for generating initial population

4 Creating Jump Sequences

DSOMA operates by calculating the number of discrete jump steps that each individual has to circumnavigate. In DSOMA, the parameter of minimum jumps ($J_{\min}$) is used in lieu of PathLength, which states the minimum number of individuals or sampling between the two individuals.

Taking two individuals in the population, one as the incumbent (X_i^t) and the other as the leader (X_L^t), the possible number of jump individuals $J_{\max}$ is the mode of the difference between the adjacent values of the elements in the individual (3). A vector J of size N is created to store the difference between the adjacent elements in the individuals. The *mode()* function obtains the most common number in J and designates it as $J_{\max}$.

$$\begin{aligned} J_j &= \left| x_{i,j}^{t-1} - x_{L,j}^{t-1} \right|, \quad j = 1, \ldots, N \\ J_{\max} &= \begin{cases} mode(J) & \textit{if } mode(J) > 0 \\ 1 & \textit{otherwise} \end{cases} \end{aligned} \tag{3}$$

The step size (s), can now be calculated as the integer fraction between the required jumps and possible jumps (4).

$$s = \begin{cases} \left\lfloor \frac{J_{\max}}{J_{\min}} \right\rfloor & \text{if } J_{\max} \geq J_{\min} \\ 1 & \text{otherwise} \end{cases} \tag{4}$$

Create a jump matrix $\mathbf{G}$, which contains all the possible jump positions, that can be calculated as:

$$\begin{aligned} \mathbf{G}_{l,j} &= \begin{cases} x_{i,j}^{t-1} + s \cdot l & \text{if } x_{i,j}^{t-1} + s \cdot l < x_{L,j}^{t-1} \text{ and } x_{i,j}^{t-1} < x_{L,j}^{t-1} \\ x_{i,j}^{t-1} - s \cdot l & \text{if } x_{i,j}^{t-1} + s \cdot l < x_{L,j}^{t-1} \text{ and } x_{i,j}^{t-1} > x_{L,j}^{t-1} \\ 0 & \text{otherwise} \end{cases} \\ & j = 1, \ldots, N; \ l = 1, \ldots, J_{\min} \end{aligned} \tag{5}$$

The pseudocode for creating jump sequences is given in Fig. 5.

5 Constructing Trial Individuals

For each jump sequence of two individuals, a total of $J_{\min}$ new individuals can now be constructed from the jump positions. Taking a new temporary population $H(H = \{Y_1, \ldots, Y_{J_{\min}}\})$, in which each new individual $Y_w(w = 1, \ldots, J_{\min})$, is constructed piecewise from $\mathbf{G}$. Each element $y_{w,j}$ $\left(Y_w = \left\{y_{w,j}, \ldots, y_{w,N}\right\}, \ j = 1, 2, \ldots, N\right)$ in the individual, indexes its values from the corresponding jth column

Pseudocode for Creating Jump Sequences

Take the population P with its associated fitness C. The number of jumps (PathLength) is given as J_{min} and the step size as s. Assume an empty schedule for storing the possible jump sequences as J and a temporary jump matrix to store the calculated individuals as G, whose size is J_{min} by N, were N is the size of the individual.

1. For $i = 1, 2, \ldots, \beta$ do the following:
   a. For $j = 1, 2, \ldots, N$ do the following:
      i. $J_j = \left| x_{L,j} - x_{i,j} \right|$
   b. $J_{max} =$ mode (J)
   c. **IF** $J_{\max} \geq J_{\min}$

      $s = \left\lfloor \frac{J_{\max}}{J_{\min}} \right\rfloor$

      ELSE

      $s = 1$
   d. For $j = 1, 2, \ldots, N$ do the following:
      i. For $l = 1, 2, \ldots, J_{min}$ do the following:
         A. **IF** $x_{i,j} < x_{L,j}$
            $\mathbf{G}_{l,j} = x_{i,j} + s \cdot l$
            ELSE IF $x_{i,j} > x_{L,j}$
            $\mathbf{G}_{l,j} = x_{i,j} - s \cdot l$
            ELSE
            $\mathbf{G}_{l,j} = 0$
      ii. Create New Trial Individual as given in Figure 2.6.

Fig. 5 Pseudocode for creating jump sequences

in $\mathbf{G}$. Each (lth $l = 1, \ldots, J_{\min}$) position for a specific element is sequentially checked in $\mathbf{G}_{l,j}$ to ascertain if it already exists in the current individual Y_w. If this is a new element, it is then accepted in the individual, and the corresponding lth value is set to zero as $\mathbf{G}_{l,j} = 0$. This iterative procedure can be given as in Eq. (6) and the pseudocode for constructing trial individual is represented in Fig. 6.

$$y_{w,j} = \begin{cases} \mathbf{G}_{l,j} \begin{cases} \text{if} \quad \mathbf{G}_{l,j} \notin \left\{ y_{w,1}, \ldots, y_{w,j-1} \right\} \text{ and } \mathbf{G}_{l,j} \neq 0; \\ \text{then } \mathbf{G}_{l,j} = 0; \end{cases} \\ 0 \quad \text{otherwise} \end{cases} \tag{6}$$
$$l = 1, \ldots, J_{\min};\; j = 1, \ldots, N;\; w = 1, \ldots, J_{\min}$$

6 Repairing Trial Individuals

Some individuals may exist, which may not contain a permutative schedule. The jump individuals $Y_w (w = 1, 2, \ldots, J_{\min})$, are constructed in such a way, that each infeasible element $y_{w,j}$ is indexed by 0.

Pseudocode for Constructing Trial Individuals

Take the temporary jump matrix G, of size is J_{min} by N, which is now populated with calculated jump sequence elements. Create a temporary population $H = \{Y_1, \ldots, Y_{J_{\min}}\}$, where $Y = \{y_1, \ldots, y_N\}$

1. For $w = 1, 2, \ldots, J_{min}$ do the following:
    a. For $j = 1, 2, \ldots, N$ do the following:
        i. For $l = 1, 2, \ldots, J_{min}$ do the following:
            A. **IF** $\left(\mathbf{G}_{l,j} \notin \{y_{w,1}, \ldots, y_{w,y-1}\} \text{ AND } \mathbf{G}_{l,j} \neq 0\right)$
            $y_{w,j} = \mathbf{G}_{l,j}$
            $\mathbf{G}_{l,j} = 0$
        ii. **IF** $y_{w,j} == \text{NULL}$
            $y_{w,j} = 0$
2. Repair the temporary population Y in Figure 2.7.

Fig. 6 Algorithm for constructing trial individuals

Taking each jump individual Y_w iteratively from H, the following set of procedures can be applied recursively.

Take A and B, where A is initialised to the permutative schedule $A = \{1, 2, \ldots, N\}$ and B is the complement of individual Y_w relative to A as given in Eq. (7).

$$B = A \backslash Y_w \tag{7}$$

If after the complement operation, B is an empty set without any elements; $B = \{\}$, then the individual is correct with a proper permutative schedule and does not require any repairment.

However, if B contains values, then these values are the missing elements in individual Y_w. The repairment procedure is now outlined. The first process is to randomise the positions of the elements in set B. Then, iterating through the elements $y_{w,j}$ $(j = 1, \ldots, N)$ in the individual Y_w, each position, where the element $y_{w,j} = 0$ is replaced by the value in B. Assigning B_{size} as the total number of elements present in B (and hence missing from the individual Y_w), the repairment procedure can be given as in Eq. (8).

$$y_{w,j} = \begin{cases} B_h & \text{if } y_{w,j} = 0 \\ y_{w,j} & \text{otherwise} \end{cases} \tag{8}$$
$$h = 1, \ldots, B_{size};\ j = 1, \ldots, N$$

Pseudocode for Repairing Trial Individuals

Take the temporary population H and its associated fitness array γ of size J_{min}. Assume two schedules A and B of maximum size N, where A is initialised to $A = \{1, 2, \ldots, N\}$. The best new trial individual is represented as Y_{best}, with its fitness $\Im\,(Y_{best})$.

1. For $w = 1, 2, \ldots, J_{min}$ do the following:
   a. $B = A \backslash Y_w$
   b. **IF** $B = \{\}$
      Randomise the elements in B.
      i. For $j = 1, 2, \ldots, N$ do the following:
         A. SET index $h = 1$
         B. **IF** $y_{w,j} == 0$
            $y_{w,j} = B_h$
            $h = h + 1$
   c. Evaluate the fitness of the new trial individual as:
      $\gamma_w = \Im(Y_w)$
   d. **IF** $w == 1$
      $Y_{best} = Y_w$
   e. **ELSE IF** $\gamma_w < Y_{best}$
      $Y_{best} = Y_w$

Fig. 7 Pseudocode for repairing trial individuals

After each individual is repaired in H, it is then evaluated for its fitness value as in Eq. (9) and stored in γ, the fitness array of size $J_{\min}$.

$$\gamma_w = \Im(Y_w), \quad w = 1, \ldots, J_{\min} \tag{9}$$

The pseudocode for repairing trial individuals is given in Fig. 7.

7 Population Update

2 Opt local search is applied to the best individual Y_{best} obtained with the minimum fitness value $(\min(\gamma_w))$. After the local search routine, the new individual is compared with the fitness of the incumbent individual X_i^{t-1}, and if it improves on the fitness, then the new individual is accepted in the population (10).

$$X_i^t = \begin{cases} Y_{best} & \text{if } \Im(Y_{best}) < C_i^{t-1} \\ X_i^{t-1} & \text{otherwise} \end{cases} \tag{10}$$

If this individual improves on the overall best individual in the population, it then replaces the best individual in the population (11).

$$X_{best}^{t} = \begin{cases} Y_{best} & \text{if } \Im(Y_{best}) < C_{best}^{t} \\ X_{best}^{t-1} & \text{otherwise} \end{cases} \tag{11}$$

8 Iteration

Sequentially, incrementing i, the population counter by 1, another individual X_{i+1}^{t-1} is selected from the population, and it begins its own sampling towards the designated leader X_{L}^{t-1} from Sects. 4–7. It should be noted that the leader does not change during the evaluation of one migration.

9 Migrations

Once all the individuals have executed their sampling towards the designated leader, the migration counter t is incremented by 1. The individual iterator i is reset to 1 (the beginning of the population) and the loop in Sects. 4–8 is re-initiated.

10 2 Opt Local Search

The local search utilised in DSOMA is the 2 Opt local search algorithm. The reason as to why the 2 Opt local search was chosen, is that it is the simplest in the k-opt class of routines. As the DSOMA sampling occurs between two individuals in k-dimension, the local search refines the individual. This in turn provides a better probability to find a new leader after each jump sequence. The placement of the local search was refined heuristically during experimentation.

The complexity of this local search is $O(n^2)$. As local search makes up the majority of the complexity time of DSOMA, the overall computational complexity of DSOMA for a single migration is $O(n^3)$.

A schematic of the DSOMA routine is given in Fig. 8, which graphically outlines the procedure for creating jump sequence between two individuals, and constructing trial individuals.

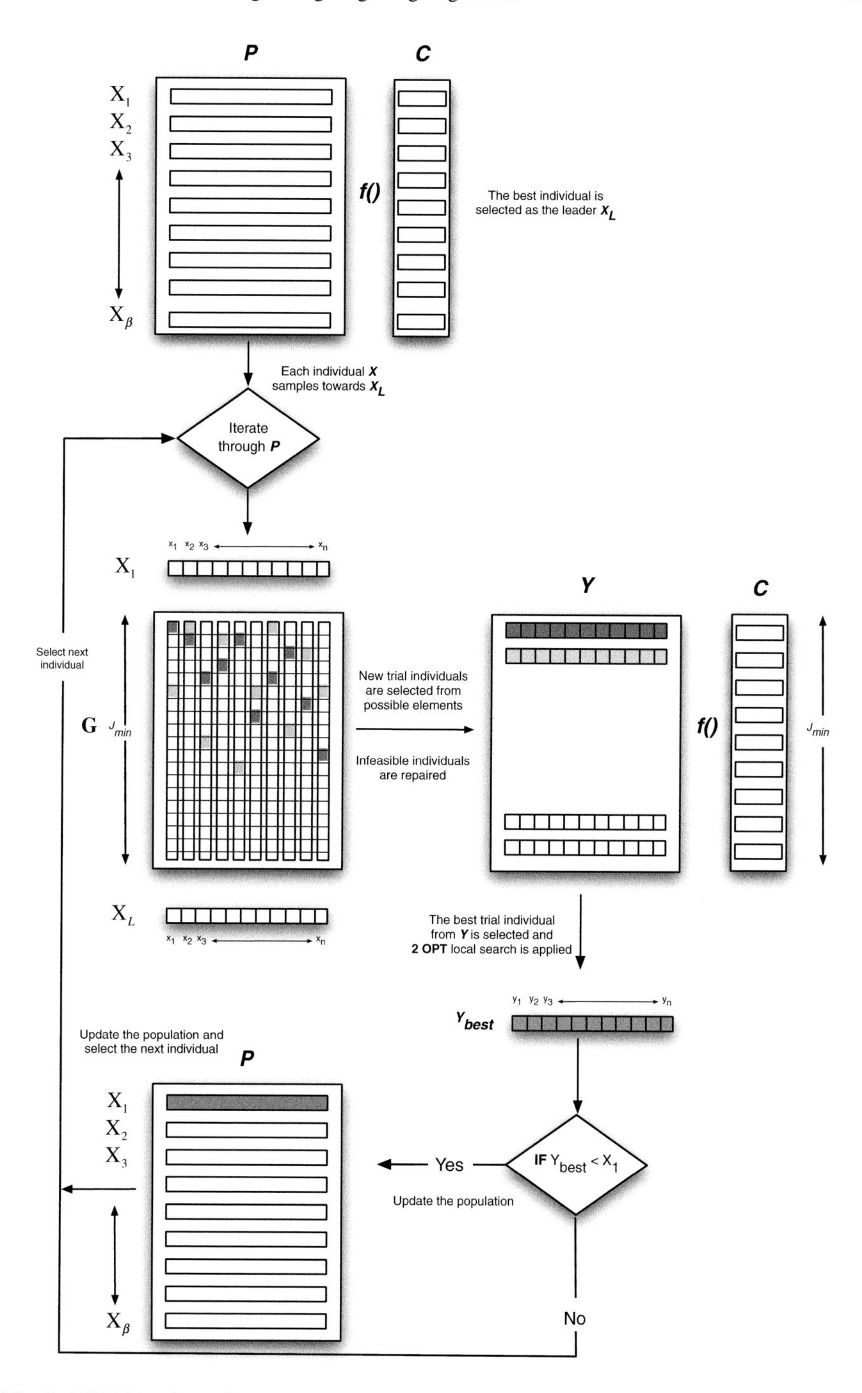

Fig. 8 DSOMA schematic

11 Conclusion

The discrete Self Organising Migrating Algorithm is described in this chapter. Using specific sampling of individuals in the search space, new individuals are constructed which assists in driving the population towards the leader.

DSOMA has proven successful in solving problem in the combinatorial optimisation domain. As a relatively new algorithm, it has a lot of scope for further development and refinement, especially in individual sampling and parallelisation aspects.

Acknowledgments The following grants are acknowledged for the financial support provided for this research: Grant Agency of the Czech Republic—GACR P103/15/06700S, VSB SGS grants of SP2015/141 and SP2015/142, research project NPU I No. MSMT-7778/2014 by the Ministry of Education of the Czech Republic, European Regional Development Fund under the Project CEBIA-Tech No. CZ.1.05/2.1.00/03.0089 and partially by the Internal Grant Agency of Tomas Bata University under the project No. IGA/FAI/2015/057.

References

1. Askarzadeh, A.: A discrete chaotic harmony search-based simulated annealing algorithm for optimum design of pv/wind hybrid system. Sol. Energy **97**(0), 93–101 (2013)
2. Davendra, D., Zelinka, I.: Optimization of quadratic assignment problem using self-organising migrating algorithm. Comput. Inform. **28**, 169–180 (2009)
3. Davendra, D., Zelinka, I., Bialic-Davendra, M., Senkerik, R., Jasek, R.: Discrete self-organising migrating algorithm for flow-shop scheduling with no-wait makespan. Math. Comput. Modell. **57**(12), 100–110 (2013) (Mathematical and Computer Modelling in Power Control and Optimization)
4. Drezne, Z.: A new genetic algorithm for the quadratic assignment problem. INFORMS J. Comput. **115**, 320–330 (2003)
5. Gambardella, L., Taillard, E., Dorigo, M.: Ant colonies for the quadratic assignment problem. Int. J. Oper. Res. **50**, 167–176 (1999)
6. Gao, K., Suganthan, P., Pan, Q., Chua, T., Cai, T., Chong, C.: Pareto-based grouping discrete harmony search algorithm for multi-objective flexible job shop scheduling. Inf. Sci. **289**, 76–90 (2014)
7. Institute, C.M.: Millennium problems (2015). http://www.claymath.org/millennium-problems
8. Lin, F., Kao, C.: Hsu: applying the genetic approach to simulated annealing in solving np-hard problems. IEEE Trans. Syst. Man Cybern. B Cybern. **23**, 1752–1767 (1993)
9. Onwubolu, G., Davendra, D.: Scheduling flow shops using differential evolution algorithm. Euro. J. Oper. Res. **171**, 674–679 (2006)
10. Onwubolu, G., Davendra, D.: Differential evolution: a handbook for global permutation-based combinatorial optimization. Springer, Germany (2009)
11. Pan, Q., Tasgetiren, M., Liang, Y.: A discrete particle swarm optimization algorithm for the no-wait flowshop scheduling problem. Comput. Oper. Res. **35**(9), 2807–2839 (2008)
12. Pan, Q.K., Wang, L., Li, J.Q., Duan, J.H.: A novel discrete artificial bee colony algorithm for the hybrid flowshop scheduling problem with makespan minimisation. Omega **45**, 42–56 (2014)
13. Reeves, C.: A genetic algorithm for flowshop sequencing. Comput. Oper. Res. **22**, 5–13 (1995)

14. Taillard, E.: Robust taboo search for the quadratic assignment problem. Parallel Comput. **17**, 443–455 (1991)
15. Tasgetiren, M., Sevkli, M., Liang, Y.C., Gencyilmaz, G.: Particle swamp optimization algorithm for permutative flowshops sequencing problems. In: 4th International Workshops on Ant Algorithms and Swamp Intelligence, pp. 389–390. Brussel, Belgium (2004)
16. Tasgetiren, M.F., Pan, Q.K., Suganthan, P., Chen, A.H.L.: A discrete artificial bee colony algorithm for the total flowtime minimization in permutation flow shops. Inf. Sci. **181**(16), 3459–3475 (2011)

Part II
Implementation

SOMA and Strange Dynamics

Ivan Zelinka

Abstract This chapter discusses the basic relations between of Self-Organizing Migrating Algorithm and complex systems that are sources of the so called strange dynamics. Interaction between SOMA and complex systems is considered from two pints of views. In the first one we are focused on chaos control, synthesis and identification. In the second one is used chaotic dynamics instead of pseudorandom number generator in order to improve SOMA performance. A few experiments with fractals, that are a part of complex systems, are introduced here too. All mentioned SOMA use is fully referenced for detailed reading and further research.

1 Introduction

Deterministic chaos and evolutionary algorithms seem to be two different areas of research that are not joined together, but this is not in fact, true. Both areas are tightly joined, as it is discussed in this chapter. Because discussed topics are wide enough and it is not possible to discuss all in the limited space of this chapter, only main ideas are reported here. For detailed study it is recommended to use references provided in the chapter. To better understand background and consequently discussed mutual relations between chaos and evolutionary dynamics, that is in fact discrete dynamical system, a brief overview of evolutionary algorithms is done here. For more detailed text about evolutionary techniques it is recommended to read [1, 2].

I. Zelinka (✉)
Department of Computer Science, Faculty of Electrical Engineering and Computer Science, VSB-Technical University of Ostrava, 17. listopadu 15, 708 33 Ostrava-Poruba, Czech Republic
e-mail: ivan.zelinka@vsb.cz

D. Davendra and I. Zelinka (eds.), *Self-Organizing Migrating Algorithm*,
Studies in Computational Intelligence 626, DOI 10.1007/978-3-319-28161-2_3

2 SOMA and Chaos

Deterministic chaos, discovered by Lorenz [3] is a fairly active area of research in last few decades. The Lorenz system produces one of the well-known canonical chaotic attractors in a simple three-dimensional autonomous system of ordinary differential equations [3, 4]. For discrete chaos, there is another famous chaotic system, called logistic equation [5]. Logistic equation is based on a predator-prey model showing chaotic behavior. This simple model is widely used in the study of chaos, where other similar models exist (canonical logistic equation [6] and 1D or 2D coupled map lattices [7]) based on it. Since then, a large set of nonlinear systems that can produce chaotic behavior have been observed and analyzed. Chaotic systems thus have become a vitally important part of science and engineering in theoretical as well as practical levels of research. The most interesting and applicable notions are, for example, chaos control and chaos synchronization related to secure communication, amongst others. Recently, the study of chaos is focused not only along the traditional trends but also on the understanding and analyzing principles, with the new intention of controlling and utilizing chaos as demonstrated in [8, 9]. The term chaos control was first coined by Ott, Grebogi and Yorke in 1990. It represents a process in which a control law is derived and used so that the original chaotic behavior can be stabilized on a constant level of output value or a n-periodic cycle. Since the first experiment of chaos control, many control methods have been developed and some are based on the first approach [10], including pole placement [11, 12] and delay feedback [13, 14]. Another research has been done on CML control by [15], special feedback methods for controlling spatio-temporal on-off intermittency has been used there and [15]. In [7] is summarized control of CML systems while in the [2, 16, 17] use of EAs on CML control. Many methods were adapted for the so-called spatiotemporal chaos represented by coupled map lattices (CML). Control laws derived for CML are usually based on existing system structures [7], or using an external observer [18]. Evolutionary approach for control was also successfully developed in, for example [16, 19, 20]. Many published methods of deterministic chaos control (DCC) were (originally developed for classic DCC) adapted for so called spatiotemporal chaos represented by CML. Models of this kind are based on set of spatiotemporal (for 1D, Fig. 5 and its fitness landscape Fig. 6) or spatial cells which represents appropriate state of system elements. Typical example is CML based on so called logistic equation [5, 18, 21], which is used to simulate behavior of system which consists of n mutually joined cells via nonlinear coupling, usually noted like ε, see [7].

2.1 Chaos Synthesis

In recent years, interests in softcomputing methods are increasing, including in particular evolutionary algorithms. These algorithms are based on similar principles of biological evolution in the real world as already mentioned. One of the typical EAs use is to solve computationally hard problems which are too complex to be solved by conventional methods. In its canonical form, EAs can be used only for numerical estimation of parameters (usually, arguments of a given cost function). Together with EAs in the canonical form, another modification allows to use EAs as symbolic “constructors”, i.e., a processor, for synthesizing complex structures in a symbolic way, based on some predefined simple elements (mathematical operators or electronic elements like diode, transistor, etc.). The term “symbolic way” specifies that mathematical structures and equations, electronic systems, etc., are generated from those simple elements just mentioned.

Given the above background in [22–24], the main motivation of investigation was the question “*Is it possible to synthesize the mathematical description of a new chaotic system, based on simple and elementary mathematical objects, by means of evolutionary computation?*” This question was also based partially on the fact that in engineering applications, it is very often vitally important to know not only when chaos can be generated but also how to generate it [8, 25]. This is extremely important in cryptography, for example, where chaotic systems are often used in the design. From a mathematical point of view, it is quite clear that there are some classes of chaotic systems which can be represented by one canonical form (one class—one canonical form) [6]. However, generally speaking, it is not so easy to exactly synthesize a chaotic system with specified features by means of classical mathematical methods. A positive answer to the question mentioned above would open possibilities to synthesize not only a set of not-yet-described chaotic systems, but also some chaotic systems with predefined features. It is believed that such possibilities would have an important impact on engineering design of various complex nonlinear systems, especially chaotic systems. The main ides of evolutionary synthesis is that in the evolutionary process are used like individuals basic building blocks (i.e. mathematical objects like $+, -, /, x, \ldots$) that are used in chaotic systems. Evolution then synthesize under user defined criteria systems, that exhibit chaos. Selected examples of evolutionary synthesized chaotic systems from [2] are depicted in Figs. 1 and 2.

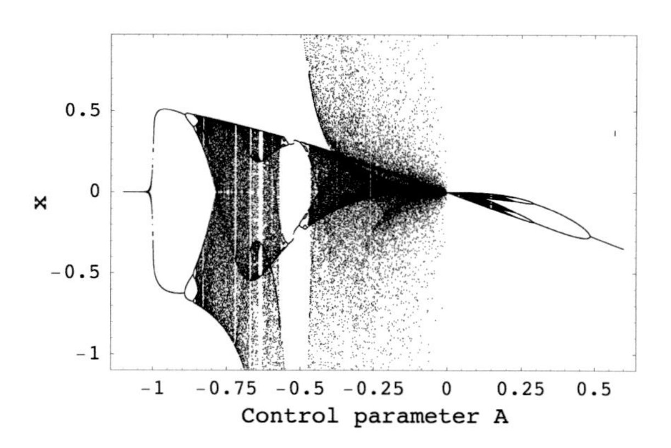

Fig. 1 Bifurcation diagram of $x_{n+1} = \frac{x}{\frac{x^3}{2A^3(x-A)(A+x)} + A}$

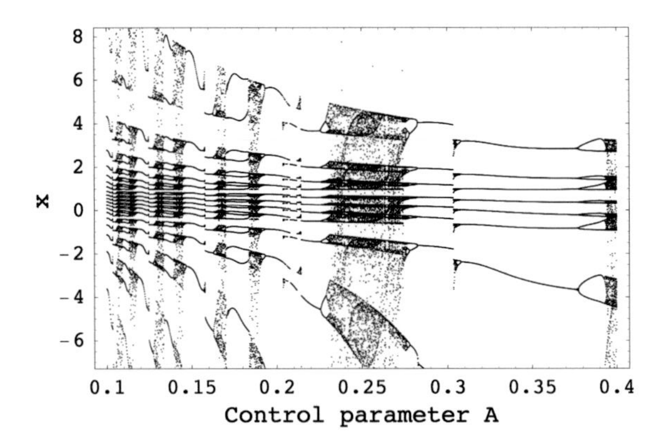

Fig. 2 Bifurcation diagram of $x_{n+1} = \frac{\frac{A^2(A-x)}{x(A(-x)+A+2x)} - 2A + x - 1}{x}$

2.2 *Chaos Control*

Since now, it has outlined how can be used and/or used chaos inside evolutionary algorithms. Opposite side of topic discussed above is use of evolutions on chaotic systems. Few important selected topics are discussed here such as the evolutionary control of deterministic chaos, evolutionary control of CML systems, evolutionary synthesis of chaotic systems, evolutionary identification and synchronization of chaotic systems. At the end evolutionary algorithms as a discrete systems with complex dynamics are discussed.

Generally said the evolutionary algorithms (EA) are known as powerful tool for almost any difficult and complex optimization problem. But the quality of optimization process results mostly depends on proper design of used cost function, especially when the EAs are used for optimization of chaos control. The results of numerous simulations lend weight to the argument that deterministic chaos in general and also any technique to control of chaos are sensitive to proper parameter set up, initial conditions and in the case of optimization they are also extremely sensitive to the construction of used cost function.

The main aim of chaos control by EAs is focused on the EA implementation to methods for chaos control for the purpose of obtaining better results, which means faster reaching of desired stable state and superior stabilization, which could be robust and effective to optimize difficult problems in the real world. In other words use EAs on control deals with an investigation on the optimization of the control of chaos by means of EA and constructing of the cost function securing the improvement of system behavior and faster stabilization to desired periodic orbits. The control law can be based on various methods as for example on two Pyragas methods [2, 26–29]: Delay feedback control—TDAS and Extended delay feedback control—ETDAS. As models of deterministic chaotic systems, one dimensional Logistic equation and two dimensional Henon map were used. The evolutionary algorithm SOMA and DE were used. Also the comparison with classical control technique—OGY is presented.

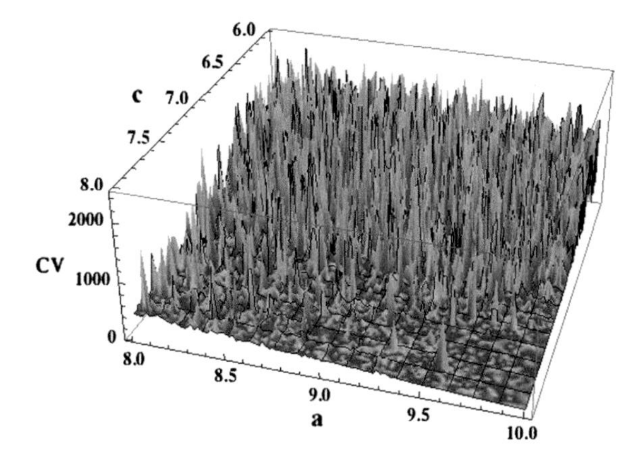

Fig. 3 Cost-function surface representing problem of synchronization

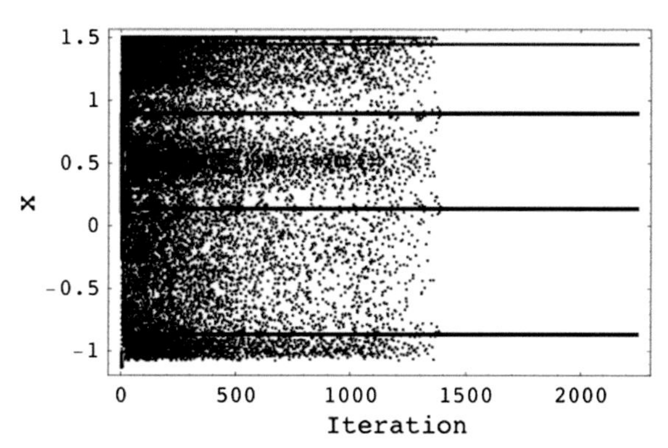

Fig. 4 Best individual solution, p-4 orbit

Some research in this field has been recently done using the evolutionary algorithms for optimization of local control of chaos based on a Lyapunov approach [19, 30]. But the approach described here is unique and novel and up to date it was not used or mentioned anywhere. We use EA to search for optimal setting of adjustable parameters of arbitrary control method to reach desired state or behavior of chaotic system. The complexity of such process is visible in Fig. 3 in which space of all possible solutions (x, y) and their quality (z) are depicted. As an example of successful chaos control is in Fig. 4, see for example [29]. In this case has been stabilized chaotic system (Henon) in four periodic orbit. In the figure is captured 100 repeated experiments and as it is visible, EAs successfully stabilized system after 1400 iterations (in the worst case). Usually solution discovered by EAs has been faster that classical control techniques [2] (Figs. 5 and 6).

2.3 Chaos Identification

Evolutionary algorithms can be also used for identification (reconstruction) of mathematical description of chaotic systems. This topic is widely discussed in [2,

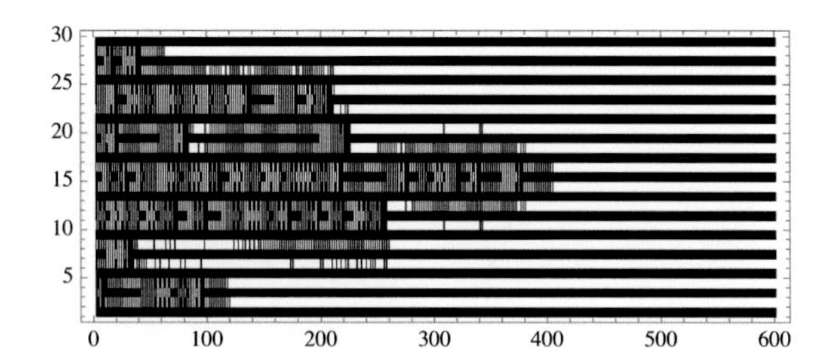

Fig. 5 CML T_1S_2 in configuration 30 × 600—stabilization after 400 iterations is visible

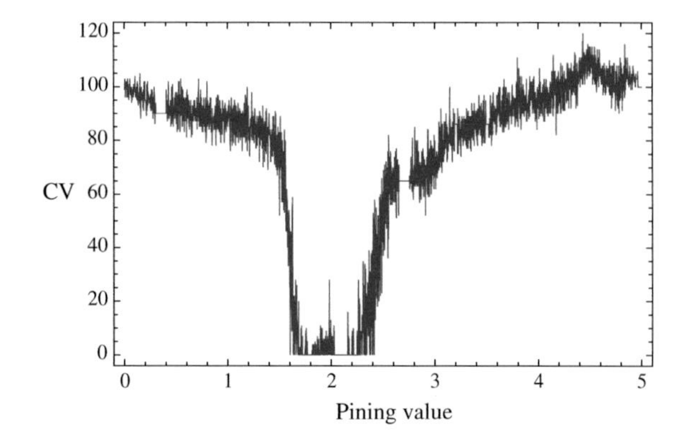

Fig. 6 CML landscape for T_1S_2 in configuration 30 × 600. Comparing with another landscapes like with T_1S_1 is this much more complex

31]. In this paper we discuss the possibility of using evolutionary algorithms for the reconstruction of chaotic systems. The main aim of this work is to show that evolutionary algorithms are capable of the reconstruction of chaotic systems without any partial knowledge of internal structure, i.e. based only on measured data and predefined set of basic mathematical objects. Algorithm such as SOMA [32] and differential evolution [33] were used in reported experiments here. Systems selected for numerical experiments in [31] are the well-known Lorenz system, Simplest Quadratic Flow, Double Scroll, Damped Driven Pendulum and Nosé—Hoover Oscilator. According to obtained results it can be stated that evolutionary reconstruction is an alternative and promising way as to how to identify chaotic systems. Identification of various dynamical systems is vitally important in the case of practical applications as well as in theory. A rich set of various methods for dynamical system identification has been developed in the past. In the case of chaotic dynamics, it is for example the well-known reconstruction of a chaotic attractor based on research of [34] which has shown that, after the transients have died out, one can reconstruct the trajectory of the attractor from the measurement of a single component. Since the entire trajectory contains a large amount of information, the series of papers by [35, 36] is introduced to show a set of averaged coordinate invariant numbers (generalized dimensions, entropies, and scaling indices) by which

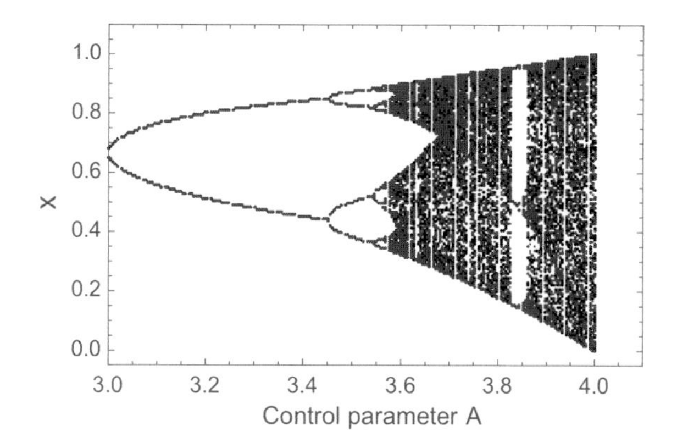

Fig. 7 Cost-function surface representing problem of synchronization

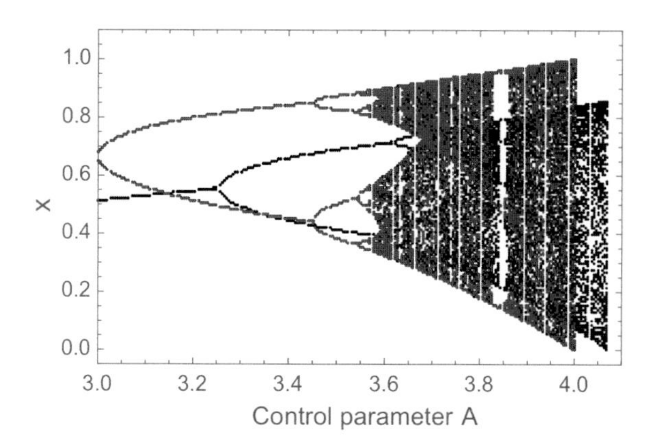

Fig. 8 Best individual solution, p-4 orbit

different strange attractors can be distinguished. The method presented in this research is based on Evolutionary Algorithm's (EAs), see [1], which allows the reconstruction not only of chaotic attractors as a geometrical objects, but also their mathematical description, based on methods of symbolic regression [22–24]. It is recommended to study those papers like [31], see Figs. 7, 8, 9 and 10.

2.4 *SOMA Powered by Pseudorandom, Chaos and Deterministic Dynamics*

It is well known that evolutionary algorithms use pseudorandom numbers generators. They need them for example to generate the first population, they are necessary in crossing or perturbation process etc. In paper [37] chaos attractors Arnold Cat Map and Sinai are used as chaotic numbers generators. The main goal was to investigate if they are usable instead of pseudorandom number generators and their

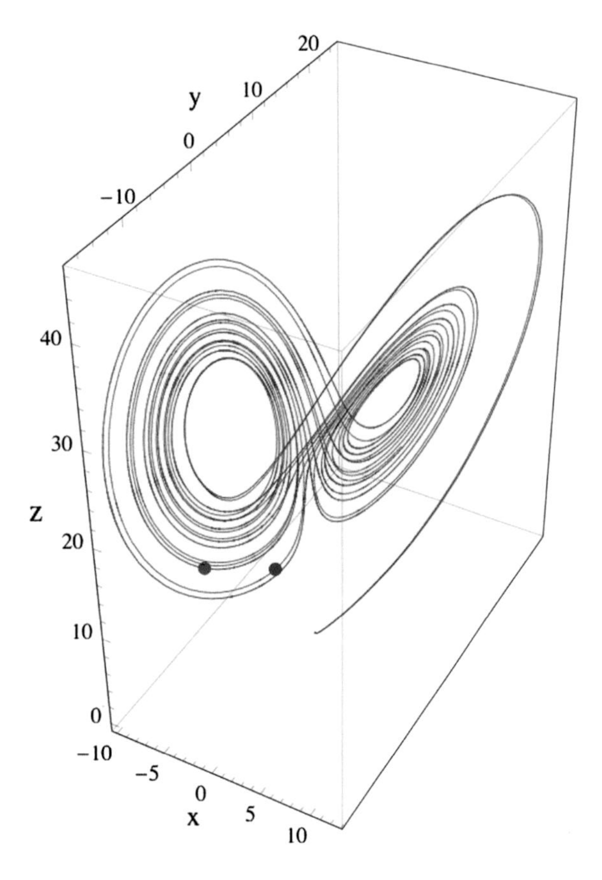

Fig. 9 SOMA identification of the Lorenz attractor

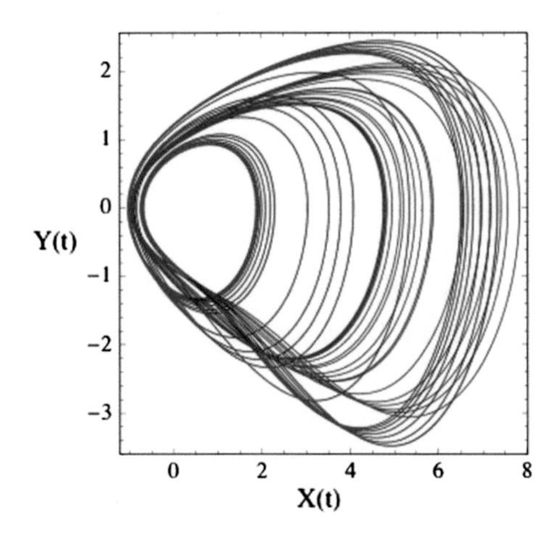

Fig. 10 SOMA identification of the Rössler attractor

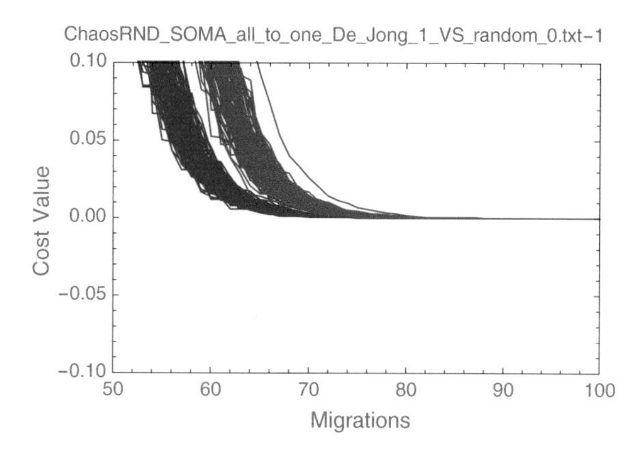

Fig. 11 SOMA powered by chaotic dynamics (*left* flow) and by MT PRNGs (*right* flow); 100× repeated solutions). The 1st DeJong's function

influence on the algorithm convergence's speed. In [37] were used together SOMA and DE for mutual comparison.

Another future research is reported in [38]. It is focused on what are relations between chaos level (measured by Lyapunov exponent) and its impact on EAs performance. In this paper was control parameter A of logistic equation varied and logistic equation then used instead of PRNG under this setting. It was found that for some settings EAs gave much more better results than with Mersenne twister generator, see Figs. 11 and 12.

Figures 11 and 12 show comparing of results of MT and Chaos random number generators. It is known that if $a = 4$ logistic equation will generate numbers from all interval [0, 1]. This fact is important for evolution process. As it is obvious, evolutions where chaos pseudo-random number generator has been used, converge faster than evolutions, where MT has been used. On the other hand when $a = 3.58$ and 1st de Jong's and Schwefel's functions have been testing function, see Table 1.

As already mentioned, inherent part of evolutionary algorithms that are based on Darwin theory of evolution and Mendel theory of genetic heritage, are random

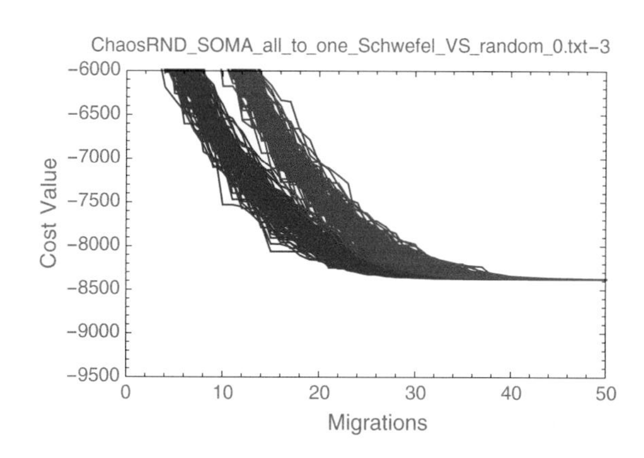

Fig. 12 SOMA powered by chaotic dynamics (*left* flow) and by MT PRNGs (*right* flow); 100× repeated solutions). The Schwefel's function

Table 1 Minimum, maximu and average fitness value of SOMA experiments for 1st de Jong's, Ranna's and Schwefel's function with settings mentioned in [38]

	Function	Min. fitness value	Max. fitness value	Average fitness value
Chaos $a = 3.58$	1st de Jong's	0.000	0.000	0.000
	Ranna's	−8588.349	−6946.222	−7716.781
	Schwefel's	−8379.658	−6543.776	−7277.547
Chaos $a = 3.828$	1st de Jong's	0.000	0.000	0.000
	Ranna's	−9180.587	−8415.158	−8857.963
	Schwefel's	−8379.658	−8379.658	−8379.658
Chaos $a = 3.855$	1st de Jong's	0.000	0.000	0.000
	Ranna's	−9110.511	−8435.843	−8830.062
	Schwefel's	−8379.658	−8379.658	−8379.658
Chaos $a = 3.8567$	1st de Jong's	0.000	0.000	0.000
	Ranna's	−9264.553	−8535.640	−8887.881
	Schwefel's	−8379.658	−8379.657	−8379.658
Chaos $a = 4$	1st de Jong's	0.000	0.000	0.000
	Ranna's	−9258.219	−8947.286	−9090.472
	Schwefel's	−8379.658	−8379.658	−8379.658
MT	1st de Jong's	0.000	0.000	0.000
	Ranna's	−9139.584	−8667.654	−8866.831
	Schwefel's	−8379.658	8379.658	8379.658

processes since genetic algorithms and evolutionary strategies use. In [39] is presented extended experiments of selected evolutionary algorithms and test functions showing whether random processes really are needed in evolutionary algorithms. In this experiments were used differential evolution and SOMA algorithms with functions 2nd DeJong, Ackley, Griewangk, Rastrigin, SineWave and StretchedSineWave. Use n periodical deterministic processes (based on deterministic chaos principles) instead of pseudorandom number generators has been done there and then compared performance of evolutionary algorithms powered by those processes and by pseudorandom number generators. Results presented are numerical demonstration rather than mathematical proofs. In [39] is proposed hypothesis that certain class of deterministic processes can be used instead of pseudorandom number generators without lowering the performance of evolutionary algorithms.

An advantage of the proposed use of deterministic processes is the fact that in such case it is possible to fully repeat runs of given algorithm, analyze its behavior deterministically, including its full path on searched fitness landscape. We also believe that mathematical proofs can be in such case more easily constructed for such class of evolutionary algorithms.

Despite the widely presumed fact that pseudorandom number generators (also for evolutionary algorithms use) has to have as big period as possible (for example Mersenne twister with $2^{19937} - 1$) and such as the 2^{32} common in many software packages, in [39] was shown here that deterministic periodical processes with period 10-35200 is enough for successful experiments reported there.

3 SOMA and Fractal Geometry

Inverse fractal problem is process during which are identified so called coefficients of affined transformations or coefficients of escape time algorithm (ETA) algorithm, which is complementary to IFS algorithm [40] as was mentioned before. Fractal reconstruction is one of well-known problems of fractal geometry and was solved not only by evolutionary algorithms. Inverse fractal problem is not only artificial problem, but its solution can help in time series processing (see fractal interpolation of time series in [40]) or in part of artificial intelligence, so called computer vision. In the computer vision fractal geometry can be used in object description. Such kind of description is not big in point of view of data size and is exact—each pixel position is given by coefficients of affined transformations [40–43].

Algorithms like differential evolution [33] and SOMA [32] were used for solution of IFP in [44]. For these purposes here was defined cost function whose minimization produced coefficients of affined transformations. For IFP was chosen ETA algorithm in grid resolution 100×100 cells. Rule of color association of original ETA was modified here so that recalculated cell remained black if color was calculated like black otherwise the color was white. In this way usually colored fractals were transformed into black/white versions. Thanks to this transformation black cells were associated to 1 and white cells to 0. It represents original and actually identified elements of matrix (cells of fractal picture). This formula calculate simply sum of cells in which both fractals differs. Minimization of this cost function (in ideal case 0) has lead to the optimal coefficients, which should be the same like coefficients of original fractal. For each combination of coefficients the cost function return total sum—scalar.

Identification of fractal object was done by differential evolution and SOMA so that there was generated population of individuals of size = number of searched coefficients [44]. For Mandelbrot set individuals of size = 2 (a, b see Eq. 1). Also another fractals has been identified like the Vortex and Spider, see [44]. In the case

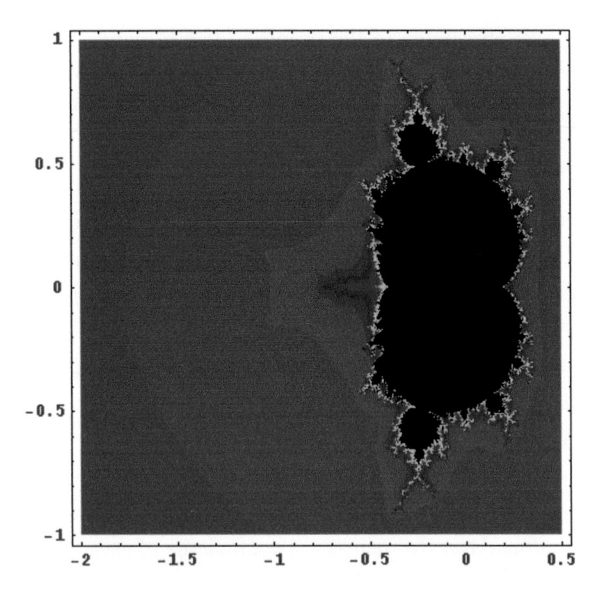

Fig. 13 Inverse fractal problem of Mandelbrot set, the best fractal estimation on the start

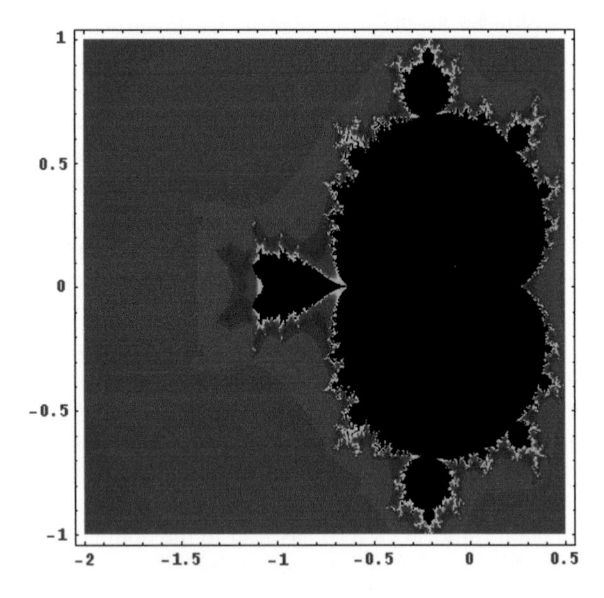

Fig. 14 Inverse fractal problem of Mandelbrot set, the best fractal estimation during the evolution

of Mandelbrot set, graphically visualizes evolution of fractals, leading to the IFP solution is depicted in Figs. 13, 14, 15 and 16.

For full report and detail, it is recommended to read [44].

$$z_{n+1} = az_n^b + c \quad where\ z\ and\ c \in\ C\ and\ a, b \in \langle 0, 3\rangle \tag{1}$$

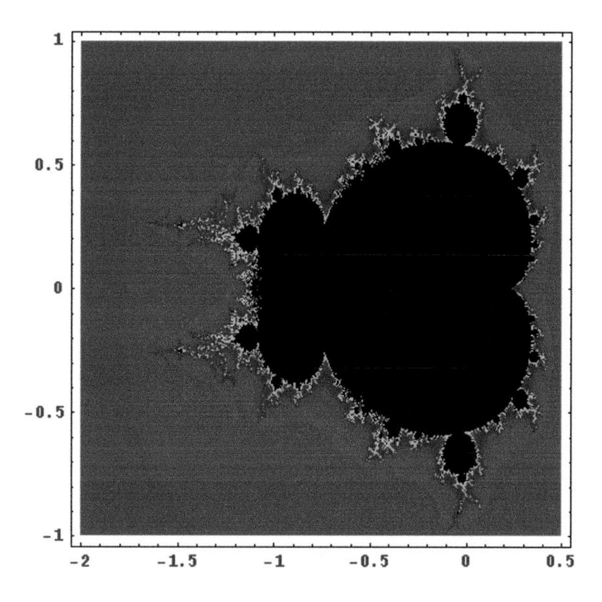

Fig. 15 Inverse fractal problem of Mandelbrot set, the best fractal estimation during the evolution

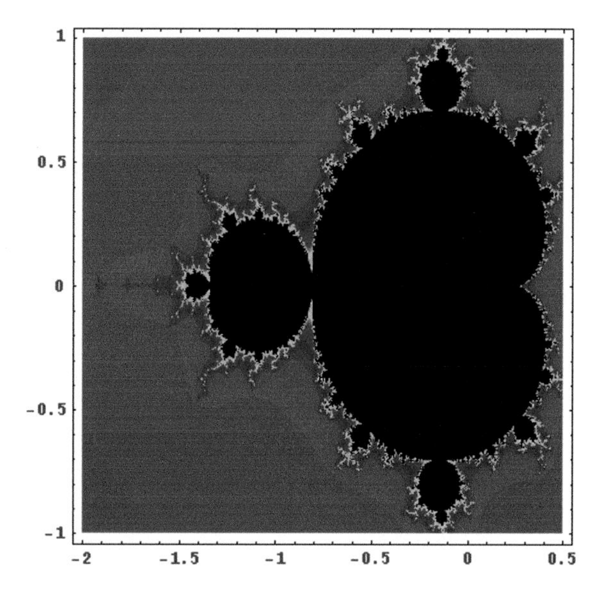

Fig. 16 Inverse fractal problem of Mandelbrot set, the best fractal estimation at the end

4 SOMA Dynamics as a Complex Networks

The main aim of this SOMA research reported in [45–52], is to show how to convert EAs dynamics into complex network then interpret tools of complex networks analysis on complex networks that are given by evolutionary dynamics. Reason for this is that today various techniques for analysis and control of complex networks exist. If complex network structure is hidden behind EA dynamics, then we believe and we have numerically demonstrated, that for example above mentioned control techniques could be used to improve dynamics and performance of used EAs. The first steps (i.e. conversion of the EA dynamics to the complex network and to CML system) have been done in the [45–52]. Now we propose how standard tools of complex networks analysis can be understood for EAs dynamics purposes. It has been successfully reported in [45–52] where we discuss conversion of selected EA to complex network and consequently to CML, its analysis and EA performance improvement based on network attributes like degree centrality etc.

5 Conclusion

In this chapter were discussed mutual relations between SOMA algorithm and its dynamics and nontrivial dynamics as chaos for example is. It has been introduced our results on the field of chaos control, chaos synthesis, chaos identification, as well as on chaos use instead of PRNGs inside SOMA with constant setting and also varying of chaos level with measurement of Lyapunov exponent. At the end was also mentioned so called inverse fractal problem, solved by SOMA. All reported results here show, that all those applications and hybridizations are useful and promising.

Acknowledgment The following grants are acknowledged for the financial support provided for this research: Grant Agency of the Czech Republic—GACR P103/15/06700S, by the SP2015/142.

References

1. Back, T., Fogel, D.B., Michalewicz, Z.: Handbook of Evolutionary Computation. Institute of Physics, London (1997)
2. Zelinka, I., Celikovsky, S., Richter, H., Chen, G.: Evolutionary Algorithms and Chaotic systems. Springer, Berlin (2010)
3. Lorenz, E.N.: Deterministic nonperiodic flow. J. Atmos. Sci. **20**(2), 130–141 (1963)
4. Stewart, I.: The Lorenz attractor exists. Nature **406**, 948–949 (2000)
5. May, R.: Simple mathematical model with very complicated dynamics. Nature **261**, 45–67 (1976)
6. Gilmore, R., Lefranc, M.: The Topology of Chaos: Alice in Stretch and Squeezeland. Wiley, New York (2002)
7. Schuster, H.G.: Handbook of Chaos Control. Wiley, New York (1999)
8. Chen, G., Dong, X.: From Chaos to Order: Methodologies, Perspectives and Applications. World Scientific, Singapore (1998)
9. Wang, X., Chen, G.: Chaotification via arbitrarily small feedback controls: theory, method, and applications. Int. J. Bifur. Chaos **10**, 549–570 (2000)
10. Ott, E., Grebogi, C., Yorke, J.A.: Controlling chaos. Phys. Rev. Lett. **64**, 1196–1199 (1990)
11. Grebogi, C., Lai, Y.C.: Controlling chaos. In: Schuster, H.G. (ed.) Handbook of Chaos Control. Wiley, New York (1999)

12. Zou, Y.-L., Luo, X.-S., Chen, G.: Pole placement method of controlling chaos in DC–DC buck converters. Chinese Phys. **15**, 1719–1724 (2006)
13. Just, W.: Principles of time delayed feedback control. In: Schuster, H.G. (ed.) Handbook of Chaos Control. Wiley, New York (1999)
14. Just, W., Benner, H., Reibold, E.: Theoretical and experimental aspects of chaos control by time-delayed feedback. Chaos **13**, 259–266 (2003)
15. Deilami, M.Z., Rahmani, C.Z., Motlagh, M.R.J.: Control of spatio-temporal on–off intermittency in random driving diffusively coupled map lattices. Chaos, Solitons, Fractals (2007). Available online 21 Dec 2007
16. Zelinka, I.: Investigation on realtime deterministic chaos control by means of evolutionary algorithms. In: Proceedings of First IFAC Conference on Analysis and Control of Chaotic Systems, pp. 211–217. Reims, France (2006)
17. Zelinka, I.: Real-time deterministic chaos control by means of selected evolutionary algorithms. Eng. Appl. Artif. Intell. (2008). doi:10.1016/j.engappai.2008.07.008
18. Chen, G.: Controlling Chaos and Bifurcations in Engineering Systems. CRC Press, Boca Raton (2000)
19. Richter, H., Reinschke, K.J.: Optimization of local control of chaos by an evolutionary algorithm. Physica D **144**, 309–334 (2000)
20. Richter, H.: An evolutionary algorithm for controlling chaos: the use of multi-objective fitness functions. In: Merelo Guervs, J.J., Panagiotis, A., Beyer, H.G., Fernndez Villacanas, J.L., Schwefel, H.P. (eds.) Parallel Problem Solving from Nature-PPSN VII, vol. 2439, pp. 308–317. Lecture Notes in Computer Science, Springer, Berlin, Heidelberg, New York (2002)
21. Hilborn, R.C.: Chaos and Nonlinear Dynamics. Oxford University Press, Oxford (1994). ISBN 0-19-508816-8
22. Zelinka, I., Davendra, D., Senkerik, R., Oplatkova, Z., Jasek, R.: Analytical programming-a novel approach for evolutionary synthesis of symbolic structures. In: Evolutionary Algorithms, InTech (2011). ISBN 978-953-307-171-8
23. Koza, J.R., Bennet, F.H., Andre, D., Keane, M.: Genetic Programming III. Morgan Kaufmann Publishers, Burlington (1999). ISBN 1-55860-543-6
24. O'Neill, M., Ryan, C.: Grammatical Evolution. Evolutionary Automatic Programming in an Arbitrary Language. Kluwer Academic Publishers, Berlin (2002). ISBN 1402074441
25. Perruquetti, W., Barbot, J.P.: Chaos in Automatic Control. CRC, Bota Raton (2005)
26. Senkerik, R., Zelinka, I., Davendra, D., Oplatkova, Z.: Utilization of SOMA and differential evolution for robust stabilization of chaotic logistic equation. Comput. Math Appl. **60**(4), 1026–1037 (2010)
27. Pyragas, K.: Control of chaos via extended delay feedback. Phys. Lett. **206**, 323–330 (1995)
28. Pyragas, K.: Continuous control of chaos by self-controlling feedback. Phys. Lett. **170**, 421–428 (1992)
29. Zelinka, I., Senkerik, R., Navratil, E.: Investigation on evolutionary optimitazion of chaos control. Chaos, Solitons Fractals (2007). doi:10.1016/j.chaos.2007.07.045
30. Richter, H.: An evolutionary algorithm for controlling chaos: the use of multi—objective fitness function, vol. 2439, pp. 308–320. Lecture Notes in Computer Science (2002)
31. Zelinka, I., Chadli, M., Davendra, D., Senkerik, R., Jasek, R.: An investigation on evolutionary reconstruction of continuous chaotic systems. Math. Comput. Model. **57**(1–2), 2–15 (2013). doi:10.1016/j.mcm.2011.06.034
32. Zelinka, I.: SOMA—Self organizing migrating algorithm. In: Onwubolu, G.C., Babu, B. (eds.) New Optimization Techniques in Engineering, pp. 167–218. Springer, New York (2004). ISBN 3-540-20167X
33. Price, K.: An introduction to differential evolution. In: Corne, D., Dorigo, M., Glover, F. (eds.) New Ideas in Optimisation, pp. 79–108. McGraw Hill, International (UK) (1999)
34. Takens, F.: Detecting strange attractors in turbulence, pp. 366–381. Lecture Notes in Mathematics (1981)
35. Halsey, T.C, Jensen, M.H., Kadanoff, L.P., Pro-caccia, I., Schraiman, B.I.: Fractal measures and their singularities: the characterization of strange sets. Phys. Rev. A **33**, 1141 (1986)

36. Eckmann, J.P., Procaccia, I.: Fluctuation of dynamical scaling indices in non-linear systems. Phys. Rev. A **34**, 659 (1986)
37. Skanderova, L., Zelinka, I.: Arnold cat map and sinai as chaotic numbers generators in evolutionary algorithms. In: AETA 2013: Recent Advances in Electrical Engineering and Related Sciences, vol. 282, pp. 381–389. Lecture Notes in Electrical Engineering (2014)
38. Skanderova, L., Zelinka, I., Saloun, P.: Chaos powered selected evolutionary algorithms. In: Proceedings of Nostradamus 2013: International Conference Prediction, Modeling and Analysis of Complex Systems, Springer Series: Advances in Intelligent Systems and Computing, vol. 210, pp. 111–124 (2013)
39. Zelinka, I., Lampinen, J., Senkerik, R., Pluhacek, M.: Investigation on evolutionary algorithms powered by nonrandom processes. Soft Comput. (2015). doi:10.1007/s00500-015-1689-2
40. Barnsley, M.F.: Fractals Everywhere. Academic Press Professional, New York (1993). ISBN 0-12 079061-0
41. Bunde, A., Shlomo, H.: Fractals and Disordered Systems. Springer, Berlin (1996). ISBN 3-540-56219-2
42. Hastings, H.M., Sugihara, G.: Fractals a User's Guide For The Natural Sciences. Oxford University Press, Oxford (1993). ISBN 0-19-854597-5
43. Peitgen, H.O., Jurgens, H., Saupe, D.: Chaos and Fractals, New Frontiers of Science. Springer, Berlin (1992). ISBN 3-540-97903-4
44. Zelinka, I.: Inverse fractal problem. In: Price, K., Storn, R.M., Lampinen, J.A. (eds.) Differential Evolution a Practical Approach to Global Optimization, Natural Computing Series, pp. 479–498. Springer, Berlin (2005)
45. Zelinka, I., Davendra, D., Chadli, M., Senkerik, R., Dao, T.T., Skanderova, L.: Evolutionary dynamics and complex networks. In: Zelinka, I., Snasel, V., Ajith, A. (eds.) Handbook of Optimization, p. 1100s. Springer, Germany (2012)
46. Zelinka, I., Davendra, D., Senkerik, R., Jasek, R.: Do evolutionary algorithm dynamics create complex network structures? Complex Syst. **20**(2), 127–140, 0891–2513 (2011)
47. Zelinka, I.: Mutual relations of evolutionary dynamics, deterministic chaos and complexity. In: Tutorial at IEEE Congress on Evolutionary Computation 2013, Mexico (2013)
48. Zelinka, I., Davendra, D., Lampinen, J., Senkerik, R., Pluhacek, M.: Evolutionary algorithms dynamics and its hidden complex network structures. In: Congress on Evolutionary Computation (CEC), 2014 IEEE Congress, pp. 3246–3251, 6–11 July 2014. doi:10.1109/CEC.2014.6900441
49. Metlicka, M., Davendra, D.: Chaos-driven discrete artificial bee colony. In: IEEE Congress on Evolutionary Computation, pp. 2947–2954 (2014)
50. Davendra, D., Zelinka, I., Metlicka, M., Senkerik, R., Pluhacek, M.: Complex network analysis of differential evolution algorithm applied to flowshop with no-wait problem. In: IEEE Symposium on Differential Evolution, pp. 65–72. Orlando, FL, USA, 9–12 Dec 2014
51. Davendra, D., Metlicka, M.: Ensemble centralities based adaptive artificial bee algorithm. In: IEEE Congress on Evolutionary Computation (2015)
52. Zelinka, I.: Evolutionary algorithms as a complex dynamical systems. In: Tutorial at IEEE Congress on Evolutionary Computation, Sendai (2015)

Multi-objective Self-organizing Migrating Algorithm

Petr Kadlec and Zbyněk Raida

Abstract Almost every optimization problem can be viewed as multi-objective one. Multi-objective problems with conflicting objectives lead to so called Pareto front which expresses trade-off among the objectives. Multi-objective techniques yield better understanding of the solved problem because resulting Pareto front expresses the balance between different objectives. In this chapter, fundamentals of multi-objective optimization are reviewed. Then, multi-objective optimization technique based on principle of self-organizing migration is described. The proposed method is able to solve unconstrained, constrained problems having any number of variables and objectives. The method is designed to find so called non-dominated set that covers the true Pareto front uniformly.

1 Introduction to Multi-objective Optimization

Almost every optimization problem can be viewed as a multi-objective one: an engineer can view the design from different angles. An individual fitness (or objective) function can be formulated for every objective. The considered design objectives can be either corresponding or conflicting. In the first case, the solved optimization task leads to a unique solution that achieves the best value of defined objective function. On the contrary, when two or more objectives are conflicting some trade-off solution has to be found. Set of all possible trade-off solutions is usually called as Pareto front. Vilfredo Pareto (1848–1923) was an Italian economist. He studied conflicting objectives in his works about economic efficiency and redistribution of incomes. So called Pareto efficiency expresses that improvement of the proposed solution in one objective leads to deterioration of at least one other

P. Kadlec (✉) · Z. Raida
Brno University of Technology, Antonínská 1, 602 00 Brno, Czech Republic
e-mail: kadlecp@feec.vutbr.cz

Z. Raida
e-mail: raida@feec.vutbr.cz

D. Davendra and I. Zelinka (eds.), *Self-Organizing Migrating Algorithm*,
Studies in Computational Intelligence 626, DOI 10.1007/978-3-319-28161-2_4

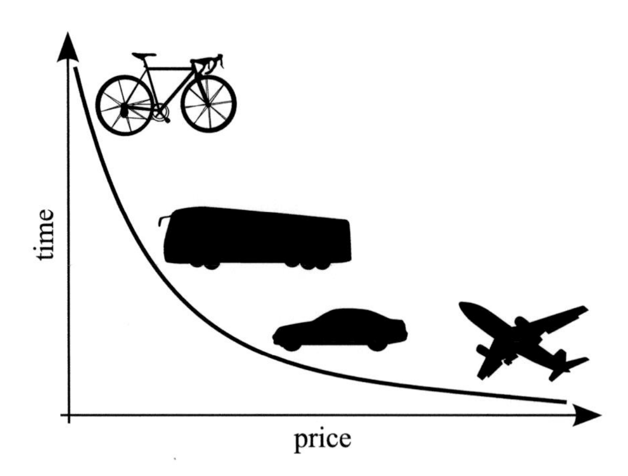

Fig. 1 Example of multi-objective problem: choosing a traveling option [12]

objective. Let us consider an example of choosing traveling method. We have four options: a plane, a car, a bus and a bike. Every option brings different traveling time and different price. If we want to optimize our traveling method from both these objectives at the same time, there is no obvious winner as indicated in Fig. 1. None of the options is better than the others in both the objectives. The cheapest variant is to go by bike but we have to accept that our trip will take the longest time. On the contrary, flying with a plane is the fastest but the most expensive variant. We can make some compromise and choose some trade-off solution: a car or a bus. In that case, we do not get the best possible performance in any of the objectives but probably satisfactory performance in all of them.

There are two options how to solve a multi-objective optimization problem. First, we can estimate the balance between objectives and assign individual objectives importance a priori. This means that one function aggregating all objectives with relative weight specified by the designer is composed and solved by a single-objective optimization technique. Second approach is based on searching for the Pareto front. Once it is known, the designer can see the trade off between all objectives and can choose the final solution with better understanding of the solved problem. Generally, MOOP (Multi-Objective Optimization Problem) is composed of a finite set of objective functions defined for common variables. All of these objectives are either maximization or minimization problems. Since maximization problem can be converted to minimization one by simply multiplying the objective function by −1, just the minimization problems will be considered in the rest of this text. The general definition of a MOOP is as follows:

$$\begin{array}{ll} \text{Minimize} & F_m(\vec{x}), \quad m = 1, 2, \ldots, M. \\ \text{subject to} & g_j(\vec{x}) \geq 0, \quad j = 1, 2, \ldots, J. \\ & x_{\min}^{(n)} \leq x^{(n)} \leq x_{\max}^{(n)}. \end{array} \tag{1}$$

where M denotes number of objectives, $\vec{x}$ stands for vector of decision variables for individual solutions and $x_{\min}^{(n)}$ and $x_{\max}^{(n)}$ are lower and upper bounds of the n-th

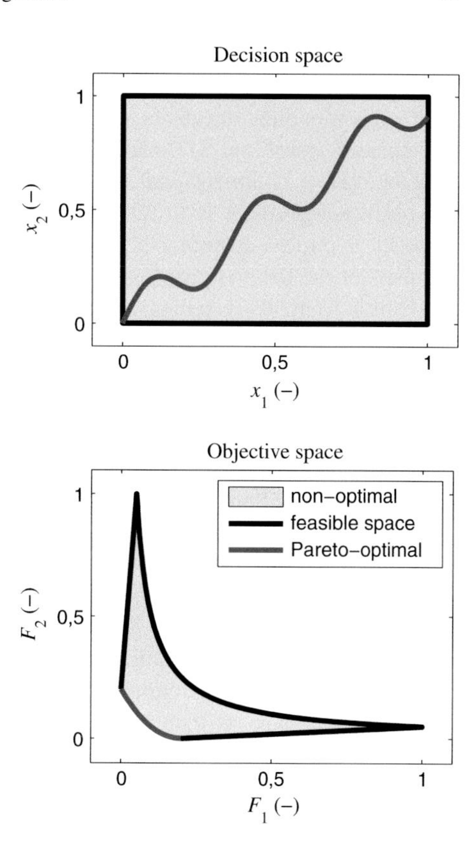

Fig. 2 Pareto optimal solutions of multi-objective problem depicted in decision space and objective space [13]

decision space variable respectively, J stands for number of constraints g. As can be seen from (1) a multi-objective problem takes place in two multidimensional spaces: the N-dimensional decision space formed by input variables and M-dimensional objective space formed by values of defined objective functions. The Pareto optimal solutions can be then depicted in both of them (see Fig. 2).

These two spaces are connected with objective functions. Every objective function "evaluates" a state vector $\vec{x}$ and places it in the objective space. Most often, these objective functions are non-linear. Therefore, the properties in these two spaces are not the same. For example two state vectors $\vec{x}$ having minimal distance in the decision space can be far from each other in the objective space.

Generally speaking, the search for Pareto optimal solutions takes place in the decision space and is controlled by information from the objective space. Thanks to the non-linearity of the solved problems this process can be very difficult. Moreover, the mechanism of control has to be independent on the nature of solved problem so the optimization technique can be used for various problems. The Pareto optimal solutions (state vectors $\vec{x}$) build a set which will be denoted here as P^*. Every member of set P^* can be depicted in both the decision and objective spaces (see red lines in Fig. 2). In general, size of the set P^* can be from interval

$|P^*| \in \langle 1; \infty \rangle$. Size of the set depends on the relation and nature of the objectives. If all the objectives are conflicting then the size of P^* is larger than 1. On the other hand, corresponding objectives lead to the degenerate solution having size $|P^*| = 1$.

Generally speaking, M-dimensional problem leads to a Pareto front which is built by $(M-1)$-dimensional hypervolume. For example the Pareto front of two-objective problem is a curved line while for the three-objective problem it becomes a curved surface etc. In the real world, it is not possible to find all members of the Pareto optimal set. However, it is very important to find all parts of the Pareto front. This brings another goal for the multi-objective optimization. Beside the request to find the solutions lying exactly on the Pareto front we need to cover the whole Pareto front uniformly with finite number of solutions. The example of Pareto-optimal sets with uniform and poor coverage of the Pareto front are depicted in Fig. 3. As can be seen, members of the set with good distribution are spaced on the Pareto front equidistantly. On the contrary, the other set covers just few parts of the Pareto front. Therefore, some information about the problem remains covered to the designer. The strategy applied for search in the decision space has to consider the necessity to find solutions with uniform spread.

Most of the pure multi-objective optimizers apply the strategy of so called *principle of dominance*. This principle compares two solutions in the objective space and states if one dominates the other or are mutually non-dominated. Definition of principle of dominance states that [1]: Solution $\vec{x}_1$ is said to dominate the other solution $\vec{x}_2$, if both conditions 1 and 2 are true:

1. Solution $\vec{x}_1$ is no worse than $\vec{x}_2$ in all objectives.
2. Solution $\vec{x}_1$ is strictly better than $\vec{x}_2$ in at least one objective.

To understand the principle of dominance well see Fig. 4. In this figure five solutions of the two-objective problem are depicted in the objective space. The

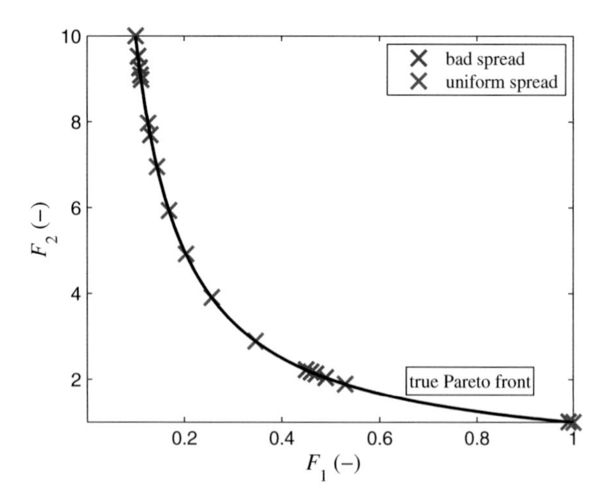

Fig. 3 Pareto optimal sets with poor and good distribution over the true Pareto front [7]

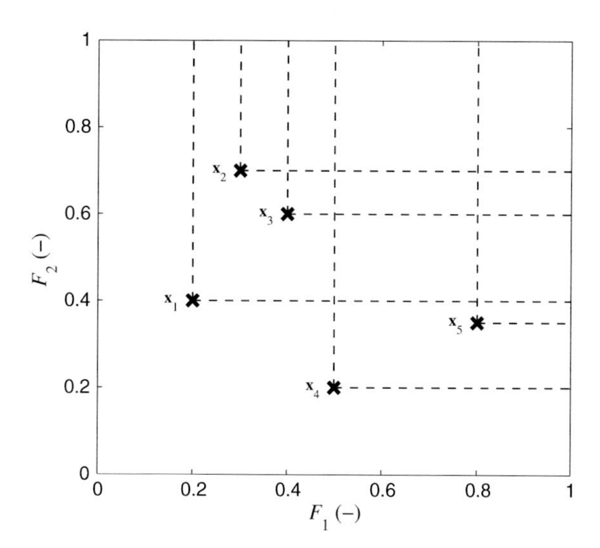

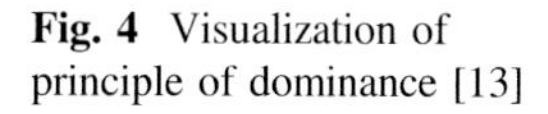
Fig. 4 Visualization of principle of dominance [13]

dashed lines mark out parts of the objective space that are dominated by the corresponding solution. As can be seen from this example set: solution $\vec{x}_1$ dominates solutions $\vec{x}_2$ and $\vec{x}_3$ while solution $\vec{x}_5$ is dominated by solution $\vec{x}_4$. Sets $\vec{x}_1$ and $\vec{x}_4$ are so called non-dominated solutions. If we search in the figure carefully, we can see that also solutions $\vec{x}_2$, $\vec{x}_3$ and $\vec{x}_5$ are non-dominated. we can state that there are two non-dominated subsets: $\vec{x}_1$ and $\vec{x}_4$ and $\vec{x}_2, \vec{x}_3, \vec{x}_5$ building advancing fronts. It is obvious, that non-dominated solutions of any subset of the solved problem are the solutions closes to the true Pareto front and therefore seem to be very good candidates for the final solution.

It is obvious that e.g. solution $\vec{x}_1$ is more suitable for our problem than $\vec{x}_3$ because it is better in all objectives. But for non-dominated solutions, it is impossible to decide which solution is better since. It is good to notice here, that any solution cannot dominate itself. It is ensured by the second condition of the dominance principle definition. Therefore, when designing multi-objective optimizer any solution cannot be in the non-dominated subset more than once. Also, the principle of dominance is transitive. It means that if $\vec{x}_1$ dominates $\vec{x}_2$ and $\vec{x}_2$ dominates $\vec{x}_3$, then $\vec{x}_3$ has to be dominated also by $\vec{x}_1$. This can speed up the process of search for non-dominated solutions.

The procedure searching for non-dominated solutions is called *non-dominated sorting*. There are several approaches in the literature. The problem is to find non-dominated subset P from the set Q consisting of all solutions. Clearly, any member of set P can be dominated by any other solution from Q. If set Q covers the entire objective space, then P becomes the true Pareto optimal set P^*. In other words, P has to cover the whole Pareto front because its members dominates all other possible solutions of the problem.

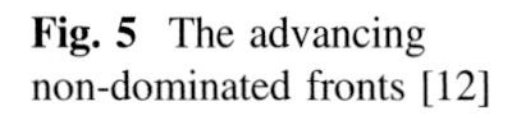

Fig. 5 The advancing non-dominated fronts [12]

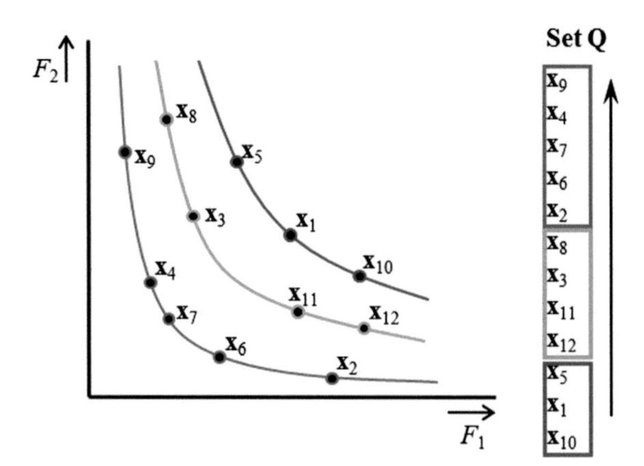

As mentioned before, there are so called advancing fronts of non-dominated solutions. First front is built by the non-dominated solutions P. All the members of the second front are dominated by members of the first front and dominate all solutions from advancing fronts (third and higher). The advancing fronts of two-objective problem are depicted in Fig. 5. Brief overview of the strategies how to find non-dominated sets can be found in [1]. If any of the strategies is repeated, the whole set Q can be sorted into consecutive non-dominated fronts.

Non-dominated sets of higher levels can be important to enhance diversity of final non-dominated set P found by optimizer. As mentioned before, it is necessary to cover the Pareto front uniformly. Therefore, the strategy applied in the optimizer cannot focuse just on members with best values of particular objective functions. Also, the contribution of a particular solution to the diversity of the non-dominated set P has to be considered when evaluating quality of particular solutions.

The so called *crowding* [2] can be applied to enhance diversity of the non-dominated set P. These methods try to emphasize the solutions from current population Q that are less crowded in the objective space to preserve the diversity among the resulting non-dominated set found by the optimizer.

The method *crowding distance* introduced by Deb et al. in [3] can be used without need to define any extra parameter. First, members within one front are sorted according to one objective. Then, the density of solutions surrounding every member of the front is computed from the Euclidean distance to the neighboring solutions. For the boundary solutions (extreme in sense of the objective functions values) large crowding distance value is set to save the found extreme solutions automatically. This technique gives preferential treatment to the less crowded solutions.

The crowding distance c is measured in the objective space. First of all, the members of one front are sorted according to all the objectives F_m. The vectors of sorted indices $\vec{I}_m$ are found. The crowding distance c for each member of the front can be computed using the following equations [1]:

$$c(I_m(i)) = \sum_{m=1}^{M} c_m(I_m(i)) \tag{2}$$

where

$$c_m(I_m(i)) = \frac{f_m(I_m(i+1)) - f_m(I_m(i-1))}{f_{m,\max} - f_{m,\min}} \tag{3}$$

where $I_m(i)$ is the i-th index from the m-th vector of indices, $f_{m,\max}$ and $f_{m,\min}$ are the maximal and minimal values of the m-th objective in the current front, respectively. The value c_m for these two extreme solutions is set to infinity (or very large value considering minimization problems). The crowding distance is the average side length of the hyper-cuboid defined by solutions surrounding a particular solution (see Eq. 6). The less crowded solutions (with a higher value of c) are preferred in the rest of the algorithm (Fig. 6).

Next term from theory of multi-objective optimization that should be explained here is so called *external archive* [4]. The external archive is formed by fixed number of non-dominated solutions. It can be viewed as form of *elitism* for multi-objective swarm optimization techniques. Members of the swarm are influenced by the solutions that are present in the external archive. Different methods for including and removing solutions from external archive can be used.

Similarly, as in case of single-objective optimization, elitism brings the danger of premature convergence. If all members of the eternal archive lie in the region of local optimum (e.g. non-dominated front of the second level), all other members of set Q can be attracted towards these local optima. While the global optimum can be in the decision space placed very far from solutions in current external archive. it is good to notice here, that distances between solutions in the decision space and objective space are usually not in proportion.

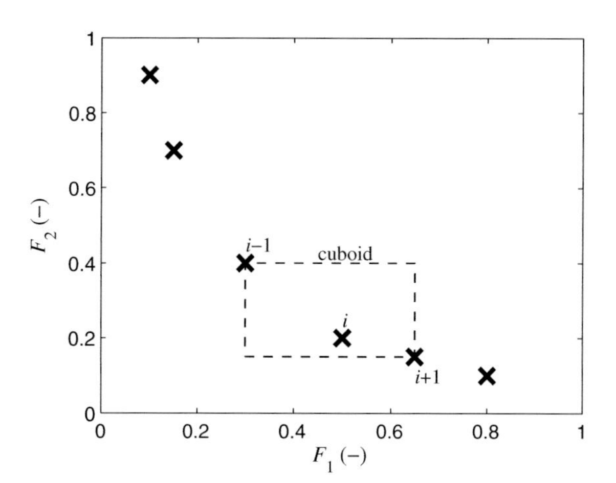

Fig. 6 Explanation about crowding distance computation [1]

2 MOSOMA

Self-organizing Migrating Algorithm (SOMA) has proven very good performance on both the benchmark and real-life optimization problems [5]. It has been also successfully used to solve real-life problems having multiple objectives. Nevertheless, it was used in connection with so called aggregation methods that transfers MOOP to SOOP (for overview see [6]). As shown in [7] these methods are not efficient for searching of Pareto front of the pure multi-objective problem. These methods lack in the diversity of final non-dominated set and usually have problems to find solutions on concave parts of the Pareto front.

The algorithm MOSOMA (Multi-Objective Self-Organizing Migrating Algorithm) was derived in [8] and [9]. This algorithm is able to solve both the constrained and unconstrained MOOPs having any number of decision space variables (discrete or continuous) and objective functions. It is suitable for solving problems with different types of Pareto front: convex, non-convex or discontinuous. The MATLAB code of MOSOMA can be downloaded at website [10].

MOSOMA makes profit of two basic principles mentioned above: (i) exploration of the decision space employing techniques that original SOMA uses and (ii) and search for the non-dominated set in the objective space from current population. For illustration see Fig. 7. The run of the algorithm can be described by following steps:

1. Defining controlling parameters of the algorithm.
2. Generating the initial population, evaluating objective functions.
3. Choosing external archive from the current population.
4. Migrating agents to members of external archive. Evaluating objective functions for new positions. Updating the external archive. Selecting migrating agents for next migration loop.

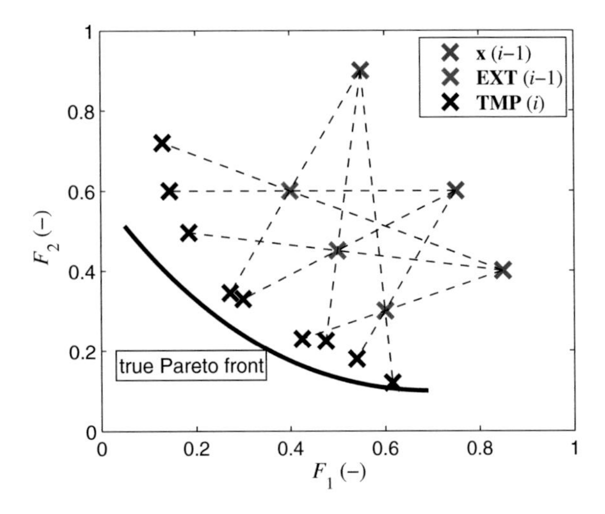

Fig. 7 Main idea of MOSOMA [13]

```
Start
    Define initial population P(1)
    Compute objective functions
    Find external archive EXT
    Select T migrating agents
    While i < I | FFC < N_f,max | |EXT(i)| < N_ex,max
        For q = 1 : T
            x_q migrates to all members of EXT(i - 1)
            Compute objective functions in TMP
        End
        Find EXT(i) from TMP ∪ EXT(i - 1)
        While |EXT(i)| < N_ex,min
            Find advancing front and crowding distance
            Fill EXT_i with best agents from advancing front
        End
        Choose T migrating agents for loop i+1
        i++
    End
    Chose final set P from EXT(i)
End
```

Fig. 8 Pseudo-code of MOSOMA [13]

5. Testing for stopping condition. If no stopping condition is accomplished, go back to 3.
6. Choosing final non-dominated set from the current external archive.

The pseudo-code of MOSOMA can be found in Fig. 8. In the following sub-sections the key parts of the algorithm will be discussed more in detail.

2.1 Controlling Parameters

At the beginning of the algorithm, user has to set some controlling parameters. Most of them are took over from the single-objective SOMA (see Part I, section "SOMA Principles and Control Parameters" of this book) but few of them had to be added. They can be summarized as follows:

FFC Number of fitness function computations.
$|P(1)|$ Initial population size.
T Number of migrating agents.
ST Number of steps in one migration.
PL Relative length of path for one migration.
PRT Probability of perturbation.
$N_{\mathrm{ex,min}}$ Minimal size of the external archive.

Initial population size strongly influences the convergence rate of the algorithm. According to our experience, it is better to generate larger initial population, because it is expedient to research whole decision space. The higher value of the parameter reduces speed of MOSOMA just for initial iteration because the convergence rate of the algorithm is controlled especially by number of migrating agents T and actual size of external archive $|\vec{EXT}|$.

The number of migrating agents in every migration loop is specified by parameter T. Its integer value is restricted just by the size of initial population $(T < |P(1)|)$. With higher number of T, research of a decision space is more precise on one hand, but the speed of MOSOMA decreases on the other hand. It is appropriate to choose this value similar to size of final non-dominated set $|P|$. The strategy for choosing the migrating agents influences efficiency of the algorithm more significantly.

In MOSOMA, the meaning of parameters path length and steps is slightly different than is their meaning in classical SOMA. Path length (*PL*) is the multiple of length between two migrating agents. And *ST* is number of steps that is the migration path divided to. These two parameters should not be set independently. They should be set so that the migration does not go through previously visited places. The condition can be expressed as:

$$s \cdot \frac{PL}{ST} \neq 1, \quad \forall s = 1, 2, \ldots, ST \tag{4}$$

As indicated in Fig. 7, MOSOMA assumes that solutions closer to the true Pareto front lie close to positions of agents currently saved in *EXT*. Some of s-multiples of ratio between *PL* and *ST* should acquire values slightly higher or lower than one. The other steps of the migration are also necessary, because they should avoid the bottleneck of MOSOMA in region of local optimum.

Objective functions values are evaluated in every step of the migration. After that, a new external archive members $EXT(i)$ are selected by the non-dominated sorting with crowding the union between two sets: previous external archive $EXT(i-1)$ and set of all temporary positions $\vec{TMP}$ defined during last migration. Within i-th migration loop, the number of fitness function computations $FFC(i)$ is given by number of steps for one migration, number of migrating agents and external archive size in the previous migration loop:

$$FFC(i) = |EXT(i-1)| \cdot T \cdot ST \tag{5}$$

The minimal size of external archive is denoted by symbol $N_{ex,\min}$. This parameter is important especially for more complex problems when MOSOMA is not able to find sufficient number of non-dominated solutions during last two migration loops. It is usually better to fill the external archive with solutions from advancing fronts to keep efficiency of those migration loops rather than let the agents to make less migrations during current iteration. Parameter $N_{ex,\min}$ cannot exceed size of the initial population $|P(1)|$ and should be similar to the size of the final non-dominated set specified by user $|P|$. As discussed above, the number of objective function computations *FFC* increases with growing external archive size.

Table 1 Recommended intervals for MOSOMA controlling parameters [11]

Parameter	Recommended interval
$\lvert P(1)\rvert/N$	$\langle 5; 12\rangle$
T/N	$\langle 5; 10\rangle$
ST	$\langle 2; 5\rangle$
PL	$\langle 1.1; 1.7\rangle$
PRT	$\langle 0.1; 0.4\rangle$
$N_{ex,\min}/\lvert P(1)\rvert$	$\langle 1/3; 2/3\rangle$

Usually, MOSOMA finds larger number of non-dominated solutions. Thus, final size of *EXT* is greater than parameter $N_{ex,\min}$.

In [11], sensitivity of MOSOMA on controlling parameters has been investigated. According to the large amount of tests on benchmark problems (few of them and evaluation metrics are presented in the Appendix of this chapter), the intervals for individual parameters have been recommended. The values are summarized in Table 1.

2.2 *Migration of Agents*

MOSOMA exploits the concept of external archive where all the best solutions are stored. These solutions build usually the non-dominated set assigned as *P*. In the *i*-th migration loop, every migrating agent visits temporarily new positions in the decision space:

$$\vec{T}MP_{t,s}(i) = \vec{x}_t(i-1) + (\vec{x}_p(i-1) - \vec{x}_t(i-1)) \cdot \frac{s}{ST} \cdot PL \cdot \vec{P}RTV_{t,s} \tag{6}$$

where $\vec{T}MP_{t,s}$ is the vector specifying s-th position during the migration of agent $\vec{x}_t$ towards agent $\vec{x}_p$. Symbol ST defines the number of steps for one migration ($s = 1, 2, \ldots, ST$). Parameter PL denotes the multiple of the distance between agents $\vec{x}_t$ and $\vec{x}_p$. So called perturbation vector $\vec{P}RTV$ has the same size as a decision space vector $\vec{x}$ and consists of zeros and ones. For every migration, N random numbers from zero to one are generated and the *PRTV* is then:

$$PRTV(n) = \begin{cases} 1 & \text{if } rand(n) > PRT, \\ 0 & \text{if } rand(n) < PRT. \end{cases} \tag{7}$$

where *PRT* denotes the probability of perturbation set by user.

T migrating agents are selected within every migration loop. According to our experience, the best way is to choose the migrating set partly randomly (the premature convergence is suppressed) and partly from the members of the current *EXT* (the region of the so far found best solutions is researched carefully to speed-up the convergence rate).

At the end of every migration loop non-dominated sorting of P is performed. If size of P (number of first front members) drops to less than $N_{ex,\min}$ the remaining positions are filled with members of advancing fronts with best values of so called crowding distance according to Eq. (2).

As in case of single-objective SOMA, individuals migrate through the N-dimensional hyper-space of input variables and try to find better solutions. MOSOMA uses the strategy which should be called '*AllToMany*'. Every individual migrates towards all members of the external archive as depicted in Fig. 7. We assume that using Eq. (6) with the path length parameter PL defined so that some of steps are very close to solution from the external archive should provide new solutions, which are placed closer to the true Pareto front.

After reviewing the basic principles of MOSOMA, following three stopping conditions are combined:

- Total number of migration loops I.
- Maximal size of external archive $N_{ex,\max}$ (usually multiple of the desired number of Pareto-optimal solutions $|P|$).
- Limited number of objective functions computations FFC.

The setting of appropriate stopping conditions is very important from the computational time viewpoint, especially. Using the above described '*AllToMany*' strategy, increasing size of external archive brings more computations of objective functions, which is usually very CPU time consuming. In MOSOMA, the ratio of CPU-time devoted for two consecutive migration loops is typically greater than one. Combination of these three stopping conditions ensures that optimization process stops in an estimable time and that sufficient number of non-dominated set members (candidates of the Pareto front) are found.

2.3 Final Non-dominated Set Choice

The size of external archive grows very quickly during migration loops and at the end of the algorithm, there are more non-dominated solutions then user asked for. Therefore, novel approach to select final non-dominated set P from EXT was proposed in [8] and [9]. This approach was proposed to enhance the spread of final non-dominated set. Obviously, it can be applied only if the size of external archive is greater than desired number of Pareto optimal solutions.

First, M extreme solutions (with minimal value of m-th objective function F_m) are saved to set P. The rest of P is than filled with EXT members so that set P covers the Pareto front as uniform as possible. Two methods has been proposed: two-objective and M-objective. The latter one can be applied aalso for two objective problems, but its computational time is larger.

The previously described crowding strategy prefers less crowded solutions and leads MOSOMA to provide solutions with better spread. The performance can be further improved adding an approach based on measuring Euclidean distance among the non-dominated solutions.

This strategy works with external archive set EXT. First step is sorting of the set in ascending order according to first objective function F_1 value. Then, the length of the found non-dominated set e is computed by:

$$e = \sum_{p=2}^{|EXT|} \sqrt{\sum_{m=1}^{2} (F_m(p) - F_m(p-1))^2} \tag{8}$$

where $F_m(p)$ is m-th objective function value of the p-th solution from external archive EXT.

Knowing this, the ideal distance between two uniformly spread solutions e_u is computed:

$$e_u = \frac{e}{|P| - 1} \tag{9}$$

where $|P|$ is desired number of final non-dominated set. Now, extreme solutions are saved into P as the first and the last member. Then, j-th member is that one having the minimal distance D_j:

$$D_j = \left| e_j - (j-1) \cdot e_u \right|, \quad j = 2, 3, \ldots, |P| - 1. \tag{10}$$

where e_j is computed as e using Eq. 8 where j replaces p.

Resulting Pareto front after applying this procedure is depicted in Fig. 9. Five non-dominated solutions (marked by plus symbol) arc chosen from external archive containing seven solutions (denoted by cross symbol).

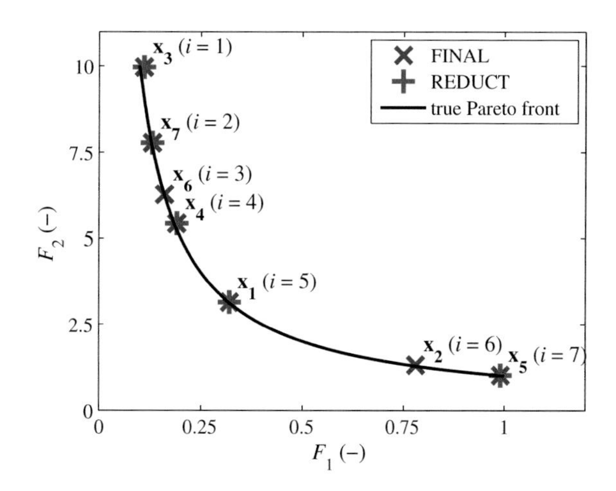

Fig. 9 Choice of the final non-dominated set from external archive—two-objective variant [13]

Unfortunately, the previously described two-objective approach cannot be directly extended for Pareto fronts having more than two objectives. In fact, when sorting the solutions of a multi-objective ($M > 2$) non-dominated set according to one of the objectives, the sorted solutions are not neighboring in sense of a topology [12]. Also, various curved shapes of multi-objective Pareto fronts can make the choice of the final set P from *EXT* impossible according to regular elements of a particular hypervolume (e.g. elements of a surface for $M = 3$, elements of a volume for $M = 4$, etc.).

Therefore, we have proposed a new method for the choice of the final non-dominated set P from the external archive *EXT* in [9]. We measure the Euclidean distance among the found solutions. First, M solutions with minimal values of particular F_m are saved into P. Particular solutions are then successively saved into P until P does not contain the desired number of solutions. Always, the q-th solution from *EXT* having the maximal distance D (the sum of Euclidian distances towards all members of the current P) is chosen:

$$D = \sum_{p=1}^{|P|} \sqrt{\sum_{m=1}^{M} (F_m(p) - F_m(q))^2} \tag{11}$$

An example result of this procedure can be found in Fig. 10. A part of the surface of the paraboloid builds the Pareto front. Red crosses mark the chosen solutions in set P. It is obvious, that the chosen solutions are spread along the whole Pareto front uniformly.

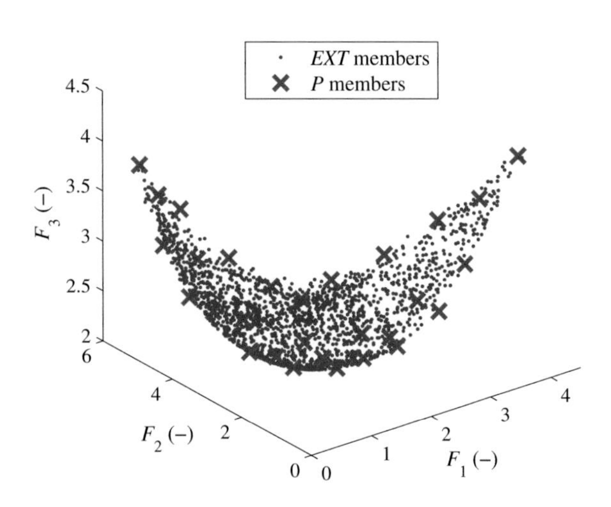

Fig. 10 Choice of the final non-dominated set from external archive—*M*-objective variant [9]

3 Appendix I—Evaluation Metrics

The quality of multi-objective optimizers can be measured by several metrics on benchmark problems with known Pareto front. Here, few evaluation metrics are briefly reviewed, for further information see [1] or [12].

Generational distance

The generational distance (GD) evaluates the accuracy of found non-dominated set P from a set of 500 uniformly spread true Pareto-optimal solutions $P*$:

$$GD = \frac{\sqrt{\sum_{p=1}^{|P|} d_p^2}}{|P|} \tag{12}$$

where d_p stands for the Euclidean distance measured in the objective space between the p-th solution from the set P and the closest member of $P*$:

$$d_p = \min_{k=1}^{|P^*|} \sqrt{\sum_{m=1}^{M} (f_{m,p} - f_{m,k}^*)^2} \tag{13}$$

where k denotes the index of the solution in set $P*$.

Spread

Spread (Δ) measures the uniformity of found non-dominated set:

$$\Delta = \frac{\sum_{m=1}^{M} d_{e,m} + \sum_{p=1}^{|MST|} \left|d_p - d_{\text{avg}}\right|}{\sum_{m=1}^{M} d_{e,m} + |MST| d_{\text{avg}}} \tag{14}$$

where d_p denotes the Euclidean distance between the p-th and $(p+1)$-st solution from $P, d_{e,m}$ denotes the distance from computed extreme solutions to the true ones and d_{avg} is the average distance among all computed solutions. Symbol MST stands for so called Minimum spanning tree of the non-dominated set.

Hit rate

Hit rate HR is expressing efficiency of the search. It is simply ratio between number of found non-dominated solutions and number of objective function evaluations:

$$HR = \frac{|P|}{FFC} 100\,\% \tag{15}$$

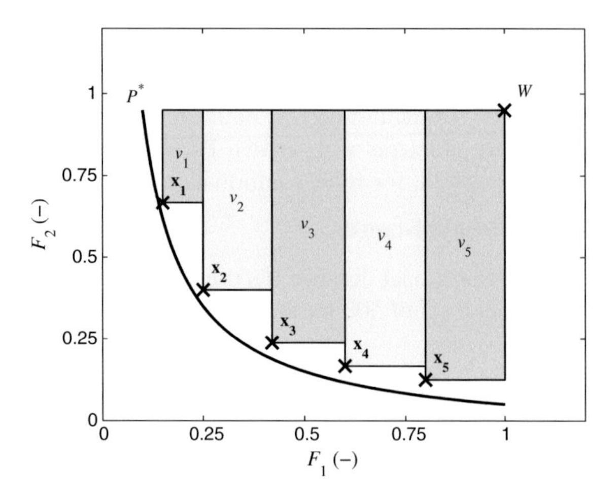

Fig. 11 Hypervolume metric explanation [12]

Hypervolume

Hypervolume HV evaluates accuracy and uniformity of found non-dominated set at the same time. It measures hypervolume in the objective space dominated by the found non-dominated set:

$$HV = \bigcup_{p=1}^{|P|} V_p \tag{16}$$

where V_p is hypervolume defined by p-th point from set P and the reference point W (see Fig. 11). Sometimes, relative hypervolume HVR as ratio between HV for found non-dominated set P and HV for 500 true Pareto front members P^* is used:

$$HVR = \frac{HV(P)}{HV(P^*)} \tag{17}$$

4 Appendix II—Benchmark Problems

The quality of multi-objective optimizers can be measured by several metrics on benchmark problems with known Pareto front. Here, few benchmark problems are briefly reviewed, for further information see [1] or [12]. For every problem, objective function definition, variable bounds and Pareto front visualization are derived here.

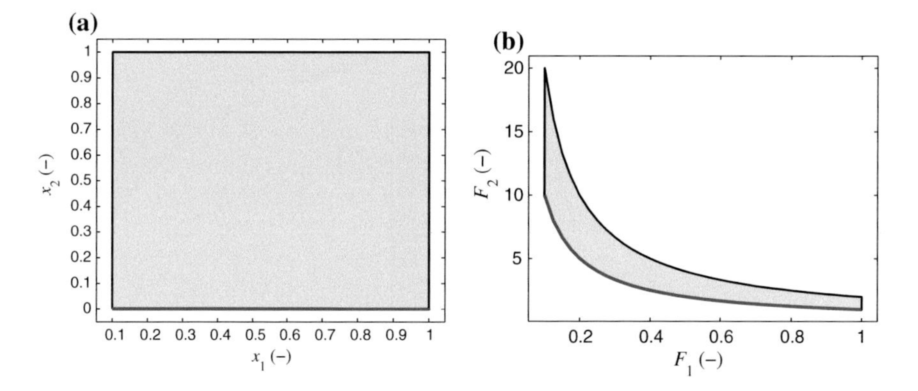

Fig. 12 Benchmark problem SC1 [12]. **a** Decision space cut, **b** objective space

Simple convex SC1 (Fig. 12)

$$F_1 = x_1$$
$$F_2 = \frac{1+x_2}{x_1}$$
$$0.1 \le x_1 \le 1$$
$$0 \le x_2 \le 1$$
$$P^* : x_2^* = 0, \quad x_1^* \in \langle 0.1; 1 \rangle$$

Schwefel SCH1 (Fig. 13)

$$F_1 = x^2$$
$$F_2 = (x-2)^2$$
$$-5 \le x \le 5$$
$$P^* : x^* \in \langle 0; 2 \rangle$$

Fonseca FON (Fig. 14)

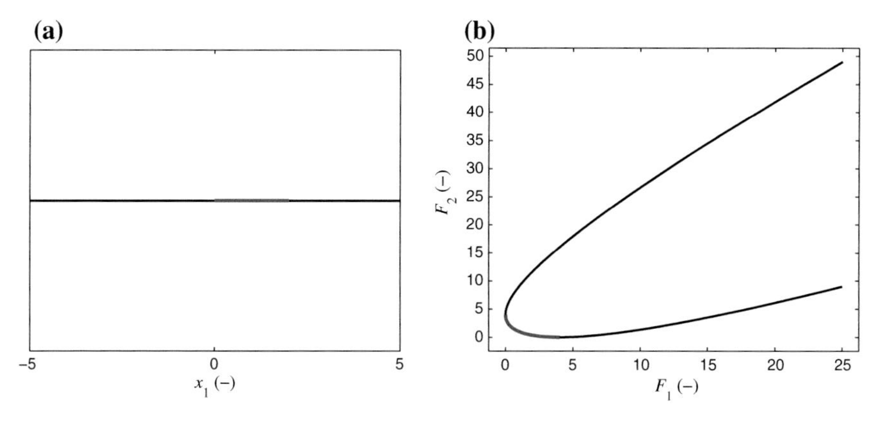

Fig. 13 Benchmark problem Schwefel [12]. **a** Decision space cut, **b** objective space

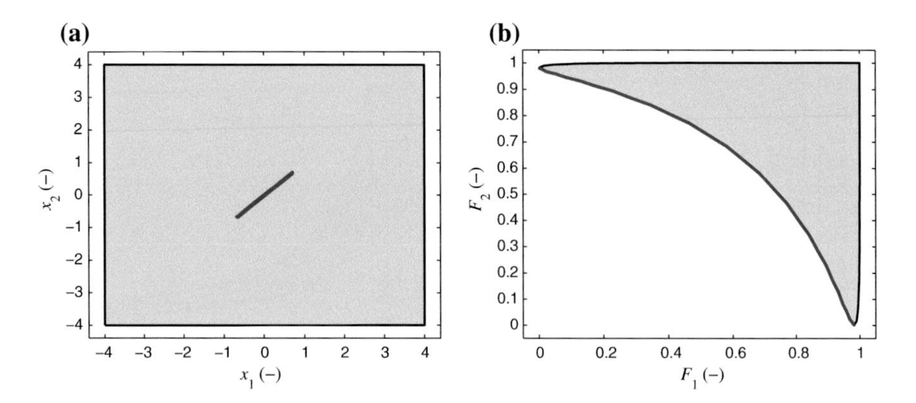

Fig. 14 Benchmark problem Fonseca [12]. **a** Decision space cut, **b** objective space

$$F_1 = 1 - \exp\left[-\sum_{n=1}^{N}(x_n - \tfrac{1}{\sqrt{N}})^2\right]$$
$$F_2 = 1 - \exp\left[-\sum_{n=1}^{N}(x_n + \tfrac{1}{\sqrt{N}})^2\right]$$
$$-4 \le x_n \le 4, \quad n = 1, 2, \ldots, N$$

$$P^* : F_2^* = 1 - \exp\{-[2 - \sqrt{\ln(1 - f_1^*)}]\}, \quad 0 \le F_1^* \le 1 - \exp(-4)$$

Poloni POL (Fig. 15)

$$F_1 = 1 + (A_1 - B - 1)^2 + (A_2 - B_2)^2$$
$$F_2 = (x_1 + 3)^2 + (x_2 + 3)^2$$
$$A_1 = 0.5\sin(1) - 2\cos(1) + \sin(2) - 1.5\cos(2)$$
$$A_2 = 1.5\sin(1) - \cos(1) + 2\sin(2) - 0.5\cos(2)$$
$$B_1 = 0.5\sin(x_1) - 2\cos(x_1) + \sin(x_2) - 1.5\cos(x_2)$$
$$B_2 = 1.5\sin(x_1) - \cos(x_1) + 2\sin(x_2) - 0.5\cos(x_2)$$
$$-\pi \le x_n \le \pi, \quad n = 1, 2$$

P^*: analytically not available

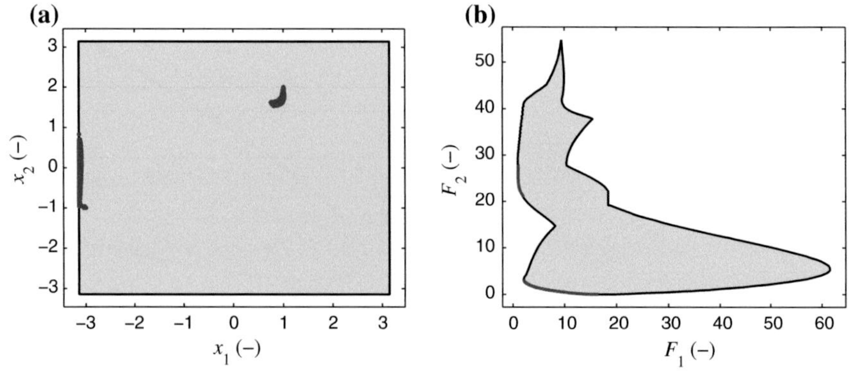

Fig. 15 Benchmark problem Poloni [12]. **a** Decision space cut, **b** objective space

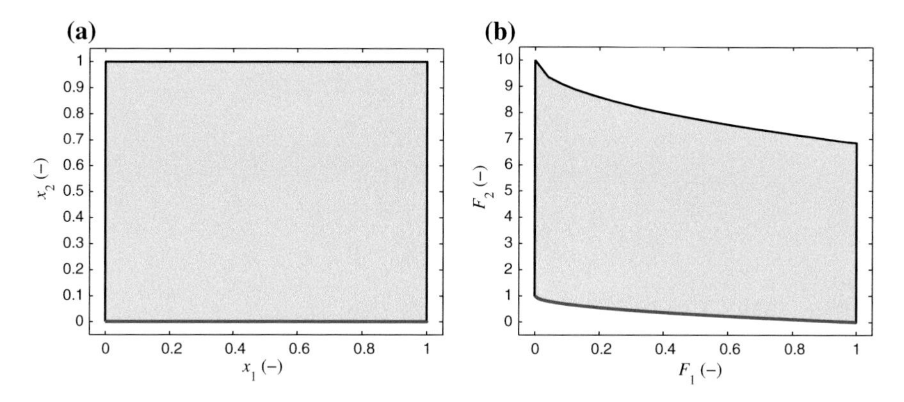

Fig. 16 Benchmark problem Zitzler, Deb, Thiele 1 [12]. **a** Decision space cut, **b** objective space

Zitzler, Deb, Thiele 1 ZDT1 (Fig. 16)

$$\begin{aligned} F_1 &= x_1 \\ F_2 &= gh \\ g(\vec{x}) &= 1 + \tfrac{9}{N-1}\sum_{n=2}^{N} x_n \\ h(F_1, g) &= 1 - \sqrt{\tfrac{F_1}{g}} \\ 0 \le x_n &\le 1, \quad n = 1, 2, \ldots, N \\ P^* : x_1^* &\le 1, \quad x_n^* = 0 \end{aligned}$$

Zitzler, Deb, Thiele 2 ZDT2 (Fig. 17)

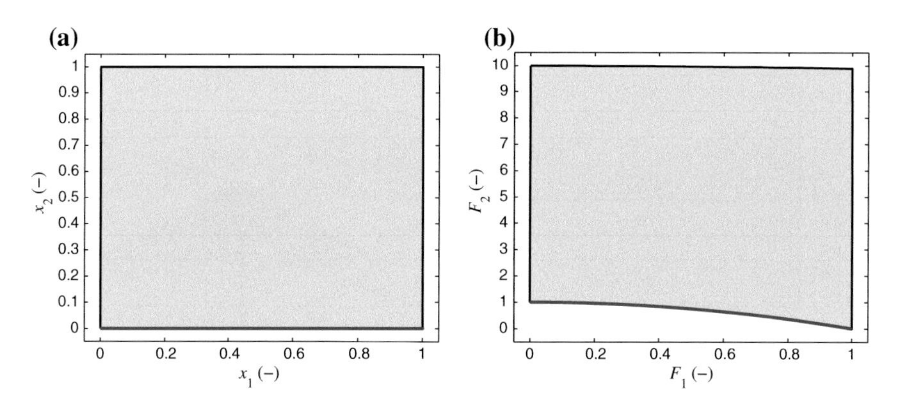

Fig. 17 Benchmark problem Zitzler, Deb, Thiele 2 [12]. **a** Decision space cut, **b** objective space

$$
\begin{aligned}
F_1 &= x_1 \\
F_2 &= gh \\
g(\vec{x}) &= 1 + \tfrac{9}{N-1} \sum_{n=2}^{N} x_n \\
h(F_1, g) &= 1 - \left(\tfrac{F_1}{g}\right)^2 \\
0 \leq x_n \leq 1, &\quad n = 1, 2, \ldots, N \\
P^* : x_1^* \leq 1, &\quad x_n^* = 0
\end{aligned}
$$

Deb, Thiele, Laumanns, Zitzler 1 DTLZ1 (Fig. 18)

$$
\begin{aligned}
F_1 &= \tfrac{1}{2} x_1 x_2 [1 + g(\vec{x}_K)] \\
F_2 &= \tfrac{1}{2} (1 - x_2)[1 + g(\vec{x}_K)] \\
F_3 &= \tfrac{1}{2} (1 - x_1)[1 + g(\vec{x}_K)] \\
g(\vec{x}) &= 100 \left\{ |\vec{x}_K| + \sum_{k=1}^{|\vec{x}_K|} \left[(x_k - 0.5)^2 - \cos(20\pi(x_k - 0.5)) \right] \right\} \\
P^* : x_2^* &= \tfrac{1}{2} - x_1^*, \quad x_k^* = 0.5, \quad k = 3, 4, \ldots, N
\end{aligned}
$$

Deb, Thiele, Laumanns, Zitzler 2 DTLZ2 (Fig. 19)

$$
\begin{aligned}
F_1 &= [1 + g(\vec{x}_K)] \cos\left(x_1 \tfrac{\pi}{2}\right) \cos\left(x_2 \tfrac{\pi}{2}\right) \\
F_2 &= [1 + g(\vec{x}_K)] \cos\left(x_1 \tfrac{\pi}{2}\right) \sin\left(x_2 \tfrac{\pi}{2}\right) \\
F_3 &= [1 + g(\vec{x}_K)] \sin\left(x_1 \tfrac{\pi}{2}\right) \\
g(\vec{x}) &= \sum_{k=1}^{|\vec{x}_K|} (x_k - 0.5)^2 \\
P^* : x_2^* &= \sqrt{1 - x_1^* 2}, \quad x_k^* = 0.5, \quad k = 3, 4, \ldots, N
\end{aligned}
$$

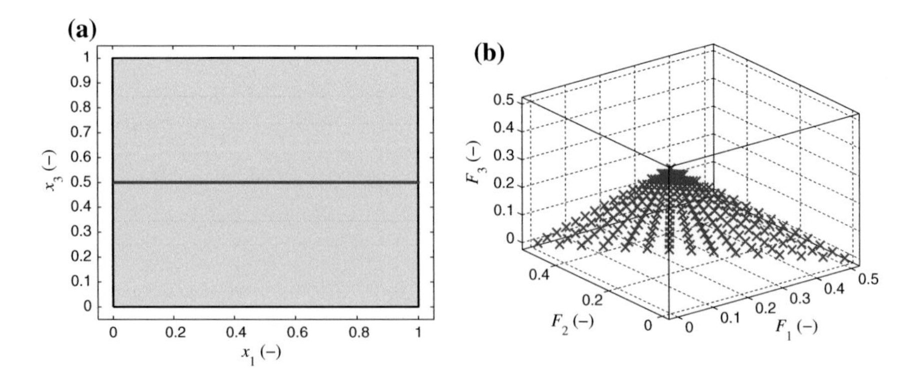

Fig. 18 Benchmark problem Deb, Thiele, Laumanns, Zitzler 2 [12]. **a** Decision space cut, **b** objective space

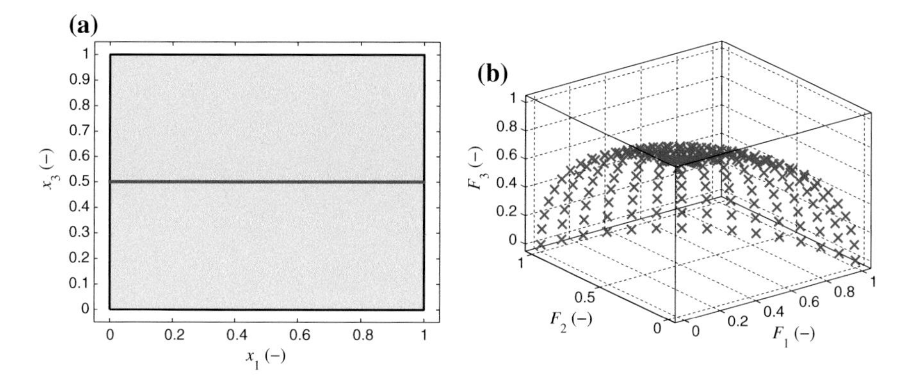

Fig. 19 Benchmark problem Deb, Thiele, Laumanns, Zitzler 2 [12]. **a** Decision space cut, **b** objective space

Acknowledgements Research described in this chapter was financially supported by Czech Science Foundation under grant no. P102/12/1274. Support of projects SIX CZ.1.05/2.1.00/03.0072 is also gratefully acknowledged.

References

1. Deb, K.: Multi-objective Optimization Using Evolutionary Algorithms. Wiley, Chichester (2001)
2. DeJong, K.A.: An analysis of the behavior of a class of genetic adaptive systems. Ph.D. thesis, University of Michigan, Ann Arbor (1975)
3. Deb, K., Pratap, A., Agarwal, S., Meyarivan, T.: A fast and elitist multiobjective genetic algorithm: NSGA-II. IEEE Evol. Comput. **6**, 182–197 (2002)
4. Bartz-Beilstein, T., et al.: Particle swarm optimizers for pareto optimization with enhanced archiving techniques. Congress on Evolutionary Computation (2003)
5. Zelinka, I.: SOMA—self-organizing migrating algorithm. In: Onwubolu, G.C., Babu, B.V. (eds.) New Optimization Techniques in Engineering, pp. 167–217. Springer, Berlin (2004)
6. Miettinen, K.: Nonlinear Multiobjective Optimization. Kluwer, Boston (1999)
7. Kadlec, P., Raida, Z.: Comparison of novel multi-objective self organizing migrating algorithm with conventional methods. In: Radioelektronika, 2011 21st International Conference (2011). doi:10.1109/RADIOELEK.2011.5936395
8. Kadlec, P., Raida, Z.: A novel multi-objective self-organizing migrating algorithm. Radioengineering **20**, 77–90 (2011)
9. Kadlec, P., Raida, Z.: Self-organizing migrating algorithm for optimization with general number of objectives. In: Radioelektronika, 2012 22nd International Conference (2012)
10. Kadlec, P.: MOSOMA: Matlab codes. Brno University of Technology (2013) Available via DIALOG. http://www.urel.feec.vutbr.cz/kadlec/userfiles/downloads/MOSOMA/mosoma.zip. Cited 20 Oct 2014
11. Kadlec, P., Raida, Z., Døínovský, J.: Multi-objective self-organizing migrating algorithm: sensitivity on controlling parameters. Radioengineering **22**, 296–308 (2013)
12. Kadlec, P.: Multiobjective optimization of EM structures based on self-organizing migration. Ph.D. thesis, Brno University of Technology, Czech republic (2012)
13. Kadlec, P., Raida, Z.: Multi-objective self-organizing migrating algorithm applied to the design of electromagnetic components. IEEE Ant. Propag. Mag. **55**, 50–68 (2013)

Multi-objective Design of EM Components

Petr Kadlec and Zbyněk Raida

Abstract Design of EM components is usually very demanding task. It comprises setting of large number of variables. With increasing number of variables, the number of possible combinations increases exponentially. Therefore, the use of global stochastic optimizers became essential. Use of multi-objective optimizers such as MOSOMA (Multi-Objective Self-Organizing Migrating Algorithm) gives the user extra knowledge about the solved problem and its contradictory requirements. In this chapter, applications of MOSOMA for solution of problems from electromagnetics are first briefly reviewed. Then, three applications are discussed more in detail: design of Yagi-Uda antenna array, design of dielectric layered filter and control of adaptive beamforming in time domain of slotted antenna array. Results of MOSOMA are compared with previously published solutions. The possibility how to treat problems having discrete decision space is discussed here.

Multi-objective optimization can be used to solve problems from any field of human activities: e.g. civil engineering, economics, mechanical engineering etc. Also MOSOMA (Multi-Objective Optimization Algorithm) is derived as general tool and therefore can be used relatively easily to solve any type of problem.

As discussed in previous chapter, solution of the multi-objective problem leads to so called Pareto front. The advantage of multi-objective approach is that designer acquires from the shape of the Pareto front extra information about trade-off between particular objectives. The optimizer offers him a set of non-dominated solutions and he can choose the final one, which meets best his requirements.

This chapter is divided into two parts: first one discusses application of MOSOMA to benchmark problems and compares MOSOMA with other

P. Kadlec (✉) · Z. Raida
Brno Univesrity of Technology, Antonínská 1, 602 00 Brno, Czech Republic
e-mail: kadlecp@feec.vutbr.cz

Z. Raida
e-mail: raida@feec.vutbr.cz

D. Davendra and I. Zelinka (eds.), *Self-Organizing Migrating Algorithm*,
Studies in Computational Intelligence 626, DOI 10.1007/978-3-319-28161-2_5

multi-objective algorithms from the point of view of convergence properties. Second part shows some real-life problems which has been successfully solved by MOSOMA.

1 Design of EM Components

Design of EM (ElectroMagnetic) components e.g. antennas, filters, guiding structures etc. implies the large number of degrees of freedom or variables. With increasing number of variables, the number of combinations increases exponentially which makes the parametric analysis of the problem impossible. Therefore, use of global optimization methods has long tradition in the field of electromagnetics.

MOSOMA is relatively new algorithm, since it was introduced in 2011 [1]. It has been successfully applied to design various EM components since then. In [2] MOSOMA was used to control adaptive beamforming of array of slot antennas in time domain. In [3] a digital filter coefficients were found using MOSOMA so that the filter is able to truncate computational domain of the FDTD (Finite Difference Frequency Domain) method with no reflections. In [4] traditional EM components parameters—Yagi-Uda antenna and dielectric layered filter—are designed by MOSOMA. In [5, 6] MOSOMA helps to design so called filtenna—an antenna with filtering properties. In the following subsections, selected applications will be discussed more in detail.

1.1 Yagi-Uda Antenna Design

Yagi-Uda antenna is well known concept of wired antenna array. It was named after two Japanese physicists: Yagi [7] introduced this concept in 1928 and Uda [8] who continued in Yagi's work in 1954. Venkatarayalu and Ray [9] and Kuwahara [10] have defined design of Yagi-Uda antenna as multi-objective optimization problem.

The antenna consisting of N elements is depicted in Fig. 1. The first element is usually the longest one and acts as reflector. Second element is driven by a source current. Rest of the elements are so called directors. In the figure, d_n denotes the total length of the nth element. Symbol s_n stands for the spacing between nth and $(n + 1)$st element.

The design of Yagi-Uda array consists in searching for lengths of N elements and spacing between them. To keep the optimization task as general as possible, all parameters are defined with respect to the operating wavelength λ. The optimization task is then to compute $2N-1$ parameters. The length of every element can be chosen from interval $d_n/\lambda \in \langle 0.30; 0.70 \rangle$ and the spacing from interval $s_n/\lambda \in \langle 0.10; 0.35 \rangle$.

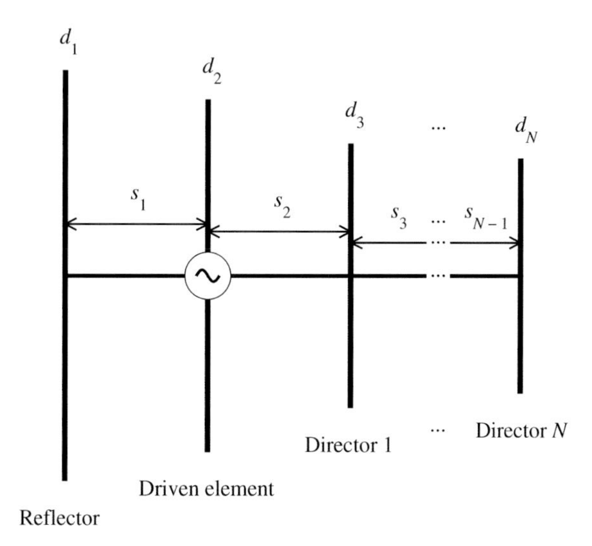

Fig. 1 Description of Yagi-Uda antenna [4]

Two objectives are considered in this formulation of the Yagi-Uda design problem—maximization of gain G and minimization of relative side lobe levels SLL:

$$F_1 = -G \tag{1}$$

$$F_2 = SLL \tag{2}$$

Antenna parameter gain expresses how much power is transmitted in the direction of peak radiation in comparison with isotropic source. Side lobe level is parameter, that describes maximal level of far-field radiation pattern, that is not the main lobe. Usually, side lobes should be suppressed as much as possible. Please notice that gain is multiplied by −1 in the first objective function. The codes of MOSOMA [11] presume just minimization objective functions. Two constraint functions are added to ensure a proper impedance matching of the antenna:

$$|50 - \Re(Z_{in})| = 5 \tag{3}$$

$$|50 - \Im(Z_{in})| = 10 \tag{4}$$

where Z_{in} denotes the input impedance of the antenna. The analysis of the antenna array necessary for the computation of objectives and constraints can be performed using the freeware software 4NEC2 based on Method of Moments [12]. The detailed description how to run 4NEC2 software directly from Matlab can be found in [4].

The settings of MOSOMA was made so that it computes objective functions maximal 36,000-times. Then, our results can be compared with results published in

Table 1 MOSOMA settings for design of Yagi-Uda antenna

Parameter	*FFC*	*PL*	*ST*	$\lvert P1 \rvert$	*T*	$N_{ex,\min}$
	36,000	1.3	5	30	20	15

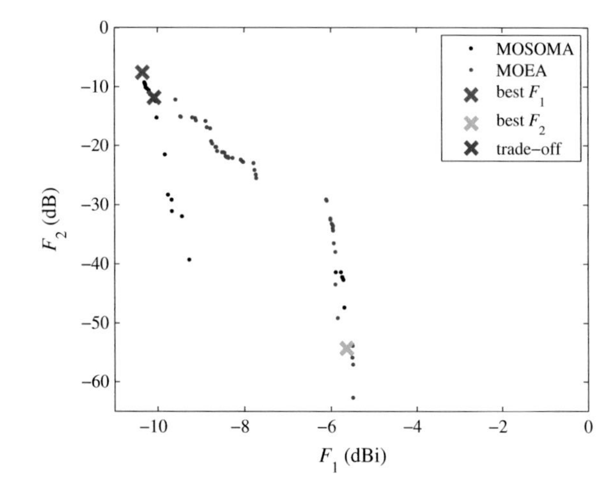

Fig. 2 Pareto optimal solutions found by MOSOMA [4] and by MOEA [9] for 4-element Yagi-Uda antenna

[9, 10]. The MOSOMA controlling parameters set for the run are summarized in Table 1.

For simplicity, we will consider the Yagi-Uda antenna having $N = 4$ elements. MOSOMA has been run ten-times and the best results are published here. Resulting Pareto front is depicted in Fig. 2. Here, results of the MOSOMA can be directly compared with results of algorithm MOEA (Multi-Objective Evolutionary Algorithm) published in [9]. MOSOMA outperforms MOEA, because most of the solutions proposed by MOEA are dominated by solutions proposed by MOSOMA.

In Fig. 2, three selected solutions are highlighted: the best solution according to objective F_1 (maximal gain $G = 10.35$ dBi), the best solution according to objective F_2 (minimal side lobe level $SLL = -54.29$ dB) and the trade-off solution which shows compromise between the gain ($G = 10.08$ dBi) and relative side lobe level ($SLL = -11.81$ dB). Radiation patterns for all these solutions are depicted in Fig. 3. The lengths and spacing between individual elements can be found in Table 2.

1.2 Dielectric Layered Filter Design

Design of a dielectric filter can be formulated as constrained two-objective problem. It involves an optimization of a relative permittivity and a width of individual layers to perform desired filtering properties in microwave frequency band. Venkatarayalu used his MOEA to solve this problem in [13]. Goudos used for its solution a variant

Table 2 Yagi-Uda antenna parameters designed by MOSOMA ([4]) and MOEA ([9])

Algorithm	MOSOMA						MOEA [9]			
	Best F_1		Best F_2		Trade-off		Best F_1		Best F_2	
	$d(\lambda)$	$s(\lambda)$	$d(\lambda)$	$s(\lambda)$	$d(\lambda)$	$s(\lambda)$	$d(\lambda)$	$s(\lambda)$	$d(\lambda)$	$s(\lambda)$
1	0.473	0.313	0.494	0.245	0.474	0.295	0.480	0.270	0.628	0.204
2	0.445	0.350	0.469	0.196	0.463	0.260	0.474	0.186	0.488	0.195
3	0.439	0.288	0.402	0.272	0.440	0.249	0.436	0.274	0.436	0.114
4	0.434	–	0.651	–	0.433	–	0.434	–	0.582	–
G (dBi)	10.35		5.64		10.08		9.60		5.50	
SLL (dB)	−7.60		−54.29		−11.81		−12.14		−62.60	
$Z_in(\Omega)$	45.18 + 6.96i		45.32 − 1.70i		45.36 + 3.65i		47.59 − 5.67i		45.51 + 7.83i	
$VSWR(-)$	1.19		1.11		1.13		1.13		1.21	

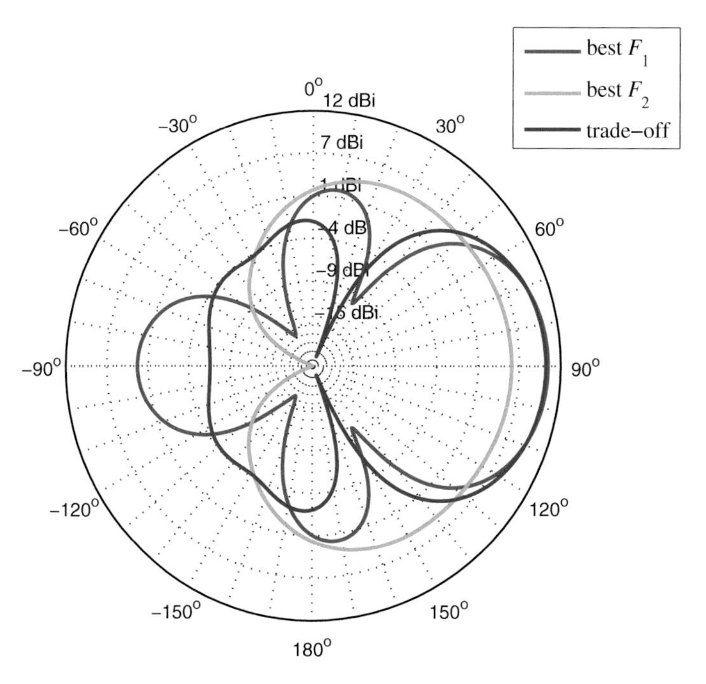

Fig. 3 H-plane radiation patterns of chosen solutions from Pareto front obtained by MOSOMA for four-element Yagi-Uda antenna [4]

of MOPSO (Multi-Objective Particle Swarm Optimization) to solve the same problem in [14].

If we want to design a filter having N layers, $2N$ parameters are set by the optimizer. The geometry of layered medium is depicted in Fig. 4. The theory of diffraction solves the problem of wave propagating through layered medium [15].

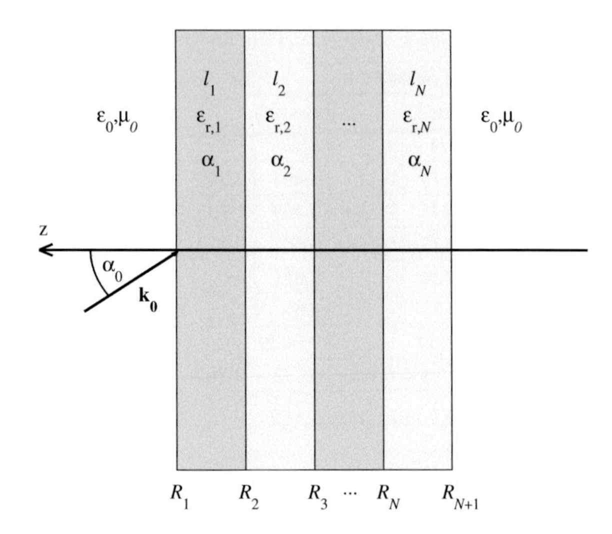

Fig. 4 Dielectric filter composed of N layers [4]

In Fig. 4, $\boldsymbol{k}_0$ stands for the wave vector of the impinging wave, l_n denotes width of nth layer, $\varepsilon_{r,n}$ denotes relative permittivity of nth layer, α_n is the incident angle for nth interface and R_n is the reflection coefficient of nth interface. Interface between the first and second dielectric layer is denoted by R_2.

In theory of diffraction, reflection coefficient is the complex number which expresses how much of the waves energy is reflected from the interface of two dielectric media. For this example, homogeneous, lossless, nonmagnetic materials are considered. Then, generalized recursive reflection coefficient R_n for the nth interface is derived [15]:

$$R_n = \frac{r_n + R_{n+1} \exp(2i\boldsymbol{k}_n l_n)}{1 + r_n R_{n+1} \exp(2i\boldsymbol{k}_n l_n)} \tag{5}$$

The wave vector $\boldsymbol{k}_n$ valid in nth layer is computed using:

$$\boldsymbol{k}_n = \frac{2\pi f}{c} \sqrt{\varepsilon_{r,n}} \tag{6}$$

In theory, two modes can propagate through the layered medium: transversally electric (TE) and magnetic (TM). The two reflection coefficients are then:

$$r_{n,\mathrm{TE}} = \frac{\sqrt{\varepsilon_{r,n-1}(1 - \sin^2 \alpha_{n-1})} - \sqrt{\varepsilon_{r,n}(1 - \sin^2 \alpha_n)}}{\sqrt{\varepsilon_{r,n-1}(1 - \sin^2 \alpha_{n-1})} + \sqrt{\varepsilon_{r,n}(1 - \sin^2 \alpha_n)}} \tag{7}$$

and

$$r_{n,\mathrm{TM}} = \frac{\varepsilon_{r,n}\sqrt{\varepsilon_{r,n-1}(1-\sin^2\alpha_{n-1})} - \varepsilon_{r,n-1}\sqrt{\varepsilon_{r,n}(1-\sin^2\alpha_n)}}{\varepsilon_{r,n}\sqrt{\varepsilon_{r,n-1}(1-\sin^2\alpha_{n-1})} + \varepsilon_{r,n-1}\sqrt{\varepsilon_{r,n}(1-\sin^2\alpha_n)}} \tag{8}$$

The nth layer incidence angle is:

$$\alpha_n = \sin^{-1}\left(\frac{\sqrt{\varepsilon_{r,n-1}}}{\sqrt{\varepsilon_{r,n}}}\alpha_{n-1}\right) \tag{9}$$

Considering the whole medium having N layers, reflection coefficient $R_{1,(TE)}$ and $R_{1,(TM)}$ expresse the reflection properties of the whole filter. Then, two-objective functions can be defined [4]:

$$F_1 = \frac{1}{P}\sum_{p=1}^{P}\left[|R_{1,\mathrm{TE}}(f_p)|^2 + |R_{1,\mathrm{TM}}(f_p)|^2\right] \tag{10}$$

$$F_2 = \frac{1}{S}\sum_{s=1}^{S}\left[2 - |R_{1,\mathrm{TE}}(f_s)|^2 - |R_{1,\mathrm{TM}}(f_s)|^2\right] \tag{11}$$

where f_p and f_s denote the pass and stop frequencies of the filter, respectively, and P and S stands for the number of frequencies in the frequency list. The objective function F_1 minimizes the reflection of the layered media in the pass band while the other function F_2 maximizes the reflection in the stop band. Four constraint functions are added to objectives to formulate the problem. These constraints define limits for the pass band f_{pc} and stop band f_{sc} [13]:

$$20\,\log|R_{1,\mathrm{TE}}(f_{pc})| < -10\ \mathrm{dB} \tag{12}$$

$$20\,\log|R_{1,\mathrm{TM}}(f_{pc})| < -10\ \mathrm{dB} \tag{13}$$

$$20\,\log|R_{1,\mathrm{TE}}(f_{sc})| > -5\ \mathrm{dB} \tag{14}$$

$$20\,\log|R_{1,\mathrm{TM}}(f_{sc})| > -10\ \mathrm{dB} \tag{15}$$

Here, f_{pc} and f_{sc} denote the pass and stop band frequencies considered for constraints, respectively.

For the example, number of layers was fixed to $N = 7$ as in [4, 13, 14]. The incidence angle of wave impinging the filter α_0 (see Fig. 4) was set to $\frac{\pi}{4}$. The width of every layer l_n can vary in the interval $\langle 1; 10\rangle$ mm. The relative permittivity of all layers can be chosen from commercially available dielectric materials 1.01, 2.20, 2.33, 2.50, 2.94, 3.00, 3.02, 3.27, 3.38, 4.48, 4.50, 6.00, 6.15, 9.20, 10.20 [13]. MOSOMA works in general only with continuous input variables. The problem with discrete input variables can be solved using a relatively simple approach. As we have 15 discrete values, the input variables denoting relative permittivity from the discrete set containing 15 values, are set from the interval $\langle 0; 15 \rangle$. This interval is divided uniformly into 15 subintervals, each corresponding to one value of an available dielectric permittivity (e.g. variable denoting width of the first layer $x_8 = 6.35$ corresponds to the seventh value from the list of available permittivities $\varepsilon_r = 3.02$). The variables are within the algorithm treated as continuous. Only objective functions are evaluated with the corresponding value of the relative permittivity.

This procedure brings obviously some shortcomings. If both the agents that participate on the migration have similar values of the input variable, all steps of the migration can cause, that the continuous variable does not leave the original subinterval and the same permittivity is examined again. Another problem is caused by the fact that different values of the input variable means one value of the relative permittivity. Then, the result of the migration depends on the initial value of the migrating agent in the subinterval. Let us consider Eq. (6) from Chap. 1 with no influence of perturbation and parameters PL = 1.3 and ST = 3 and two different migrating agents having only one variable x_q = 0.10 and $x_q^* = 0.90$. Now, let them migrate towards the member of the external archive x_p = 3.5. The resulting steps of the migration correspond to different dielectric materials $\boldsymbol{TMP} = 1.62, 3.13, 4.65$ and $\boldsymbol{TMP}^* = 2.51, 4.02, 5.54$. Beside all the shortcomings, the algorithm is able to solve problems with continuous and discrete variables simultaneously without any change of MOSOMA program. The only change comprises the evaluation of objective functions—discrete values are used according to subintervals of the input variable.

In the example, band-pass filter for the band from 28 to 32 GHz is designed. The frequencies for the objective functions (10) and (11) and constraint functions (12)–(15) are summarized in Table 3. The controlling parameters are set according to Table 4.

The best Pareto front obtained from ten runs of MOSOMA is depicted in Fig. 5. Again, three solutions are highlighted here: the best solution according to the first objective (red marker), second objective (green) and the trade-off solution (blue).

Table 3 Frequency bands for the band-pass filter design

Band	Lower limit (GHz)	Upper limit (GHz)
f_p	28	32
fs	24; 32	28; 36
f_{pc}	29	31
f_{sc}	24; 34	26; 36

Table 4 MOSOMA settings for design of dielectric layered filter

Parameter	FFC	PL	ST	$\lvert P1 \rvert$	T	$N_{ex,\min}$
	15,000	1.3	5	30	20	15

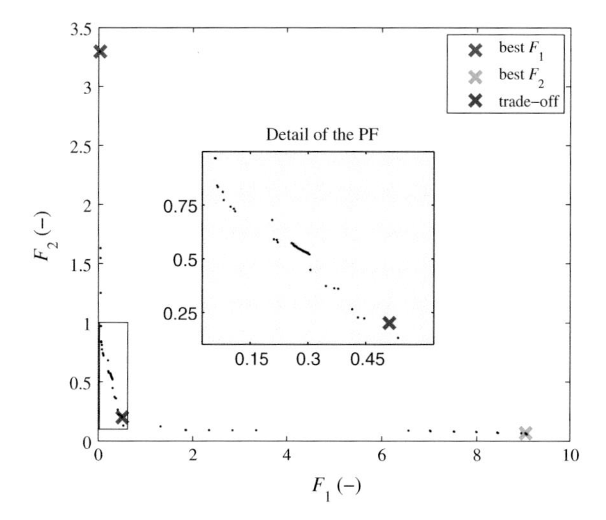

Fig. 5 Pareto front of dielectric filter design having $N = 7$ layers found by MOSOMA [4]

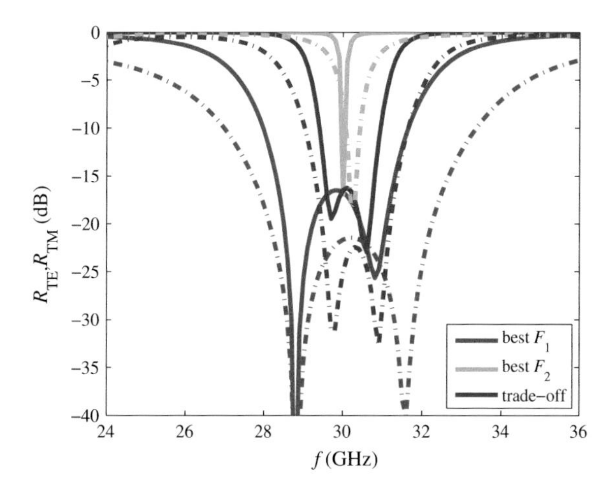

Fig. 6 Reflection properties of three selcted solutions found by MOSOMA [4]

The reflection properties according to frequency of those three solutions is depicted in Fig. 6. Here, the red solution ideally satisfies the first objective, but the last two constraints are violated. On the contrary, green solution suits the second objective very well but violates first two constraint functions. Finally, blue solution respects both the objectives and does not violate any constraint function.

The trade-off solution composed of layers having width respectively 4.686, 1.995, 4.739, 1.001, 1.003, 1.002, 8.663 mm and relative permittivities 10.20, 1.01, 10.20, 1.01, 1.01, 2.94, 2.35 was chosen as the final trade-off solution. This can be compared with solutions published in [9, 14]. Resulting reflection coefficients are compared in Fig. 7. Reflection coefficient for our solution remains below −16 dB for the TE mode and −22 dB for the TM mode in the whole operational band.

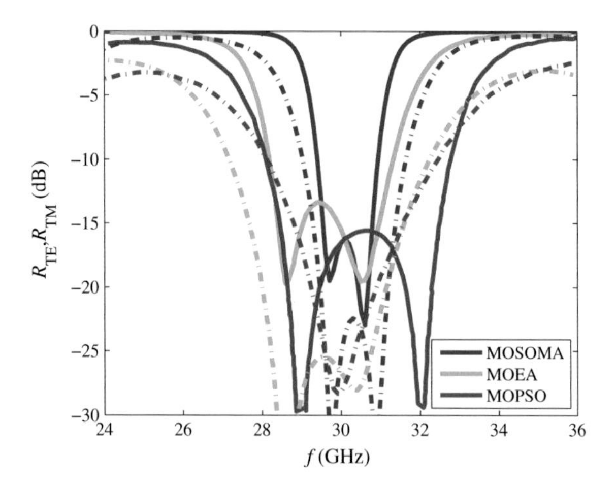

Fig. 7 Reflection properties of band pass filter designed by MOSOMA [4] compared with solution found by MOEA [9] and MOPSO [14]

Coefficients R_{TE} and R_{TM} of our proposal decrease steeper at the boundaries of the desired frequency band than for solutions from [9, 14]. Total width of our design is 23.08 mm compared to 33.44 mm [9] and 21.35 mm [14].

1.3 Adaptive Beamforming in Time Domain

Previous two examples of MOSOMA application in electromagnetics considered design of some product—filter and antenna. In this case, specific physical properties of the designed product were changed by the optimization. In addition, MOSOMA can be applied for controlling purposes, also. In [3], MOSOMA was used to control properties of theoretical approach—digital filter for purposes of the FDTD (Finite Difference Time Domain). The coefficients of filter are designed so that it truncates the computational domain as the perfectly matched layer—no energy is reflected back to the computational domain from its boundary.

Another example of MOSOMA application as controlling mechanism can be the adaptive beamfoming of slot antennas array in time domain presented in [2]. Beam forming is used for an antenna array directional transmissions. It can be applied in many fields of human interest: e.g. radar, wireless communication or biomedicine. Beamforming tries to set antenna elements so that nulls and peaks appears in desired positions of the radiation pattern. It can be achieved either in time domain—by delaying the inputs of individual elements, or in frequency domain—by changing the phase of individual elements. The amplitude of the feeding can vary for both the methods.

To this example, slot antenna array feeding is optimized so the radiated energy is focused to desired space while vanishes in another part of the irradiated body. This scenario meets the requirements of biomedicine hyperthermia [16]. The radiated

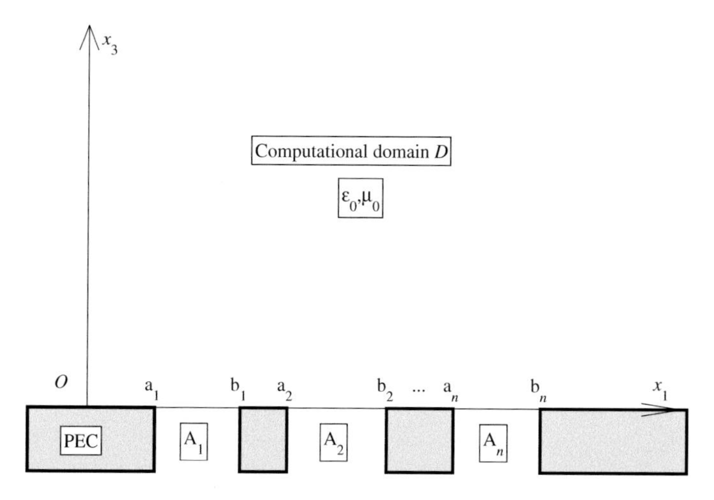

Fig. 8 The antenna array model description [2]

energy is focused to a certain part of the body to heat e.g. a tumor while another parts of the body should not be heated more than is necessary.

The computational scenario can be found in Fig. 8. Here, N non-overlapping slot antennas with a finite width in infinitesimal PEC (Perfect Electric Conductor) ground plane are depicted. The computational domain D is irradiated by the antenna array. It is filled with vacuum with relative permittivity $\varepsilon_r = 1$ and relative permeability $\mu_r = 1$. The nth slot A_n is defined as $a_n < x_1 < b_n$ with respect to $a_1 < b_1 < a_2 < \cdots < b_n$.

Every slot is excited by power exponential EM pulse [17]:

$$V(t) = V_{\max} \frac{t}{t_r}^{\nu} \exp[-\nu(\frac{t}{t_r} - 1)]H(t - T) \tag{16}$$

where $V_{\max}$ denotes pulse amplitude, t denotes time, t_r means rise time when the pulse reaches its amplitude $V_{\max}$, $\nu > 0$ is so called rising exponent, $H(.)$ assigns the Heaviside step function and T is the start time of the pulse. It is assumed that the first slot is excited at time T_1 and consecutive slots ($n = 2, 3, \ldots, N$) are excited at times $T_n > T_1$.

The closed form expressions for the EM pulsed fields radiated by a planar array of slot antennas have been derived in [18]. This time domain approach is based on so called Cagniard-DeHoop method [19]. The field in the computational domain D is the superposition of the fields radiated by individual slots. The space-time (x_1, x_3, t) EM fields can be computed according to [18]:

$$\{E_1, E_3, H_2\} = \sum_{n=1}^{N} \frac{V_n}{b_n - a_n}$$

$$* \left[\frac{1}{\pi} \left\{ \frac{x_3(x_1 - b_n)}{c^2t^2 - x_3^2}; 1; \frac{1}{\mu_0} \frac{t(x_1 - b_n)}{c^2t^2 - x_3^2} \right\} \frac{\mathrm{H}(t - T_{b,n})}{\sqrt{t^2 - T_{a,n}^2}} \right.$$

$$- \frac{1}{\pi} \left\{ \frac{x_3(x_1 - a_n)}{c^2t^2 - x_3^2}; 1; \frac{1}{\mu_0} \frac{t(x_1 - a_n)}{c^2t^2 - x_3^2} \right\} \frac{\mathrm{H}(t - T_{a,n})}{\sqrt{t^2 - T_{a,n}^2}}$$

$$\left. + \left\{ 1; 0; \sqrt{\frac{\varepsilon_0}{\mu_0}} \right\} - [\mathrm{H}(x_1 - b_n) - \mathrm{H}(x_1 - a_n)] \delta(t - \frac{x_3}{c}) \right] \tag{17}$$

where * denotes the time convolution, c denotes velocity of light in vacuum, $T_{a,n}$ denotes the time of arrival from the edge denoted a_n of the nth slot and $\delta(.)$ denotes the Dirac pulse. First two constituents in Eq. (17) are cylindrical waves from the edges of the slot, third constituent is a plane wave propagating from the slot aperture. All the components contain the time convolution integral that can be solved numerically. The field components from Eq. (17) can be used for computation of two components of the Poynting vector S:

$$\begin{aligned} S_1 &= -E_3 H_2 \\ S_3 &= -E_1 H_2 \end{aligned} \tag{18}$$

In the example, slot antenna array having $N = 5$ elements is considered. All the slots have the same width W and are in positions with spacing $w/2$. In this configuration, 14 parameters can vary during the optimization: excitation times T_n of four most right slots, amplitudes $V_{\max,n}$ of the excitation pulses for all the slots ($v = 2$ for Eq. (16)) and rice times of the excitation pulses $t_{r,n}$.

Knowing that, two objectives can be formulated. First one maximizes the energy in point $\mathrm{P}_{\max}$ while the second one minimizes it in point $\mathrm{P}_{\min}$ at time $t/t_n = 10$. Points of interest are defined as follows: $\mathrm{P}_{\max}\{x_1/w = 0, \quad x_2/w = 6.25\}$ and $\mathrm{P}_{\min}\{x_1/w = 2.25, \quad x_2/ = 6.25\}$. Then, objective F_1 is:

$$F_1 = -\sqrt{\left[\frac{S_1(\mathrm{P}_{\max})}{S_n}\right]^2 + \left[\frac{S_3(\mathrm{P}_{\max})}{S_n}\right]^2} \tag{19}$$

and second objective F_2:

$$F_1 = +\sqrt{\left[\frac{S_1(\mathrm{P}_{\min})}{S_n}\right]^2 + \left[\frac{S_3(\mathrm{P}_{\min})}{S_n}\right]^2} \tag{20}$$

Table 5 MOSOMA settings for control of adaptive beamforming of slot antennas array

Parameter	*FFC*	*PL*	*ST*	$\lvert P1 \rvert$	*T*	$N_{ex,\min}$
	45,000	1.3	5	30	50	30

where S_n denotes the Poynting vector of the TEM mode propagating in a parallel plate waveguide used for the feeding of the apertures.

The settings of MOSOMA for the experiment have been chosen with respect to $N = 14$ variables of the optimization problem. Settings for MOSOMA parameters has been set according to Table 5. The bounds for individual variables were defined as follows: $T_n \in \langle 0; 2t_n \rangle$, $V_{\max,n} \in \langle 0; 1 \rangle$ and $t_{r,n} \in \langle 0; 1.5t_n \rangle$.

The non-dominated solutions found by MOSOMA are depicted in Fig. 9. The selected trade-off solution is depicted as a green cross here. The values of corresponding variables for the trade-off solution are: $\boldsymbol{x} = \{0.50;\ 1.00;\ 2.00;\ 2.00;\ 1.00;\ 1.00;\ 1.00;\ 1.00;\ 1.00;\ 1.50;\ 1.44;\ 1.50;\ 1.50;\ 1.35\}$.

EM fields time evolution is depicted in Fig. 10. For the sake of place, just four steps for times $n = 2$, 3, 7 and 10 are depicted here. As can be seen here, at the time $n = 10$, the radiated energy is focused to the region of $P_{\max}$ (devoted in the figure as green rectangle), while the energy in point $P_{\min}$ is minimal. Thanks to the formulation of the objective functions, the radiated field needs to be computed only in two points at one time of the computational domain D, which decreases the computational demands of this approach.

Now, if we apply the feeding scheme found by MOSOMA periodically, the selected part of irradiated body will absorb much more energy than the rest of it. As a result of that, the temperature of the tissue should increase in the region of interest.

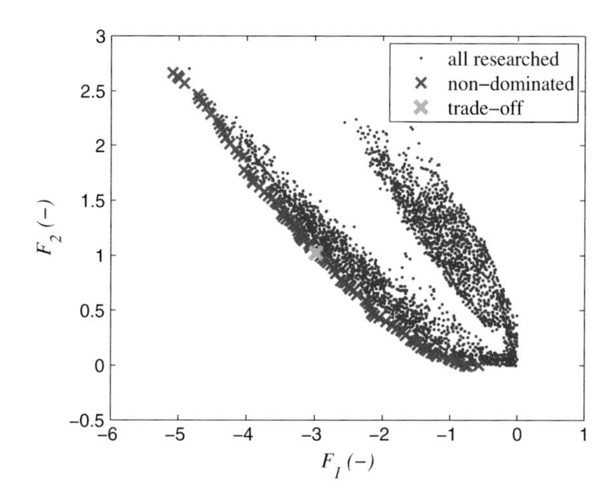

Fig. 9 The Pareto front of the adaptive beamfoming problem found by MOSOMA [2]

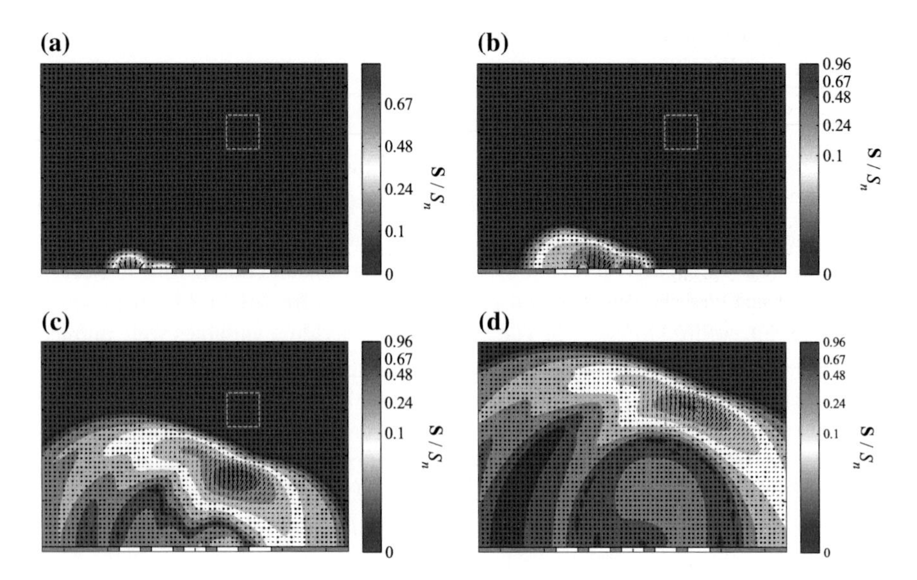

Fig. 10 Time evolution of EM fields radiated by the slot antenna array [2]. **a** Time step n = 2, **b** Time step n = 3, **c** Time step n = 7, **d** Time step n = 10

Acknowledgements Research described in this chapter was financially supported by Czech Science Foundation under grant no. P102/12/1274. Support of projects SIX CZ.1.05/2.1.00/03.0072 is also gratefully acknowledged.

References

1. Kadlec, P., Raida, Z.: A novel multi-objective self-organizing migrating algorithm. Radioengineering **20**, 77–90 (2011)
2. Kadlec, P., Štumpf, M., Raida, Z.: Adaptive beam forming in time-domain. In: 2011 International Conference on Electromagnetics in Advanced Applications (ICEAA) (2011). doi:10.1109/ICEAA.2011.6046359
3. Wiktor, M., Kadlec, P., Raida, Z.: Performance limits for low order absorbing boundary conditions in waveguides in time domain. In: 22nd International Conference on Radioelektronika, IEEE (2012)
4. Kadlec, P., Raida, Z.: Multi-objective self-organizing migrating algorithm applied to the design of electromagnetic components. IEEE Ant. Propag. Mag. **55**, 50–68 (2013)
5. Raida, Z., et al.: Communication subsystems for emerging wireless technologies. Radioengineering **21**, 1036–1049 (2012)
6. Všetula, P., Raida, Z.: Dipole antenna array with synthesized frequency dependency of gain and reflection coefficient. In: 2013 International Conference on Electromagnetics in Advanced Applications (ICEAA) (2013). doi:10.1109/ICEAA.2013.6632410
7. Yagi, H.: Beam transmission of ultra short waves. Proc. IRE **16**, 715–741 (1928)
8. Uda, S., Mushiake, Y.: Yagi-Uda Antenna. Maruzen Co. (1954)
9. Venkatarayalu, N.V., Ray, T.: Optimum design of Yagi-Uda antennas using computational intelligence. IEEE Trans. Ant. Propag. **52**, 1811–1818 (2004)

10. Kuwahara, Y.: Multiobjective optimization design of Yagi-Uda antenna. IEEE Trans. Ant. Propag. **53**, 1984–1992 (2005)
11. Kadlec, P.: MOSOMA: Matlab codes. Brno University of Technology. Available via DIALOG. http://www.urel.feec.vutbr.cz/kadlec/userfiles/downloads/MOSOMA/mosoma.zip (2013). Cited 20 Oct 2014
12. Burke, G.J., Poggio, A.J.: Numerical Electromagnetics Code (NEC)—Method of Moments. Lawrence Livermore Laboratory, CA. Report UCID18834 (1981)
13. Venkatarayalu, N.V., Ray, T., Yeow-Beng, G.: Multilayer dielectric filter design using a multiobjective evolutionary algorithm. IEEE Trans. Ant. Propag. **53**, 3625–3632 (2005)
14. Goudos, S.K., Zaharis, Z.D., Salazar-Lechuga, M., Lazaridis, P.I., Gallion, P.B.: Dielectric filter optimal design suitable for microwave communications by using multiobjective evolutionary algorithms. Microwave Opt. Technol. Lett. **49**, 2324–2329 (2007)
15. Chew, W.C.: Waves and Fields in Inhomogeneous Media. IEEE Press, Piscataway (1994)
16. Fenn, A.J., King, G.A.: Experimental investigation of an adaptive feedback algorithm for hot spot reduction in radio-frequency phased-array hyperthermia. IEEE Trans. Biomed, Eng. **45**, 273–280 (1996)
17. Quak, D.: Analysis of transient radiation of a (traveling) current pulse on a straight wire segment. In: 2001 IEEE International Symposium on Electromagnetic Compatibility, EMC (2001). doi:10.1109/ISEMC.2001.950488
18. Stumpf, M., de Hoop, A.T., Lager, I.E.: AClosed-form time-domain expressions for the 2D pulsed EM field radiated by an array of slot antennas of finite width. In: 2010 URSI International Symposium on Electromagnetic Theory (EMTS) (2010). doi:10.1109/URSI-EMTS.2010.5637269
19. de Hoop, A.T.: A modification of Cagniard's method for solving seismic pulse problems. Appl. Sci. Res. **B8**, 349–356 (1960)

Utilization of Parallel Computing for Discrete Self-organizing Migration Algorithm

Marek Běhálek, Petr Gajdoš and Donald Davendra

Abstract Evolutionary algorithms can take advantage of parallel computing, because it decreases the computational time and increases the size of processable instances. In this chapter, various options for a parallelization of DISCRETE SELF-ORGANISING MIGRATING ALGORITHM are described, with three implemented parallel variants described in greater detail. They covers the most frequently used hardware and software technologies, namely: parallel computing with threads and shared memory; general purpose programming on GPUs with CUDA; and distributed computing with MPI. The first two implementations speed up the computation, the last one moreover changes the original algorithm. It adds a new layer that simplifies its usage in the distributed environment.

1 Introduction

DISCRETE SELF-ORGANISING MIGRATING ALGORITHM (DSOMA) [7, 8] is one of the newer meta-heuristic algorithms. It has been developed to solve (NP-hard) combinatorial optimization problems. For such problems (it is believed), that there is no

M. Běhálek (✉) · P. Gajdoš · D. Davendra
FEECS, Department of Computer Science, VŠB—Technical University of Ostrava, Ostrava, Czech Republic
e-mail: marek.behalek@vsb.cz

P. Gajdoš
e-mail: petr.gajdos@vsb.cz

D. Davendra
e-mail: DonaldD@cwu.edu; donald.davendra@vsb.cz

M. Běhálek · P. Gajdoš
IT4Innovations National Supercomputing Center, VŠB—Technical University of Ostrava, Ostrava, Czech Republic

D. Davendra
Department of Computer Science, Central Washington University, 400 E. University Way, Ellensburg, WA 98926-7520, USA

D. Davendra and I. Zelinka (eds.), *Self-Organizing Migrating Algorithm*,
Studies in Computational Intelligence 626, DOI 10.1007/978-3-319-28161-2_6

exact method to get a solution in a reasonable period of time. Still, computer scientists and programmers frequently encounter such problems and they are often addressed by heuristic methods or approximation algorithms. But even if we use these algorithms, the solution remains still very computationally demanding. Parallel/distributed computing in this aspect can be a great asset. It can decrease the computational time or increase the size of processable instances.

In this chapter, different options how to take advantage of the parallel computing for DSOMA are explored. Firstly, approaches that are frequently used in this area are described. Then various options how to parallelize DSOMA are analyzed. Finally, three selected variants that were implemented are introduced. The last section briefly summarizes results of various experiments that were performed.

2 Levels of Parallelization

Various meta-heuristics have been introduced in the last years. There are articles (for example [1, 21]) that try to categorize the most frequently used approaches, how to extend these algorithms to take advantage from the parallel/distributed computing architecture. DSOMA is a population based meta-heuristic, that iteratively searches for better solutions. Iterations in DSOMA are called migrations. The DSOMA population composes of individuals. In each migration, selected pairs of individuals are used to produce new trial individuals. If a better solution is found among the trial individuals, then the successful trial individual replaces the original individual in the population.

From the algorithm point of view, we can define different levels, where we can apply parallel computing.

- *Fitness function*—usually the most time consuming function is the computation of the fitness function. It is frequently used in every step of DSOMA. First option how to speed up the computation is to speed up this function. On the other hand, used fitness function depends on the solved problem and some fitness functions may be easier to parallelize then others.
- *Constructing trial individuals*—Trial individuals in DSOMA are computed based on two individuals from the population. After the trial individuals are generated, they are repaired and their fitness values are computed. These last two operations are more time consuming then generating the trial individuals, moreover for a given set of trial individuals they are independent and thus suitable to be performed in parallel.
- *Computing a migration*—a common strategy is to select pairs of individuals and construct trial individuals by combining the current best individual in the population with the remaining individuals. Computing the trial individuals for a given pair does not depend on other pairs and it can be performed in parallel.

To improve the obtained results, DSOMA (similarly to other evolutionary algorithms) utilizes some local search algorithm. In [8], the *2-Opt* local search is utilised. Because an application of such local search algorithm on a single individual can be even more time consuming then the whole DSOMA computations, its parallelization can be crucial, if we want to speed up the computation.

Moreover, similarly to other population based evolution algorithms, we can add some new layers to improve the overall parallel behaviour. For example, if we want to use a computer (cluster) with the distributed memory architecture (they represent a majority between current supercomputers) then none of the previously described levels may be appropriate. The reason is the communication bottleneck, which can outweigh the speed up achieved by the parallel execution.

The most popular and simplest model in the distributed environment is the so called *island model* [1, 2]. In this model, the population is partitioned in a small set of sub-populations (islands, colonies). These islands execute the original algorithm and then some (sparse) individuals exchange algorithm is applied to exchange information between such islands. It was shown, that sparse information exchange between such islands brings diversity into the population and thus prevents convergence in local optima. Moreover, the overall parallel behaviour is affected by the used communication topology or exchange data rate [14, 22].

In practical applications, some hybrid approaches that combine more than one level of parallelization are used (see Sect. 4.2). In general, the higher level of parallelization is coarse-grain implementation (for example the mentioned set of islands) and then each of these islands integrates other parallel model (or models).

3 Hardware and Software Options for Parallelization

In Sect. 2, different levels of parallelization for DSOMA were mentioned. But the *appropriate* level (or levels) of parallelization depends also on the target hardware and used software technologies.

The following paragraphs summarize the most distinguishing options that are state-of-the art in the world of high performance computing (HPC). The first such choice, that greatly affects the solution especially from a programmer's perspective, is a target device that will be used for computing. There are two main options:

- *general purpose processors*—current (super)computers used for HPC contains up to hundreds of thousands of computational cores. Even current mainstream desktop processors contain between 4 and 8 cores. Considering the usage of general purpose processors, the parallelization usually implements a model—*Multiple Program–Multiple Data* (MPMD). In this model, each core runs independently of others and can perform a different program with its own data.
- *many-core (or massively multi-core) coprocessors*—the most common devices from this category are graphics processing units (GPUs) that allows general purpose programming. However, recently Intel has also introduced their Xeon

PHI,[1] a coprocessor for HPC. These coprocessors usually contain many (tens or hundreds) lightweight cores and frequently implement Single Instruction—Multiple Data (SIMD) model. It describes devices with multiple processing elements that perform the same operation on multiple data simultaneously.

Considering the many-core coprocessors, usually the computation is divided between general purpose processor and many-core coprocessor. Such SIMD coprocessors successfully exploit data level parallelism, but not concurrency. A common approach is to use the coprocessor when appropriate and for remaining computations use the general purpose processors. Moreover, most applications do not use the coprocessor and CPU at the same time [13, 16]. Some effort has been made to exploit the full computation power of CPUs and coprocessors at the same time for evolutionary algorithms [26], but this idea is not explored in this chapter any further.

Current computers used in HPC frequently combine both type of devices. For example a computer named TITAN (from the *Top 500* list[2]) is a cluster composed from 18.688 nodes where each node contains a traditional general purpose processor with 16 cores and the NVIDIA TESLA K20 GPU accelerator.

Considering general purpose processors, there are two main memory architectures.

- *Shared memory architecture*—in this architecture, all processors share the same memory and this memory is used to exchange data and for synchronization. This architecture is frequently used for *smaller* HPC computers with tenths or hundreds computing cores.
- *Distributed memory architecture*—it is the main model in todays supercomputers. In this architecture, a computer (usually a cluster) composes from interconnected nodes. Each of these nodes usually contains multiple computational cores with the shared memory. Nodes can exchange some kind of messages, but they cannot directly access a memory in a different node. Thus the memory is in fact distributed among the nodes.

Considering HPC, the most widely used programming languages are C, C++ or FORTRAN. In this chapter, we will focus on C++, as our original sequential implementation of DSOMA is also in C++.

Concurrent programming with threads is the simplest way as to how to create a parallel application in the shared memory environment. Nearly every programming language supports this programming paradigm. Also C++ contains libraries for concurrent programming with threads.[3] Moreover, there are technologies that simplify the development of such applications. Like an example we can list OPEN

[1] http://www.intel.com/content/www/us/en/processors/xeon/xeon-phi-detail.html.

[2] http://www.top500.org/system/177975.

[3] http://www.cplusplus.com/reference/thread/thread/.

MULTI-PROCESSING (OPENMP),[4] CILK and CILK++.[5] For the distributed memory model, the de facto standard is MESSAGE PASSING INTERFACE (MPI).[6]

The technologies for programming on GPUs have quickly evolved in last ten years. Currently, there are various approaches and technologies. Earlier, general purpose computing on graphics processing units was primarily based on modification of graphics pipeline, e.g. in OPEN GRAPHICS LIBRARY (OPENGL) where programmable shaders are used for time consuming BASIC LINEAR ALGEBRA SUBPROGRAMS (BLAS). Nowadays, OPEN COMPUTING LANGUAGE (OPENCL) is the currently dominant open general-purpose GPU computing language. It is an open standard defined by the Khronos Group.[7] The dominant proprietary framework is NVIDIA's COMPUTE UNIFIED DEVICE ARCHITECTURE (CUDA) that was launched in 2006 as an SDK and API that allows using the C programming language to code algorithms executable on GPUs. MICROSOFT[8] also introduced its own API called DIRECTCOMPUTE that supports general-purpose computing on GPUs, this time with full support of DIRECTX.

In the following subsections, technologies that are later used for parallelization of DSOMA are introduced in more detail.

3.1 OpenMP

OPENMP is maybe the most widely used technology considering shared memory architecture. Moreover, it is also frequently used in combination with MPI for distributed memory systems. OPENMP is an API that supports multi-platform shared memory programming in C, C++, and FORTRAN. Its support is implemented on most processor architectures and operating systems. It consists of a set of compiler directives, library routines, and environment variables that influence run-time behavior.

The simplest example of the OPENMP usage is the automatic parallelization of `for` cycles. Listing 1 demonstrates a simple `for` cycle. The `pragma` directive defines, that this cycle should be parallelized by OPENMP. OPENMP is supported by a wide range of compilers. For example, its support is built in widely used GNU COMPILER COLECTION (GCC).[9] This directive is ignored by a compiler, if there is no OPENMP support or if it is switched off.

[4]http://openmp.org/.

[5]https://www.cilkplus.org/.

[6]http://www.mpi-forum.org/.

[7]http://www.khronos.org.

[8]http://www.microsoft.com.

[9]https://gcc.gnu.org/onlinedocs/libgomp/Enabling-OpenMP.html.

Listing 1: A simple for in OpenMP

```
const int N = 100000;
int a[N];

#pragma omp parallel for
for (int i = 0; i < N; i++) {
        a[i] = 2 * i;
}
```

Listing 2 shows how to enable OpenMP with GCC g++ compiler (parameter -fopenmp). Using the environment variable OMP_NUM_THREADS, it is possible to define the number of threads that are used during the execution.

Listing 2: Compilation and execution of a source code with OpenMP pragmas in g++

```
g++ -fopenm source.cpp -o result
OMP_NUM_THREADS=4 ./result
```

Even if OpenMP automatically parallelize annotated for cycles, other issues like a concurrent modification of shared data, must be solved by a programmer. OpenMP provides additional constructs controlling data sharing or synchronization.

3.2 Message Passing Interface

MPI is a language-independent communications protocol used to program parallel computers or clusters with distributed memory architecture. It defines a message-passing interface, together with protocol and semantic specifications for how its features must behave in any implementation. The MPI applications compose from independent processes where the MPI messages and constructs are the only communication mechanisms. It is used to exchange data and also to solve synchronization issues.

The primary functionality is a point-to-point and collective communication. Listing 3 shows a simple example of a point-to-point communication in MPI. Similarly to OpenMP, a user specifies the number of processes when the application starts. Each process is uniquely identified by its rank (a number from 0 to the number of processes). In the following example, two processes are expected. Process ranked as 0 sends some data and the process 1 is waiting until the data arrives.

Listing 3: A simple MPI example

```
MPI_Init (&argc , &argv );
int myrank ;
MPI_Comm_rank(MPI_COMM_WORLD, &myrank);
if( myrank == 0){
  int data[10];
  MPI_Send(data, 10, MPI_BYTE, 1, 1, MPI_COMM_WORLD);
}
if( myrank == 1){
  int data [10];
  MPI_Recv(data, MPI_BYTE, 10,
    MPI_ANY_SOURCE, MPI_ANY_TAG, MPI_COM_WORLD);
}
MPI_Finalize ();
```

Even though MPI is relatively simple to use, it represents a quite low-level interface. Hence development of applications directly in C++ with MPI can be laborious and time-consuming. Furthermore, the complexity of creating parallel applications lies also in other supportive activities like debugging and profiling. Even an experienced programmer of sequential applications can spend a lot of time learning a new set of complex tools. Therefore, for many non-experts in the area of parallel computing, it can be difficult to create, debug, and analyze their distributed applications.

To overcome the described issues, the tool KAIRA [3, 4] was used for the development of a distributed variant of DSOMA KAIRA is a complete development environment for MPI C++ applications. It provide an environment in which a user can implement and experiment with his or her ideas in a short time; create a real running program; and verify its performance, scalability, and correctness.

KAIRA is an open source application and it is freely available at: http://verif.cs.vsb.cz/kaira/.

3.2.1 Brief Introduction into KAIRA

The key aspect of KAIRA is its usage of a visual program. A user specifies *communication* and *parallel aspects* in a visual way. However, the application is not completely programmed visually. Sequential parts of the developed application are written in C++ in a textual form and they are inserted into the visual program. So for real applications, the visual program is usually relatively small. The visual representation serves also as a natural unifying element for supportive activities like debugging and performance analysis. The used visual programming language is inspired by COLORED PETRI NETS (CPNS) [10].

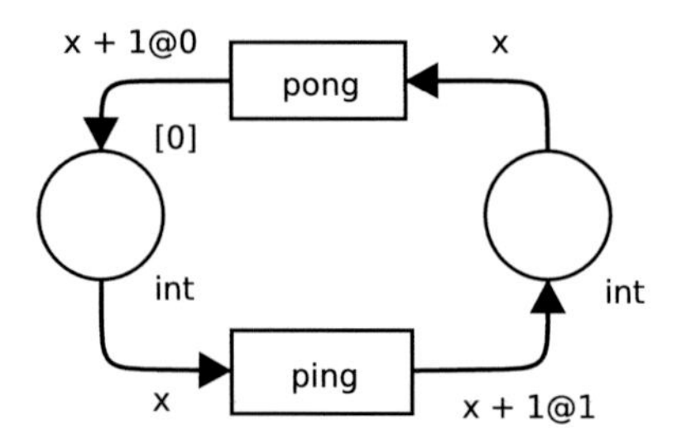

Fig. 1 The *ping-pong* example's visual program

From the combination of the visual program (that captures parallel aspects) and inserted sequential codes, Kaira is able to generate a stand-alone MPI application in a fully automatic way. Such a generated program can be run directly on a cluster computer. For debugging purposes, multi-threaded and sequential versions of the application can be generated. It is important to mention that Kaira is *not* an automatic parallelization tool. It does not discover parallelisms in applications. The user has to explicitly define them, however they are defined in a high-level way and the tool derives the implementation details.

In Sect. 4.2, a Kaira's visual program is used to present a parallel behaviour of an implemented distributed variant of DSOMA. The basic notation is the same as that of CPNs, hence circles (places) represent memory spaces and boxes (transitions) represent actions. Places have its types defined in lower right corner. In the upper right corner, there is an initial marking. Moreover, transitions can have priorities (a number in a upper right corner). Transitions with higher priorities are executed (fired) before transitions with lower priorities.

Figure 1 shows a simple example—*Ping-Pong* in Kaira.[10] This net is executed on each (MPI) process. An enriched C++ is used as an inscription language on arcs. When fired, every transition takes values (tokens) from its input places and produces tokens into its output places. The expression in the form `expr@target` means that created tokens (by evaluating `expr`) are sent to another process determined by its evaluating `target`. It allows communication between (MPI) processes.

In this example, first two processes exchange an integer token. Transition `ping` takes a value (named x in a scope of this transition), increments this value (by the expression (x + 1)) and sends it to the process 1 (`@1` on its output arc). Similarly, transition `pong` takes a value, it increments this value and sends it to the *process 0*. While the initial marking is used only for *process 0*, there is just one token `0` at *process 0* at the beginning of the computation and the only enabled transition is the transition `ping` on *process 0* (it is detonated by *green* elements in Fig. 2). After it is

[10]The example is a part of Kaira's distribution.

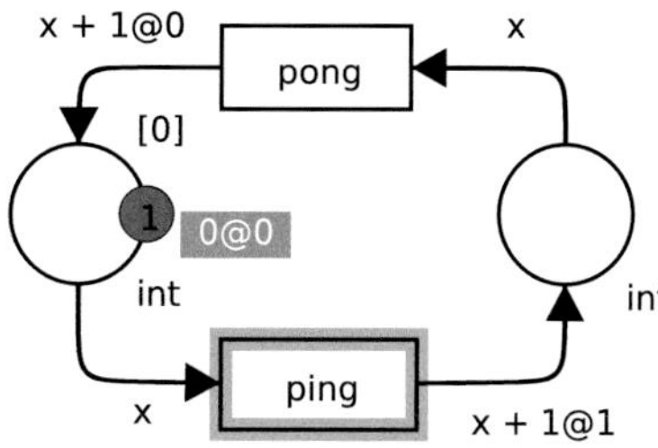

Fig. 2 Initial marking

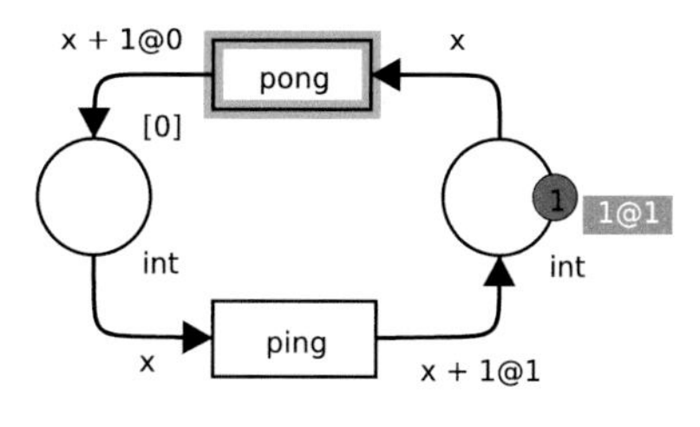

Fig. 3 First step in the computation

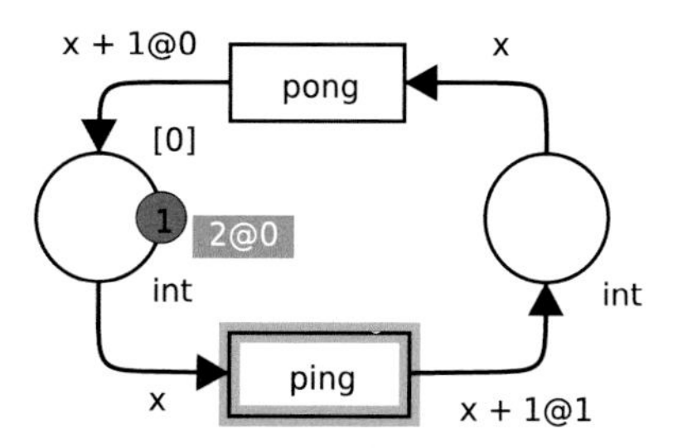

Fig. 4 Second step in computation

fired, there is just one token with a value `1` on *process 1* and only the transition `pong` is enabled (see Fig. 3). After the transition `pong` is fired, the token returns to *process 0* and the exchange can start again (Fig. 4).

The double bordered transition contains a C++ code that is executed whenever the transition is fired. This code is in fact a simple function with predefined definition. This definition is automatically derived from the net.

Figure 5 contains an initial marking as an example, where a C++ code is inserted into the transition `Compute`. This code is present in Listing 4. The structure `Vars` and the function's header are generated automatically. Arcs detonated with `[bulk]` construct take not only one token but all tokens during the transition's firing. The transition `Compute` takes all numbers from place `p1` and it multiplies them by a value from `p2`. The results are inserted into the vector `result`. The first step in computation is captured by Fig. 6.

Listing 4: The function inserted into the transition `Compute` form Figure 5

```
struct Vars {
    std::vector<int> &result;
    int &x;
    ca::TokenList<int > &y;
};

void transition_fn(ca::Context &ctx, Vars &var)
{
    ca::Token<int> *t;
    for(t=var.y.begin(); t!=NULL; t=var.y.next(t)) {
        var.result.push_back(t->value * var.x);
    }
}
```

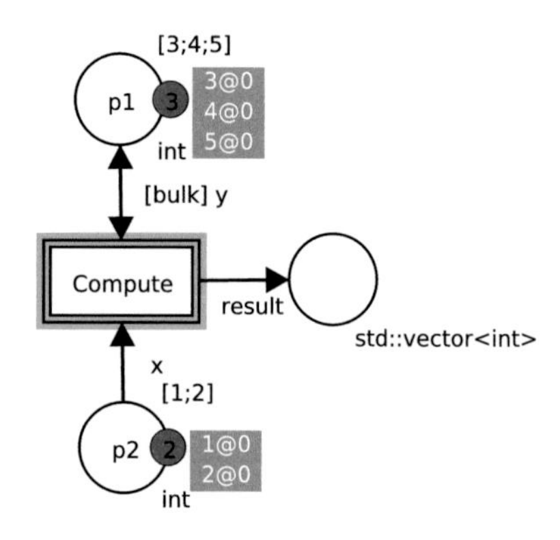

Fig. 5 An example with inserted C++ code

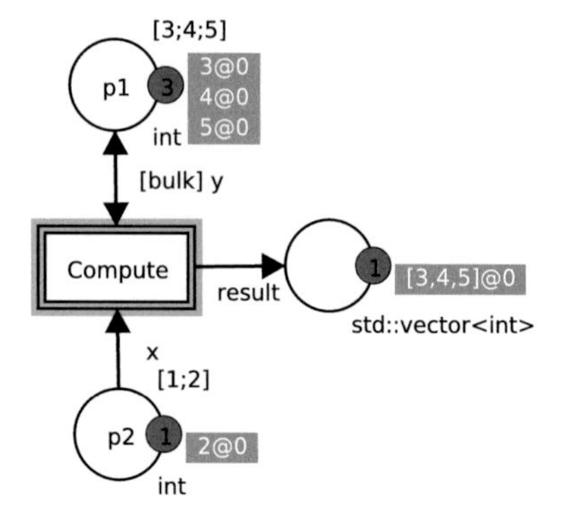

Fig. 6 A marking after the transition `compute` was fired

3.3 GPU Computing with CUDA

Compute Unified Device Architecture (CUDA) [5, 28] was introduced by NVIDIA as a general parallel programming and computing platform in 2006. Although utilization of CUDA was very limited at the beginning and the first GPU chips based on G80 Tesla Architecture were too expensive, it became very popular after a few years. NVIDIA developed their own hardware architecture that enables solving known problems in shorter time due to massive parallelism. The GPUs were well-suited to address real problems that could be expressed as data-parallel computations. Except for the gaming community that still plays a very important role for hardware producers and vendors, NVIDIA decided to focus on research and computational areas as well. Nowadays, CUDA is the most popular and supported architecture running on GPUs [6, 12, 25] and its environment allows a heterogeneous programming approach.

Simply stated, a part of program pipeline can be processed by GPU, whereas another part by CPU. Moreover, current design patterns for parallel programming [13, 15, 16, 27] strongly suggest to distinguish between CPU and GPU parts to achieve the best application performance. A good program structuring itself does not ensure expected results. Memory arrangement and data alignment belong to the most important tasks that the programmer has to solve. CUDA ready graphic cards have basically five types of memory. The biggest is a global memory, which serves as a communication point between GPU and CPU and a primary storage if the data can not be given somewhere else. Since global memory is too slow, a shared memory (shared by all threads in one thread-block) is often employed to achieve better performance. Registers represent the fastest memory. Constant and texture memory are used in such cases where the read-only data structures can be used. The most important functions performed on GPUs are called kernels. Every kernel needs its runtime configuration that consists at least of CUDA grid and block settings. Both grid and blocks can have up to three dimensions. The grid consists of blocks and every block encapsulates a set of threads. The computation is performed on several streaming multi-processors independently (Maxwell Multiprocessor SMM on Maxwell architecture). The maximum number of blocks, threads and SMMs depends on GPU specification. We refer to [5, 20, 23, 28] for more details on CUDA programming since it is out of scope of this chapter.

Internal architecture of GPUs is suitable for vector and matrix algebra operations. That leads to the wide usage of GPUs in the area of information retrieval, data mining, image processing, data compression, etc. Nowadays, programmers usually choose between OpenCL which is supported by all hardware producers, and CUDA which is supported by NVIDIA only. An important benefit of OpenCL is its platform independence; however, CUDA still sets the trends in GPU programming.

CUDA kernels are usually relatively complex and particular implementation can require suitable data arrangement and indexing. Listing 5 represents an illustrative kernel implementation of the same loop as can be seen in Listing 1.

Listing 5: A simple kernel in CUDA

```
__global void foo(const unsigned int N, int* a)
{
    unsigned int tOffset = blockIdx.x * blockDim.x + threadIdx.x;

    while (tOffset < N)
    {
        a[tOffset] = 2 * tOffset;
        tOffset = blockIdx.x * blockDim.x;
    }
}
```

4 Parallelization of DSOMA

In this section, three implemented parallel versions of DSOMA are introduced. All these variants are implemented in C++. First two solutions use general purpose processors. First is an OpenMP solution that uses the shared memory. OpenMP was chosen because it is relatively easy and it provides a meaningful speedup even on a common desktop computer. Second solution uses the distributed memory architecture. From the programmer's perspective, it combines the usage of MPI with OpenMP. This solution can meaningfully use a computational power of hundreds of processors. Finally, the last solution uses the CUDA technology for the general purpose programing on NVIDIA GPUs.

4.1 OpenMP Implementation of DSOMA

OpenMP is a relatively easy to use technology. It is usually easy to identify time consuming `for` cycles that are suitable for the parallelization with OpenMP. Still, to get a meaningful speed up and at least modes scalability, it requires additional work and a programmer's insight.

Simplified `main` method from our original sequential DSOMA implementation is captured in Listing 6. The most important class is `Soma`. The instance of this class allocates a memory space for storing the DSOMA population, information as to which individual represents the current best solution and also it contains a memory space for computing the fitness function and constructing trial individuals.

What remains are in fact two `for` cycles. The outer one is not suitable for a parallelization, because the current best solution is used during the construction of trial individuals. So, each step depends on results of the previous steps. In contrary, the inner cycle computes trial individuals for selected pairs from the population. Each of these computations is in fact independent and the trial individuals for every pair can be computed in parallel.

Listing 6: Initial simplified DSOMA `main` function

```
Problem problem = new Problem();     // a problem description
// an object storing the population and an allocated memory
// for computing the fitness function and jump solutions.
Soma soma(problem);
soma.InitializePopulation();

for(int i=0; i<soma.GetNumberOfMigrations(); i++)
{
    for(int j=0; j<problem.numberOfIndividuals; j++)
    {
        // compose trial individuals between the best solution in
        // the population and i-th solution in the population
        soma->ComputeTrialIndividuals(i);
    }
    soma.UpdateBestSolution();
}
```

So, the basic idea is to add the OpenMP `pragma` to parallelize the inner cycle. Still, there are some issues. First of all, there are some memory issues that can lead to incorrect results. To compute the fitness function or the trial individuals in parallel, each of the involved threads needs its own memory to store data for ongoing computations. Moreover, while the population changes during the computation of a migration (new better individuals are included into the population), it is easier to prepare the whole migration before its computation starts. These modifications along with OpenMP clauses are captured in Listing 7.

Listing 7: The modified DSOMA `main` function with OpenMP *pragmas*

```
Problem problem = new Problem();      // a problem description
Soma soma(problem);                   // the population
soma.InitializePopulation();

for(int i=0; i<soma.GetNumberOfMigrations(); i++)
{
    //prepared migration
    std::vector<Migration*> migration;
    soma.PrepareMigration(migration);

    #pragma omp parallel default(shared) shared(soma, migration)
    {
        //contains a memory for computing the fitness function
        Fitness fitness(problem);

        //a memory buffer for generated trial individuals
        MemoryForTrials trials(problem.numberOfJobs);

        #pragma omp for schedule(dynamic,20)
        for(int j=0; j<migrations.size(); j++)
        {
            Solution solution;
            migration.at(j)->Compute(solution, fitness, trials);
            soma.IncludeSolution(solution);
            delete migration.at(j);
        }
    }
    migration.clear();
    soma.UpdateBestSolution();
}
```

After the mentioned memory issues were removed, the parallel implementation computed correct results. Still, the resulting application worked even slower than the original sequential implementation. The reason was due to some functions from the C++ API. For an example the function `rand`[11] can be named. It is frequently called while repairing trial individuals. An access to this function is restricted to a single thread, because it modifies internal state objects and thus its usage slows the whole computation. As a solution, the function `rand` was replaced by the functions `rand_r`.

After such issues were solved, we got the desired OpenMP parallel implementation. Its performance (scalability and speedup) is summarized in Sect. 5.

[11] http://www.cplusplus.com/reference/cstdlib/rand/.

4.2 Distributed Island Model Implementation of DSOMA

As was mentioned in Sect. 3.2.1, a distributed MPI implementation of DSOMA was implemented in Kaira. The visual program capturing the communication aspects is shown in Fig. 7.

By default, initial marking is used only for *process 0*. The blue area is used for initialization in Kaira, defining processes where the initial marking is applied. In our case, it is applied to all processes (the range is defined by the expression: `ca::range(0, ctx.process_count())`). So, every process has its migration counter and an instance of a class Soma. It is the same class as in Listing 7. It stores the DSOMA population and functions necessary to compute new migrations. It represents an island in the mentioned island model.

The most important transition is `Compute migration`. It computes one migration. The same code as in Listing 7 is used. Also for this variant, the same OpenMP constructs are used to speed up the computation of a single migration. Moreover, it optionally sends the current best result to a neighbouring process. In our implementation, islands are connected in a ring configuration, where a process with rank `x` sends its results to a process ranked as `x + 1`. The last process sends results to *process 0*. Results are exchanged every *n*-th migration (it is set as a command line parameter when starting). An `index -1` is used to define *empty* solutions, which are not exchanged between processes.

After the desired number of migration is computed (or a wall-time is reached), the best results from all processes are gathered and the overall best results is obtained.

4.3 GPU Implementation

In case of GPU utilization, the emphasis is put on the partial subtask that must be solved in advance, e.g. data transfers, memory allocations, the blocks and grids

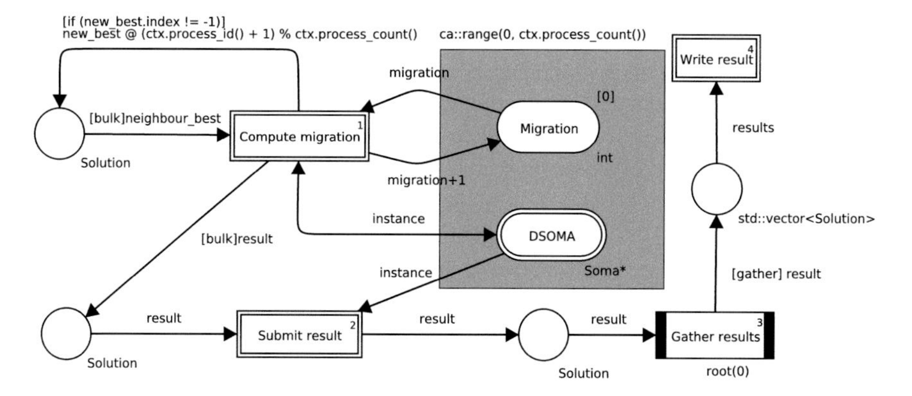

Fig. 7 A visual program in Kaira defining communication for the distributed implementation

settings, etc. Finally, all implementation must cover individual hardware specification to achieve the best performance. This can be illustrated on data alignment and arrangement of thread blocks.

4.3.1 Data Storage, Transfers and Alignment

Searching for the best parallel implementation usually starts with the design of data storage and analysis of all data transfers. There are several kinds of memory on CUDA devices, each with different scope, lifetime, and caching behaviour. The global memory, which resides in the device DRAM, is used for transfers between the host and device as well as for the data input to and output from kernels. Usually, dynamic and/or large data is stored in the global memory. However, the best approach is often based on pre-cached data model, where all static data is copied into fast memory to avoid higher latency of I/O operations. Texture memory meets these requirements. It is much more faster than the global memory due to texture cache and supports enough space for allocation. Moreover, MAXWELL architecture brings a new model that combines unified L1/texture cache. The unified L1/texture cache acts as a coalescing buffer for memory accesses, gathering up the data requested by the threads of a warp prior to delivery of that data to the warp. This function was previously served by separated L1 cache in Fermi and Kepler. Texture memory was used to store a static timetable represented by a matrix $A[M \times J]$, where M is a number of machines, J is a number of jobs and $A[m,j]$ is a time demanded by a machine $m \in \langle 0, 1, \ldots, M)$ to process a job $j \in \langle 0, 1, \ldots, J)$. Other data such as vectors of machines or jobs permutations are stored in the global memory and can be transferred into shared memory within individual CUDA kernels. This data is modified very often when DSOMA is running and the only chance to increase I/O accesses is to use some inner CUDA optimization mechanism, such as *__restricts__* pointers.

In case of DSOMA, all jobs permutations are stored in form of vectors in the global memory. Although accessing global memory is quite slow, the data alignment plays an important role as well and has a significant influence on final computation time.

Let $x \in \mathbb{N}_0^N$ represents a single jobs permutation of dimension N. An element of such permutation is represented by zero-based index of a job (jobID) in DSOMA. Then the permutation x_i is the i-th individual of the population P, where $|P| = \beta$ and $i \in \langle 0, \ldots, \beta)$. Then the i-th individual can be written as N dimensional vector $x_i = \langle x_i^0, x_i^1, x_i^2, \ldots, x_i^{N-1} \rangle$, where x_i^j is an individual's element represented by jobID, where $j \in \langle 0, \ldots, N)$.

In case of Row-Major Format (RMF), all individuals are stored as a single vector v such that the first N elements of the vector v represent the individual x_0, the second N-tuple represents the second individual x_1 etc. Then

$$v = \langle x_0^0, x_0^1, \ldots, x_0^{N-1}, x_1^0, x_1^1, \ldots, x_1^{N-1}, \ldots, x_{\beta-1}^0, x_{\beta-1}^1, \ldots, x_{\beta-1}^{N-1} \rangle.$$

In case of Column-Major Format (CMF), all individuals are stored as a single vector v such that the first β elements of the vector v represent all the first elements of all individuals $x \in P$, the second β-tuple represents all the second elements of all individual $x \in P$ etc. Then

$$v = \langle x_0^0, x_1^0, \ldots, x_{\beta-1}^0, x_0^1, x_1^1, \ldots, x_{\beta-1}^1, \ldots, x_0^{N-1}, x_1^{N-1}, \ldots, x_{\beta-1}^{N-1} \rangle.$$

The RMF or CMF is selected according to inner implementation of CUDA kernels [6]. Figures 8 and 9 illustrate the successive computation of the same schedule in three different ways. In every table (index matrix M), the header row represents a job permutation, and the header column is a vector of indices of machines. Both vectors are indexed from zero in order to simplify this example. Next, all values $M[i,j]$ represent iteration numbers in which the real schedule times $[i,j]$ were computed, e.g. $M[1,0]$ and $M[0,1]$ in Fig. 9 were computed in the second iteration indexed by 1 (zero-based indexing). Finally, Fig. 9 illustrates parallel accesses into matrices by four CUDA threads (4 threads in a block), whereas Fig. 8 shows a sequential computation processed by a single thread which will be described in more detail in the following section.

Having regard to the alignment of the memory, RMF was used in Fig. 9 which ensures required coalesced access into global memory [5, 6, 16]. Contrariwise, CMF must be used in the case of Fig. 8, because it is expected that every thread of

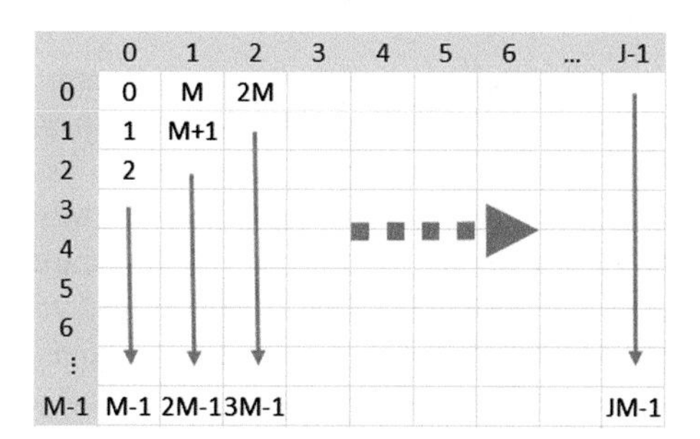

Fig. 8 Single thread computation

	0	1	2	3	4	5	6	...	J-1
0	0	1	2	3					
1	1	2	3	4					
2	2	3	4	5					
3	3	4	5	6					
4	4	5	6	⋮					
5	5	6	⋮	M-1					
6	6	⋮	M-1	M					
⋮	⋮	M-1	M	M+1					
M-1	M-1	M	M+1	M+2					

Fig. 9 Block/Warp computation

the CUDA block will process a single schedule, thus the values $M[i,j]$ of all schedules have to be aligned close to each other.

4.3.2 Data Level Prallelism

The above described memory alignment plays an important role for kernel design as well as arrangement of threads in blocks [15].

Obtaining schedules is the most time consuming part of DSOMA, especially in the case of local search, where a huge number of job permutations must be evaluated. Standard many-core architectures based on latest CPUs solve this problem by data division and distribution among several CPU cores. This is the first level of Data Parallelism (DLP) managed by SIMD architectures. In case of GPU utilization, this kind of data processing can be distributed to the level of individual computation threads. The number of active threads that run in parallel will result in the number of processed schedules. Nevertheless, modern GPUs with support of OpenCL [11, 18, 24] or CUDA [5, 28] enable deeper and more complex data decomposition. Threads within a thread block can cooperate on a single schedule processing, they can share some intermediate data and reduce global memory accesses. Finally, better data parallelism enables higher occupancy of GPU chips that leads to the better computation performance.

Two CUDA kernels that compute individual schedules will be briefly introduced in the following text.

1. Single Thread Computation
2. Block/Warp Computation

NVIDIA Kepler architecture [19] introduced several new features that can significantly decrease kernel run-time. Shuffle instructions brought another way of sharing data among threads within the same warp, in addition to shared memory utilization. Using these instructions however has its limits, e.g. at most 32 threads (=warp size) can be affected by the shuffle instruction call, the number of used registers increases, thread/lane indexing can make some code parts more complex, etc. On the other hand, shuffle instructions can reduce the amount of utilized share memory, eliminate thread synchronization barriers, or reduce total number of instructions with respect to thread data processing and warp data transfers. This can keep CUDA cores busy with memory accesses that have low latency and high bandwidth. The shuffle instructions were primarily used in case of Block/Warp Computation.

4.3.3 Single Thread Computation

In case of single thread computation (see Fig. 8), it is supposed that a single thread will subsequently process one or more schedules. CUDA blocks can be designed to fit inner limits of a device, such as an optimal number of threads with respect to the number of used registers or amount of shared memory. This implementation is suitable especially during the 2OPT search because N^2 schedules must be computed in the worst case.

4.3.4 Block/Warp Computation

Such implementation represents the next level of parallelism, where a set of threads cooperates and computes a single schedule. Figure 9 illustrates a successive evaluation of cells in the grid of a schedule marked by zero-based indices. The same indices mark the cells that are processed in parallel. Let the number of jobs be J. Then there is an inner loop of $(J+B-1)/B$ steps, where B is a block size; a block of 4 threads was used in this illustrative example. In every step, a strip of at most B columns is computed, such that all values in the last active column are called border values and are stored in the shared memory for the next step. If $t \in \{0, 1, 2, 3\}$ is a thread index and s is the zero-based step index, then a thread t computes the whole column $s * B + t$ in every step. In case of $B < 32$, where 32 is the warp size [19], threads can store intermediate data in registers and share them by shuffle instructions to achieve the best performance.

A strip of B columns is computed in a loop of $M+B-1$ iterations. If a thread computes a single value (cell of a schedule), then it moves down and evaluates a new cell in the following iteration reusing previously computed value that was shuffled to subsequent thread simultaneously. The cells marked by number 3 were evaluated in the 4-th iteration, which is the first one, where all threads run the same instructions in parallel. Until then, only $i+1$ threads were active, where $i \in \langle 0, 1, \ldots, M+B-1)$ is the zero-based iteration index. Next, all threads run parallel while $(B-1) \leq i < M$. Finally, the threads subsequently finish their computations during the last B iterations. After that, the thread block shift right and process the next column strip.

This strategy is suitable for faster processing of schedules with respects to GPU limits. As it was mentioned above, the usage of shuffle instructions needs CUDA blocks of at most 32 threads. Moreover, this number of threads limits the total number of active blocks. Although such implementation is more complicated, it finally brings significant performance improvement and it is more suitable for DSOMA in general.

5 Experiments

During the experiments, a fixed set of parameters was used. There are 1000 individuals in the DSOMA population, 300 migrations are computed, and the maximum number of trial individuals is 32. A *2-opt* variant that stops after a better solution is found is used as a local search algorithm and it is applied to the current best solution, if it is not improved in 5 consecutive migrations. These setting may not be optimal for all problems. But in our experiments, we do not evaluate the DSOMA itself, but only its parallel behavior. Moreover, while randomness is a crucial part of the DSOMA execution, runs in our experiments are repeated 10 times and average numbers are presented as results. The experiments were performed on a flow shop instances from the extended TAILLARD SETS [17].

5.1 OPENMP Experiments

In the first experiment, the performance of the OPENMP solution that was described in Sect. 4.1 is evaluated. For this experiment, problems with varying size were used as an input. The varying size is important, because the size of the solved instance can affect the overall parallel performance. The measurements were performed on a computer with 84 cores (14 times 6 core Intel Xeon E5-4610 2.40 GHz) and 1 TB of shared memory.

The average execution times for chosen problems (along with their sizes) are summarized in Table 1. The *relative speedup* and *relative efficiency* are captured in Fig. 10. As defined in [9], the relative speedup on p processes (threads in our case) is a ratio between the execution time on one processor and the execution time on p processors. Similarly relative efficiency is the relative speedup divided by the number of processes. From the experiment results, it can be observed, that the parallel implementation is suitable for larger instances, where the amount of computation overcomes the overhead introduced by parallel execution. The main parallel bottleneck in our solution is the concurrent memory access. To further optimize the memory usage, some fundamental changes to the current implementation needs to be made. Considering the used hardware, it is very hard to achieve the efficiency close to 1 with growing number of threads. Still, for larger instances, the efficiency remained close to 50 % even for 32 threads. In absolute numbers, the computational time for the largest instance was reduced by nearly 600 s from 634 s to 40 s.

Table 1 Execution times (measured in *seconds*) for the OpenMP implementation of DSOMA

Problem			Number of used OPENMP threads					
Name	Machines	Jobs	1	2	4	8	16	32
ta100	10	200	23.22	16.95	14.01	10.51	6.44	5.98
ta110	20	200	45.61	23.93	16.27	10.85	7.72	6.34
ta120	20	500	108.9	59.14	38.61	22.00	12.57	10.67
ta130	50	500	243.56	146.28	76.78	35.98	23.01	15.98
ta140	20	700	287.15	170.40	83.92	44.91	25.57	18.31
ta150	50	700	443.92	233.48	128.95	72.04	39.65	23.97
ta160	20	1000	329.37	182.93	116.36	70.80	34.25	21.24
ta170	50	1000	633.93	368.64	213.18	114.96	64.09	39.26

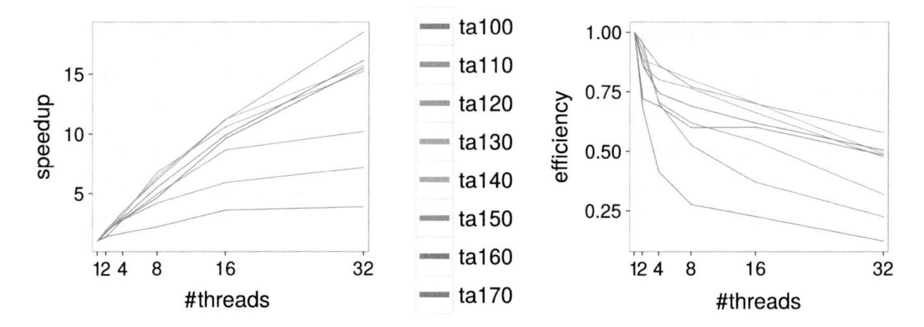

Fig. 10 Relative speedup and relative efficiency graphs

5.2 CUDA Experiments

The second set of experiments was focused on the CUDA implementation of DSOMA. The measurements were performed on a computer with the following specification: AMD FX(tm)-8150 Eight-Core Processor, 3.61 GHz, 32 GB RAM, Windows 64-bit and a single GPU NVIDIA GeForce GTX 970, 4 GB GDDR5 RAM, 13 SMx, Maxwell architecture with CUDA compute capability 5.2.

As it was aforementioned, the data alignment and Single or Warp/Block computation strategy plays an important role in the process of evaluation of time schedules. Hence the first GPU results (see Table 2) illustrate performances of different computation strategies only as a part of the whole DSOMA algorithm. 300 schedules were evaluated in parallel for every input data (row in the Table 2). According to expectations, the greater is the number of jobs, the Block/Warp computation gives better results with comparison to Single Thread Strategy. Then the Column Major Format (CMF) data alignment providing coalesced memory access enables the computation time reduction.

Table 2 Execution times (measured in *milli-seconds*) for the CUDA implementation of different schedulers

Problem			Single Thread		Block/Warp	
Name	Machines	Jobs	CMF	RMF	CMF	RMF
ta100	10	200	**1.90**	1.98	2.79	3.00
ta110	20	200	4.11	3.95	**3.51**	4.18
ta120	20	500	10.04	9.99	**8.46**	10.12
ta130	50	500	23.08	23.95	**13.37**	24.49
ta140	20	700	14.08	13.97	**11.74**	14.30
ta150	50	700	32.28	33.56	**18.53**	34.31
ta160	20	1000	20.27	19.73	**16.75**	20.06
ta170	50	1000	46.19	47.66	**16.55**	53.89

Bold indicates fastest solution for each problem instance

Table 3 Execution times (measured in *seconds*) for the CUDA implementation DSOMA

Problem			Time (s)
Name	Machines	Jobs	
ta100	10	200	3.68
ta110	20	200	4.75
ta120	20	500	10.03
ta130	50	500	12.79
ta140	20	700	13.60
ta150	50	700	16.14
ta160	20	1000	24.64
ta170	50	1000	22.68

The next experiments were focused on the total computation time with the same settings as in the case of Table 1. The Block/Warp strategy was used to evaluate all individual schedules, whereas Single Thread strategy was used during 2Opt local search phase. Nevertheless, the computation time covers the whole DSOMA runtime, i.e. generating individuals and trials, repairing trials, 2Opt search, searing for leader, etc. All partial steps were implemented with the usage of CUDA in the form of individual kernels, and no additional data transfers between host and device were needed. All runtimes are given in Table 3.

5.3 *Distributed DSOMA Experiment*

The final experiment was performed with the distributed variant of DSOMA (described in Sect. 4.2). ANSELM, a supercomputer from IT4INNOVATIONS[12] was used for this experiment. ANSLEM[13] is a cluster composed from 209 computing nodes, where each node contains two Intel Sandy Bridge E5-2665, 8-core, 2.4 GHz processors and 64 GB of physical memory.

In this experiment, 16 islands were used. On every 5-th migration, these islands exchanged their current best solutions with their neighbours. Figure 11 captures the progress in computation and shows the fitness values of the current best solution for every island in time (time is measured in seconds).

[12]http://www.it4i.cz/.

[13]https://docs.it4i.cz/anselm-cluster-documentation/hardware-overview.

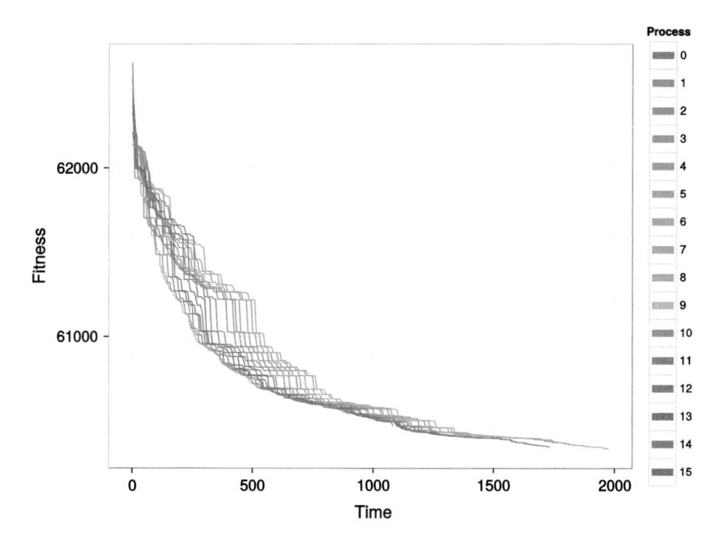

Fig. 11 Progress of fitness values of current best solutions in time in a single run

6 Conclusion

In this chapter, different approaches as to how to take advantage of the parallel computing for DSOMA were explored. From all possible variants, three distinguished options were chosen and implemented. For the first variant, OpenMP which is a relatively easy to use technology at it is suitable even for common desktop computers with a multi-core processor was used. Based on performed experiments, we can conclude that it is possible to achieve meaningful speedup, but the efficiency drops down with growing number of processors. While it was relatively easy to get the first parallel solution with this technology, it still requires a lot of tuning to get a good parallel behaviour for a larger number of processors.

The second implemented solution uses the general purpose programming on GPUs supporting CUDA. This technology can be successfully applied to speed up the computation of tasks like evolutionary algorithms. Also for DSOMA, we were able to significantly reduce the overall computational time. Still, it requires a lot of additional work to use this technology and it requires a skilled CUDA programmer. In our case, the original sequential solution was in fact completely rewritten. Furthermore, it is closely tied to the used hardware and it can be hard to use it on a different GPU.

Previous solutions implement the original DSOMA, where the parallel behaviour is added to the architecture. The last solution adds a new layer (a set of distributed islands) that makes the algorithm suitable for distributed computing. This chapter focuses on parallelization of DSOMA. It does not argue about the quality of obtained results. This is especially true for the distributed variant. It adds new *parameters* like exchange rate or a number of islands and these parameters can affect the quality of obtained results. Such solution can be executed on hundreds of processors, but to use them meaningfully, it still requires a lot of testing and tuning.

Acknowledgements This work was supported by the IT4Innovations Centre of Excellence project (CZ.1.05/1.1.00/02.0070), funded by the European Regional Development Fund and the national budget of the Czech Republic via the Research and Development for Innovations Operational Programme, as well as Czech Ministry of Education, Youth and Sports via the project Large Research, Development and Innovations Infrastructures (LM2011033). The work was also partially supported by Grants of SGS No. SP2015/146 and SP2015/123, VŠB—Technical University of Ostrava, Czech Republic.

References

1. Alba, E., Luque, G., Nesmachnow, S.: Parallel metaheuristics: recent advances and new trends. International Trans. Oper. Res. **20**(1), 1–48 (2013). doi:10.1111/j.1475-3995.2012.00862.x. URL http://dx.doi.org/10.1111/j.1475-3995.2012.00862.x
2. Araujo, L., Merelo, J.: Diversity through multiculturality: assessing migrant choice policies in an island model. IEEE Trans. Evol. Comput. **15**(4), 456–469 (2011). doi:10.1109/TEVC.2010.2064322
3. Böhm, S., Běhálek, M., Meca, O., Šurkovskỳ, M.: Visual programming of MPI applications: debugging, performance analysis, and performance prediction. Comput. Sci. Inf. Syst. **11**(4), 1315–1336 (2014)
4. Böhm, S., Běhálek, M., Meca, O., Šurkovský, M.: Kaira: development environment for mpi applications. In: Ciardo, G., Kindler, E. (eds.) Application and theory of petri nets and concurrency. Lecture Notes in Computer Science, vol. 8489, pp. 385–394. Springer International Publishing, Berlin (2014). doi:10.1007/978-3-319-07734-5_22. URL http://dx.doi.org/10.1007/978-3-319-07734-5_22
5. Cheng, J., Grossman, M., McKercher, T.: Professional CUDA C Programming, 1 edn. Wrox, Birmingham (2014)
6. Cook, S.: CUDA Programming: A Developer's Guide to Parallel Computing with GPUs (Applications of GPU Computing), 1 edn. Morgan Kaufmann, Burlington (2012)
7. Davendra, D., Senkerik, R., Zelinka, I., Pluhacek, M., Bialic-Davendra, M.: Utilising the chaos-induced discrete self organising migrating algorithm to solve the lot-streaming flowshop scheduling problem with setup time. Soft. Comput. **18**(4), 669–681 (2014)
8. Davendra, D., Zelinka, I., Bialic-Davendra, M., Senkerik, R., Jasek, R.: Discrete self-organising migrating algorithm for flow-shop scheduling with no-wait makespan. Math. Comput. Model. **57**(12), 100–110 (2013). doi:http://dx.doi.org/10.1016/j.mcm.2011.05.029. URL http://www.sciencedirect.com/science/article/pii/S0895717711002998. Mathematical and Computer Modelling in Power Control and Optimization
9. Foster, I.: Designing and Building Parallel Programs: Concepts and Tools for Parallel Software Engineering. Addison-Wesley Longman Publishing Co. Inc, Boston (1995)
10. Jensen, K., Kristensen, L.M.: Coloured Petri Nets—Modelling and Validation of Concurrent Systems. Springer, Berlin (2009)
11. Kaeli, D.R., Mistry, P., Schaa, D., Zhang, D.P.: Heterogeneous Computing with OpenCL 2.0, 1st edn. Morgan Kaufmann, Burlington (2015)
12. Kindratenko, V. (ed.): Numerical Computations with GPUs, 2014 edn. Springer, Berlin (2014)
13. Kirk, D.B., Mei, W. Hwu, W.: Programming Massively Parallel Processors: A Hands-on Approach, 2 edn. Morgan Kaufmann, Burlington (2012)

14. Kushida, J.I., Hara, A., Takahama, T., Kido, A.: Island-based differential evolution with varying subpopulation size. In: 2013 IEEE Sixth International Workshop on Computational Intelligence and Applications (IWCIA), pp. 119–124. IEEE (2013)
15. Mattson, T.G., Sanders, B.A., Massingill, B.L.: Patterns for Parallel Programming, 1 edn. Addison-Wesley Professional, Boston (2004)
16. McCool, M., Reinders, J., Robison, A.: Structured Parallel Programming: Patterns for Efficient Computation, 1st edn. Morgan Kaufmann, Burlington (2012)
17. Metlicka, M., Davendra, D., Hermann, F., Meier, M., Amann, M.: GPU accelerated NEH algorithm. In: 2014 IEEE Symposium on Computational Intelligence in Production and Logistics Systems (CIPLS), pp. 114–119 (2014). doi:10.1109/CIPLS.2014.7007169
18. Munshi, A., Gaster, B., Mattson, T.G., Fung, J., Ginsburg, D.: OpenCL Programming Guide, 1st edn. Addison-Wesley Professional, Boston (2011)
19. NVIDIA: Whitepaper: NVIDIA's Next Generation CUDA Compute Architecture: Kepler GK110. online (2012). URL http://www.nvidia.com/content/PDF/kepler/NVIDIA-kepler-GK110-Architecture-Whitepaper.pdf
20. Pacheco, P.: An Introduction to Parallel Programming, 1st edn. Morgan Kaufmann, Burlington (2011)
21. Pedemonte, M., Nesmachnow, S., Cancela, H.: A survey on parallel ant colony optimization. Appl. Soft Comput. **11**(8), 5181–5197 (2011). doi:http://dx.doi.org/10.1016/j.asoc.2011.05.042. URL http://www.sciencedirect.com/science/article/pii/S156849461100202X
22. Ruciński, M., Izzo, D., Biscani, F.: On the impact of the migration topology on the island model. Parallel Comput. **36**(10), 555–571 (2010)
23. Sanders, J., Kandrot, E.: CUDA by Example: An Introduction to General-Purpose GPU Programming, 1st edn. Addison-Wesley Professional, Boston (2010)
24. Scarpino, M.: OpenCL in Action: How to Accelerate Graphics and Computations, 1st edn. Manning Publications, Greenwich (2011)
25. Suh, J.W., Kim, Y.: Accelerating MATLAB with GPU Computing: A Primer with Examples, 1st edn. Morgan Kaufmann, Burlington (2013)
26. Taillard, E., Melab, N., Talbi, E.G., et al.: Parallelization strategies for hybrid metaheuristics using a single GPU and multi-core resources. In: Parallel Problem Solving from Nature-PPSN XII, pp. 368–377. Springer, Berlin (2012)
27. White, T.: Hadoop: The Definitive Guide, 3rd edn. Yahoo Press (2012)
28. Wilt, N.: CUDA Handbook: A Comprehensive Guide to GPU Programming, 1st edn. Addison-Wesley Professional, Boston (2013)

C-SOMAQI: Self Organizing Migrating Algorithm with Quadratic Interpolation Crossover Operator for Constrained Global Optimization

Dipti Singh, Seema Agrawal and Kusum Deep

Abstract SOMAQI is a variant of Self Organizing Migrating Algorithm (SOMA) in which SOMA is hybridized with Quadratic Interpolation crossover operator, presented by Singh et al. (Advances in intelligent and soft computing. Springer, India, pp. 225–234, 2014). The algorithm SOMAQI has been designed to solve unconstrained nonlinear optimization problems. Earlier it has been tested on several benchmark problems and the results obtained by this technique outperform the results taken by several other techniques in terms of population size and function evaluations. In this chapter SOMAQI has been extended for solving constrained nonlinear optimization problems (C-SOMAQI) by including a penalty parameter free approach to select the feasible solutions. This algorithm also works with small population size and converges very fast. A set of 10 constrained optimization problems has been used to test the performance of the proposed algorithm. These problems are varying in complexity. To validate the efficiency of the proposed algorithm results are compared with the results obtained by C-SOMGA and C-SOMA. On the basis of the comparison it has been concluded that C-SOMAQI is efficient to solve constrained nonlinear optimization problems.

Keywords Self organizing migrating algorithm · Quadratic interpolation crossover operator · Genetic algorithm · Constrained global optimization

D. Singh (✉)
Department of Applied Sciences, Gautam Buddha University, Greater Noida, India
e-mail: diptipma@rediffmail.com

S. Agrawal
Department of Mathematics, S.S.V.P.G. College, Hapur, India
e-mail: Seemagrwl7@gmail.com

K. Deep
Indian Institute of Technology Roorkee, Roorkee, India
e-mail: kusumfma@iitr.ac.in

D. Davendra and I. Zelinka (eds.), *Self-Organizing Migrating Algorithm*,
Studies in Computational Intelligence 626, DOI 10.1007/978-3-319-28161-2_7

1 Introduction

Many real life problems arising in various disciplines of engineering, economics, decision science, operations research and social system come out to be nonlinear constrained optimization problems. The main aim of the proposed work is to develop an efficient approach to find the global optimal solution of these kinds of problems.

A general single criterion constrained non linear global optimization problem can be described as follows:

$$\begin{array}{ll} \text{Minimize} & f(x) \\ \text{Subject to:} & g_i(x) \geq 0, \quad i = 1, \ldots, m \\ & h_j(x) = 0, \quad j = 1, \ldots, l \\ & x \in X \end{array}$$

$where f: R^n \rightarrow R,\ g_i: R^n \rightarrow R,\ i = 1, \ldots, m\ and\ h_j: R^n \rightarrow R,\ j = 1, \ldots, l$ are non-linear continuous functions defined on the search space $X \subseteq R^n$, $where\, X = \{x = (x_1, x_2, \ldots, x_n) | a_i \leq x_i \leq b_i, \quad i = 1, \ldots, n\}$.

A large number of design optimization problems can be modeled as single criterion nonlinear programming problems and they are usually highly constrained. In contrast to conventional optimization methods, evolutionary algorithm methods for single criterion optimization have following two advantages:

- They impose no restriction on the optimization problem. The objective function can be multimodal and noncontinuous.
- They can be used to solve any optimization models i.e., models with continuous, integer discrete and mixed continuous-integer and continuous-discrete decision variables.

A variety of constraint handling techniques has appeared in literature for solving nonlinear constrained optimization problems. The main problem in applying evolutionary algorithms to solve a constrained problem is how to deal with constraints because evolutionary operators used for manipulating chromosomes may yield infeasible solutions. Quite a large number of methods have been developed to handle constraints when evolutionary algorithms are used [1–4]). These methods can be classified as follows: (i) **Rejecting method** accept only the feasible solutions and discard the infeasible solutions during the search process. (ii) **Repairing method** Repairs infeasible solutions to feasible solutions using some repair procedure. (iii) **Modifying genetic operator method** modifies the genetic operators according to the requirement of the problem to maintain the feasibility of solutions. (iv) **Penalty function method** penalizes infeasible solutions using a penalty parameter. Before applying this strategy, first the constrained problem is transformed to an unconstrained problem in which the function to be minimized has the following form:

$$ø(x, \lambda) = f(x) + \lambda \sum_{j=1}^{l} [h_j(x)]^2 + \lambda \sum_{i=1}^{m} G_i[g_i(x)]^2, \quad (1)$$

where G_i is the Heaviside operator such that $G_i = 0\, for\, g_i(x) \geq 0\, and\, G_i = 1\, for\, g_i(x) < 0,\ and\ \lambda$ is a positive multiplier that controls the magnitude of penalty terms.

Among the above mentioned methods penalty function method is considered to be the most effective tool to produce the feasible solutions of constrained optimization problems. Many attempts have been made in literature to improve the efficiency of these penalty function methods which can be found in [5–7]. One major drawback of penalty function method is that the penalty parameter has to be fine tuned. Small value of penalty parameter leads to infeasible solution [8, 9] and on the other hand large value of penalty parameter may generate an alternate feasible solution and fails to converge to the optimal solution. To overcome the above mentioned drawbacks. Deb [10] proposed a penalty parameter free approach to handle the constraints. Coella and Mezura-Montes [11] proposed a dominance-based selection scheme to handle the constraints which does not require the fine tuning of a penalty function. Deb and Agarwal [12] developed a niched-penalty approach for constraint handling which does not require any penalty parameter. Akhtar et al. [13] proposed a socio-behavioral simulation based approach to solve engineering optimization problems. Though the results obtained by this technique are taking lesser function evaluations but the success rate was not good and also the implementation of this approach is not easy. Eskandar et al. [14, 15] proposed two algorithms, water cycle algorithm and mine blast algorithm for solving constrained engineering design optimization algorithms.

Besides this many attempt has been made in literature to hybridize evolutionary algorithms with other approaches to improve its efficiency and has been used to solve nonlinear constrained optimization problems. Millie et al. proposed a new PSO algorithm with crossover operator for finding the solution of global nonlinear optimization problems [16]. Deep and Dipti [17] proposed a self organizing migrating genetic algorithm for constrained optimization, in which Genetic Algorithm (GA) has been hybridized with Self Organizing Migrating Algorithm (SOMA). Pant et al. [18] proposed new mutation schemes for Differential Evolution Algorithm (DE) and applied them to find the relay time and plug setting arising in the optimization of directional over-current relay settings. Deep and Bansal [19] developed quadratic approximation based Particle Swarm Optimization Algorithm (PSO) for solving economic dispatch problems with valve-point effects. Deep and Das [20] proposed hybrid binary coded Genetic Algorithm for constrained optimization. Recently Singh et al. [21] proposed an algorithm SOMAQI which is hybridization of SOMA and quadratic interpolation crossover operator. SOMAQI inspired by the features of SOMA works with very less population size and outperforms standard particle swarm optimization (PSO) and SOMA in terms of population size, function evaluations, mean best and success rate. This algorithm has been proposed to solve unconstrained optimization problems only. Its performance over other methods inspire author to extend it for solving constrained optimization problems.

In this chapter a novel penalty parameter free hybrid approach C-SOMAQI has been presented to solve nonlinear constrained optimization problems. It does not require any penalty parameter to be fine tuned for constraint handling and is very easy to implement. Another advantage of this approach is that it works with very less population size. To validate the efficiency of the proposed algorithm, it is tested on ten constrained benchmark test problems taken from Deep and Dipti [17] and the results are compared with the results of constrained SOMA (C-SOMA) and constrained SOMGA (C-SOMGA) (2008).

The rest of the paper is organized as follows. In Sect. 2, preliminaries are given. In Sect. 3, the proposed Algorithm C-SOMAQI is presented. In Sect. 4, the numerical results are discussed. Finally, the paper concludes with Sect. 5 drawing the conclusions of the present study.

2 Preliminaries

The proposed algorithm (C-SOMAQI) is extended form of hybridized algorithm SOMAQI. To know more about this algorithm, the working of SOMA, QI crossover and SOMAQI has been discussed in this section.

2.1 *Self Organizing Migrating Algorithm*

The Self Organizing Migrating Algorithm is a general-purpose, population based stochastic optimization algorithm refer [22–26]. The approach is similar to that of other evolutionary algorithms but working is different. Like other evolutionary algorithms, SOMA does not create any new individuals to process the algorithm. It changes only the position of individuals from current position to better position. In the working of SOMA, first of all the population is initialized randomly using uniform distribution over the search space. In each loop called the migration loop the population is evaluated and the individual with highest fitness value is known as leader and the worst is known as active. Rather than competing with each other, the active individual proceeds in the direction of the leader and travels a certain distance (called the path length) towards the leader in n (pathlength/step size) steps of defined length (step size). This path is perturbed randomly by a parameter.

2.1.1 Mutation: Perturbation

In SOMA, mutation is replaced by perturbation. Perturbation of individuals ensures the diversity among the individuals. When the active individual moves towards the leader, its path is perturbed randomly by perturbation (PRT) parameter. Perturbation vector (PRT vector) is created before an individual proceeds towards leader. This parameter has the same effect as mutation in genetic algorithm (GA). For each individual's parameter, the algorithm generates a random number from the interval (0, 1). Then the following expression is used:

$$\begin{gathered} if\ rnd_j < PRT\ then \\ PRTVector_j = 1; \\ else \\ PRTVector_j = 0; \\ end\ if, \end{gathered}$$

The randomly generated binary perturbation vector controls the allowed dimensions for an individual of population. If an element of the perturbation vector is set to zero, then the individual is not allowed to change its position in the corresponding dimension.

2.1.2 Crossover Operator: Generation of New Positions of Individuals

In GA, new individuals are created as a result of combination of two or more parents. But SOMA does not create any new individual. Here crossover means to explore for the better solution during the movement of the active individual towards the leader. The movement of an individual is given as follows:

$$x_{i,j}^{MLnew} = x_{i,j,start}^{ML} + (x_{L,j}^{ML} - x_{i,j,start}^{ML}) \cdot t \cdot PRTVector_j \tag{2}$$

where $t \in \langle 0, by\,step\,to,\ PathLength \rangle$ *and* ML *is actual migration loop*

$x_{i,j}^{MLnew}$ *is the new positions of an individual.*
$x_{i,j,start}^{ML}$ *is the positions of active individual.*
$x_{L,j}^{ML}$ *is the positions of leader.*

The pseudo code of SOMA is given as follows:

Begin
define parameters
while termination criterion is not satisfied do
generate PRT Vector
evaluate the objective function value of all individuals of the population

select leader and active individual of the population
for active individual in population do
for k = 1 to n do
move active individual towards the position of the leader
evaluate objective function value at the new position
if objective function value at the new position is better than objective function value at the position of active individual
position of active individual = new position
end if
end for
move active individual to best new position
end for
end while
report the best individual as the final optimal solution
end

2.2 *Quadratic Interpolation (QI) Crossover Operator*

The central idea of Quadratic Approximation is to fit a quadratic curve passing through three points and to find the minimum of this curve. Quadratic Approximation works as follows:

1. Select three distinct points R_1 (with best fitness value), R_2 and R_3
2. A new trial point of minima $x' = \left(x'_1, x'_2, \ldots, x'_n\right)$ is given as

$$x' = \frac{1}{2}\frac{\left[\left(R_2^2 - R_3^2\right) * f(R_1) + \left(R_3^2 - R_1^2\right) * f(R_2) + \left(R_1^2 - R_2^2\right) * f(R_3)\right]}{\left[(R_2 - R_3) * f(R_1) + (R_3 - R_1) * f(R_2) + (R_1 - R_2) * f(R_3)\right]} \tag{3}$$

where $f(R_1)$, $f(R_2)$ and $f(R_3)$ are the objective function values at R_1, R_2 and R_3 respectively.

2.3 *Methodology of SOMAQI*

SOMAQI is the combination of SOMA and QI crossover operator. In SOMAQI both the algorithms are used in series. First SOMA is used to explore the solution search space, after that QI crossover operator is applied to exploit the search space in order to find the better optimal solution.

Methodology

First the individuals are generated randomly. At each generation the individual with highest fitness value is selected as leader and the worst one as active individual. For active individual a new population of size n is created. Where n = (Path Length/step size). This population is nothing but the new positions of the active individual, The movement of this individual is given in Eq. (2). Now the best individual of the new population is selected and the active individual is replaced with this best individual if new individual is better than the active individual. Then we again select the best and worst individual from the population. A new point is created using quadratic interpolation at the end of each generation using Eq. (3). For this we choose three particles R1, R2 and R3, where R1 is the leader and R2 and R3 are randomly chosen particles from the remaining population. This new point is accepted only if it is better than active individual and is replaced with active individual.

The computational steps of SOMAQI are given as follows:

Step 1: generate initial population;
Step 2: evaluate all individuals in the population;
Step 3: generate PRT vector for all individuals;
Step 4: sort all of them;
Step 5: select the best fitness individual as leader and worst as active;
Step 6: for active individual new positions are created using Eq. (2). Then the best position is selected. It replaces the active individual if it is better than active individual;
Step 7: sort the population and select the best fitness individual as leader and worst as active;
Step 8: create new point by QI from R1, R2 and R3 using Eq. (3);
Step 9: if new point is better than active individual replace active individual with new one;
Step 10: if termination criterion is satisfied stop else go to step 2;
Step 11: report the best individual as the optimal solution.

3 Proposed Hybrid C-SOMAQI Algorithm

Singh et al. [21] have presented a novel variant of SOMA, (SOMAQI) for unconstrained optimization which combines QI crossover operator with SOMA for creating the new solution member in the search space and maintains the diversity of the solution in the search space. This paper is an extension of SOMAQI to solve constraints optimization problems. The methodology of C-SOMAQI is given as follows:

Methodology

The constraint violation function can be evaluated as follows:

$$\psi(x) = \sum_{i=1}^{m} G_i[g_i(x)]^2 + \sum_{j=1}^{l} [h_j(x)]^2, \tag{4}$$

where G_i is the Heaviside operator such that $G_i = 0 \; for \; g_i(x) \geq 0 \; and \; G_i = 1 \; for \; g_i(x) < 0$.

Value of constraint violation function $\psi(x) \; is \; 0$ for the individuals which are feasible in the region and for the individuals which are out of feasible region the value of $\psi(x)$ indicates how far the solutions are from the feasible region. The proposed algorithm works iteratively in two phases. In the first phase, also called the global phase, the objective function is evaluated at a number of randomly sampled feasible points. In the second phase, also called the local phase, these points are manipulated by local searches to yield a possible candidate for global optima. In the global phase algorithm generates feasible points randomly. At each generation the individual with highest fitness value is selected as leader and the worst one as active individual. Now the active individual moves towards leader in n steps of defined length, where n is the ratio of path length and step size. For each active individual a new population of size n is created. This population is nothing but the new positions of the active individual. The movement of this individual is given in Eq. (2). Then this population according to the fitness value is sorted. Now starting from the best position of the new population constraint violation function is evaluated. If $\psi(x) = 0$, replace the active individual with the current position and if $\psi(x) > 0$ then move to next best position of the sorted new population. If no feasible solution is available then active individual remains the same. We again select the best and worst individual from the population. Now in the local phase a new point is created using QI crossover operator using Eq. (3). If this point satisfies constraint violation function, then this point is accepted only if it is better than active individual and is replaced with active individual. This process is continued till the termination criterion is satisfied. The computational steps of C-SOMAQI are given as follows:

Step 1: generate the initial feasible population;
Step 2: evaluate all individuals in the population;
Step 3: generate PRT vector for all individuals;
Step 4: sort all of them;
Step 5: select the best fitness individual as leader and worst as active;
Step 6: for active individual a new population of size n is created. This population is nothing but the new positions of the active individual towards the leader in n steps of defined length. The movement of this individual is given in Eq. (2);
Step 7: sort new population with respect to fitness;
Step 8: for each individual in the sorted population, check feasibility criterion;

Step 9: if feasibility criterion is satisfied replace the active individual with the new position, else move to next position in the sort order and go to step 8;
Step 10: create new point by crossover operator using Eq. (3);
Step 11: if feasibility criterion is satisfied replace the active individual with the new position, else go to step 9 until prescribed number of iterations are exhausted;
Step 12: if new point is better than active replace active with the new one;
Step 13: if termination criterion is satisfied stop else go to step 5;
Step 14: report the best individual as the final optimal solution.

4 Numerical Results on Benchmark Problems

The proposed algorithm is coded in C++ and run on a Pentium III 2.20 GHz with 2 GB RAM. The efficiency of the proposed algorithm C-SOMAQI has been tested on a set of 10 constrained test problems taken from literature which are given in appendix.

These test problems include objective functions with decision variables and equality and inequality constraints. The maximization problems were transformed into minimization ones by changing objective function $f(x)$ into $-f(x)$. All equality constraints problems were converted into inequality constraints as $|h(x)| - \epsilon \leq 0$, where ϵ is a degree of violation. Since C-SOMAQI is probabilistic technique and rely heavily on the generation of random numbers, therefore 100 trials of each are carried out, each time using a different seed for the generation of random numbers. A run is considered to be a success if the optimum solution obtained falls within 1 % accuracy of the known global optimal solution. The stopping criterion is either a run is a success or a fixed number of function calls (1, 50,000) are performed. The value of parameters related to C-SOMAQI is presented in Table 1.

The comparative performance of C-SOMAQI, C-SOMGA and C-SOMA are measured in terms of three criteria, namely accuracy, efficiency and reliability. They are described as follows:

Table 1 Parameters of C-SOMAQI

Parameter	Value
Population size	10
PRT	0.3, 0.5 and 0.9
Step	0.11
Pathlength	3
Total number of function calls allowed	150,000

Table 2 Percentage of success of C-SOMAQI, C-SOMGA and C-SOMA

Problem no.	No. of successful runs out of 100			Best amongst C-SOMAQI, C-SOMGA and C-SOMA
	C-SOMA	C-SOMGA	C-SOMAQI	
1	50	100	100	C-SOMGA and C-SOMAQI
2	10	100	100	C-SOMGA and C-SOMAQI
3	0	100	100	C-SOMGA and C-SOMAQI
4	0	87	96	C-SOMAQI
5	0	100	100	C-SOMGA and C-SOMAQI
6	0	100	97	C-SOMGA
7	50	90	99	C-SOMAQI
8	0	50	100	C-SOMAQI
9	0	70	94	C-SOMAQI
10	0	17	39	C-SOMAQI

1. *Accuracy*: Accuracy of an algorithm totally depends on its closeness to the global optimum.
2. *Efficiency*: Efficiency of an algorithm is based on average number of function calls (ANFC) required to converge to the global minima.
3. *Reliability*: Reliability of an algorithm depends on number of successful runs.

The information regarding number of successful runs, average number of function evaluations required in successful runs and mean of optimal objective function values, are given in Tables 2, 3 and 4.

The number of successful runs of a total of 100 runs, corresponding to C-SOMAQI, C-SOMGA and C-SOMA are presented in Table 2. Results show that C-SOMAQI gives 100 % success in five problems and 99 % in one problem, also the success obtained by C-SOMGA is 100 % in five problems and 90 % in one problem but C-SOMA does not give 100 % in any problem. Same results are shown

Table 3 Average number of function evaluations amongst C-SOMAQI, C-SOMGA and C-SOMA

Problem no.	Average no. of function evaluations			Best amongst C-SOMAQI, C-SOMGA and C-SOMA
	C-SOMA	C-SOMGA	C-SOMAQI	
1	2070	590	361	C-SOMAQI
2	18090	4368	133	C-SOMAQI
3	*	4115	1299	C-SOMAQI
4	*	24,592	568	C-SOMAQI
5	*	18,783	2785	C-SOMAQI
6	*	30,922	7000	C-SOMAQI
7	1598	605	186	C-SOMAQI
8	*	559	48	C-SOMAQI
9	*	3603	377	C-SOMAQI
10	*	135,468	1463	C-SOMAQI

Table 4 Mean objective function value amongst C-SOMAQI, C-SOMGA and C-SOMA

Problem no.	Mean objective function value			Best amongst C-SOMAQI, C-SOMGA and C-SOMA
	C-SOMA	C-SOMGA	C-SOMAQI	
1	−1.87228	−1.86839	−1.87264	C-SOMAQI
2	−311.601	−319.9927	−320	C-SOMAQI
3	0.26864	0.01531	0.0156381	C-SOMGA
4	−234.964	−309.150	−310	C-SOMAQI
5	6.73079	13.5960	13.5915	C-SOMAQI
6	−230.519	−14.99225	−14.99926	C-SOMAQI
7	−7.09413	−0.08816	−0.095565	C-SOMAQI
8	0.14079	0.82519	0.749792	C-SOMAQI
9	−0.23646	−0.88794	−1.000248	C-SOMAQI
10	−0.24275	−0.77542	−0.803374	C-SOMAQI

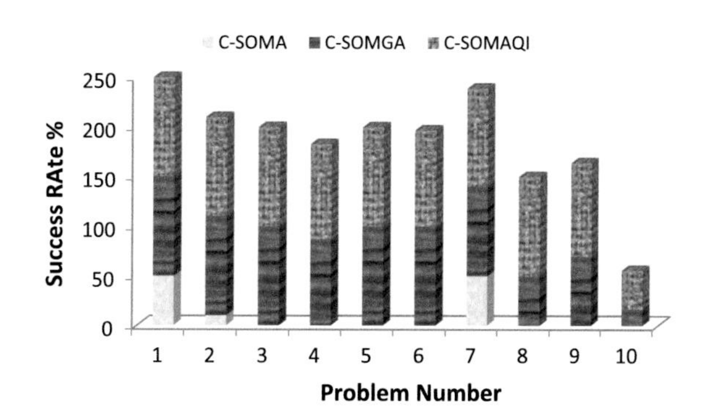

Fig. 1 Graph showing percentage of success among C-SOMA, C-SOMGA and C-SOMAQI

through bar graph in Fig. 1. C-SOMA failed to solve problems 3, 4, 5, 6, 8, 9 and 10. Based on these results ranking of all the algorithms is C-SOMA < C-SOMGA < C-SOMAQI requires. Hence C-SOMAQI is *most reliable.*

The ANFC corresponding to C-SOMAQI, C-SOMGA and C-SOMA are reported in Table 3. Results show that the ANFC taken by C-SOMA are 5–136 times than C-SOMAQI and the ANFC taken by C-SOMGA are 1.5–92 times than C-SOMAQI. The ranking of all the algorithms is C- SOMA < C-SOMGA < C-SOMAQI. C-SOMAQI requires least function evaluations among all the algorithms. Hence C-SOMAQI is *most efficient.*

The mean objective function value corresponding to C-SOMAQI, C-SOMGA and C-SOMA is given in Table 4. Results show that the optimal value obtained by C-SOMAQI is better than C-SOMA and C-SOMGA in all the 10 problems except problem 3. Ranking of all the algorithms is C-SOMA < C-SOMGA < C-SOMAQI. Hence C-SOMAQI is *most reliable.*

The problems which could not be solved by the particular algorithm is given the symbol (*) at the corresponding entries.

In order to reconfirm our results, we compare the relative performance of all the algorithms simultaneously. We use a Performance Index (PI). The relative performance of an algorithm using this modified PI is calculated in the following manner.

$$PI = \frac{1}{N_p} \sum_{i=1}^{N_p} \left(k_1 \alpha_1^i + k_2 \alpha_2^i + k_3 \alpha_3^i \right) \tag{4}$$

where

$$\alpha_1^i = \frac{Sr^i}{Tr^i},$$
$$\alpha_2^i = \begin{cases} \frac{Mo^i}{Ao^i}, & if\ Sr^i > 0 \\ 0, & if\ Sr^i = 0 \end{cases} \quad and$$
$$\alpha_3^i = \begin{cases} \frac{Mt^i}{At^i}, & if\ Sr^i > 0 \\ 0, & if\ Sr^i = 0 \end{cases} \quad and$$

where

Sr^i = Number of successful runs of ith problem
Tr^i = Total number of runs of ith problem
Ao^i = Mean objective function value obtained by an algorithm of ith problem
Mo^i = Minimum of Mean objective function value obtained by all algorithms of ith problem
At^i = Mean execution time of successful runs taken by an algorithm in obtaining the solution of ith problem
Mt^i = Minimum of mean execution time of successful runs taken by all algorithms in obtaining the solution of ith problem
N_p = Total number of problems analyzed.

$k_1, k_2\ and\ k_3\ (k_1 + k_2 + k_3 = 1\ and\ 0 \leq k_1, k_2, k_3 \leq 1)$ are the weights assigned to percentage of success, mean objective function value and mean execution time of successful runs, respectively.

From the above definition it is clear that modified PI is a function of $k_1, k_2\ and\ k_3\ since\ k_1 + k_2 + k_3 = 1,\ one\ of\ k_i, i = 1, 2, 3$ could be eliminated to reduce the number of variables from the expression of PI. But it is still difficult to analyze the behavior of this PI, because the surface of PI for all the algorithms are overlapping and it is difficult to visualize them. Hence equal weights are assigned to two terms at a time in the PI expression. This way PI becomes a function of one variable. The resultant cases are as follows:

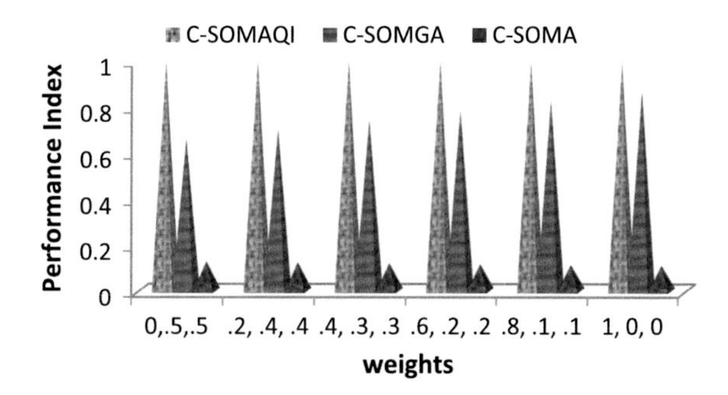

Fig. 2 PI for combination of C-SOMAQI, C-SOMGA and C-SOMA for case 1

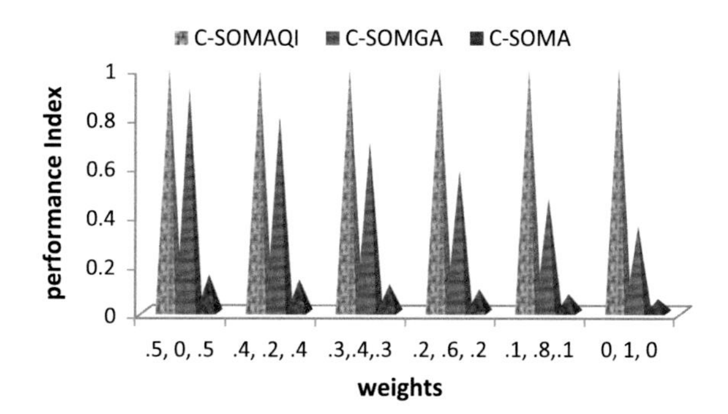

Fig. 3 PI for combination of C-SOMAQI, C-SOMGA and C-SOMA for case 2

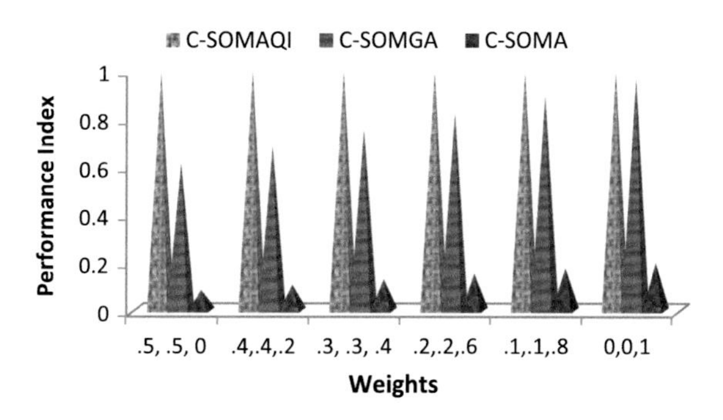

Fig. 4 PI for combination of C-SOMAQI, C-SOMGA and C-SOMA for case 3

(i) $$k_1 = w, k_2 = k_3 = \frac{1-w}{2}, \quad 0 \le w \le 1$$

(ii) $$k_2 = w, k_1 = k_3 = \frac{1-w}{2}, \quad 0 \le w \le 1$$

(iii) $$k_3 = w, k_1 = k_2 = \frac{1-w}{2}, \quad 0 \le w \le 1$$

The graphs corresponding to each of case (i), (ii) and (iii) are shown in Figs. 2, 3 and 4. The horizontal axis represents the weight w and the vertical axis represents the performance index PI.

In case (i), the mean objective function value and average no. of function evaluations of successful runs are given equal weights. PI's of C-SOMAQI, C-SOMGA and C-SOMA are superimposed in the Fig. 2. It is observed that the value of PI for C-SOMAQI is more than C-SOMGA and C-SOMA.

In case (ii), equal weights are assigned to the numbers of successful runs and mean objective function value of successful runs. PI's of C-SOMAQI, C-SOMGA and C-SOMA are superimposed in the Fig. 3. It is clear that C-SOMAQI has the highest PI.

In case (iii), equal weights are assigned to mean objective function value and average number of successful runs. PI's of C-SOMAQI, C-SOMGA and C-SOMA are superimposed in the Fig. 4. It is clear that C-SOMAQI has the highest PI.

5 Conclusions

In this chapter, a variant of self organizing migrating algorithm SOMAQI has been extended to solve constrained nonlinear optimization problems and called as C-SOMAQI. In the proposed approach a penalty parameter free approach has been used for dealing the feasibility of solutions with quadratic interpolation crossover operator. This technique requires very less population size to work with and hence requires lesser number of function evaluations. To evaluate the performance of this algorithm it has been tested on ten benchmark test problems. C-SOMAQI is a variant of SOMA, so results have been compared with the results taken by C-SOMA itself and with one more variant of SOMA that is C-SOMGA. For this purpose a performance index graph has been plotted on the basis of percentage of success, mean objective function value, average number of function evaluations. The numerical and graphical results clearly indicate that C-SOMAQI can be considered to be robust for solving constrained optimization problem.

Appendix

Problem 1

$$\min_x f(x) = -x_1 - x_2,$$

Subject to:

$$g_1(x) = \left[(x_1 - 1)^2 + (x_2 - 1)\right]\left[1/2a^2 - 1/2b^2\right] + (x_1 - 1)(x_2 - 1)\left[1/a^2 - 1/b^2\right] - 1 \geq 0.$$

where a = 2, b = 0.25.

This problem has two global minima one of which is at (1, 0.8729) with $f_{min} = -1.8729$. The feasible domain of the problem is disconnected. The bounds on the variables are $0 \leq x_1, x_2 \leq 1$.

Problem 2

$$\min_{x} f(x) = 3x_1 + x_2 + 2x_3 + x_4 - x_5,$$

Subject to:

$$g_1(x) = 25x_1 - 40x_2 + 16x_3 + 21x_4 + x_5 \le 300,$$
$$g_2(x) = x_1 + 20x_2 - 50x_3 + x_4 - x_5 \le 200,$$
$$g_3(x) = 60x_1 + x_2 - x_3 + 2x_4 + x_5 \le 600,$$
$$g_4(x) = -7x_1 + 4x_2 + 15x_3 - x_4 + 65x_5 \le 700.$$

This problem has global minima at (4, 88, 35, 150, 0) with f_{min} = 320. The bounds on the variables are $1 \le x_1 \le 4$, $80 \le x_2 \le 88$, $30 \le x_3 \le 35$, $145 \le x_4 \le 150$, $0 \le x_5 \le 2$.

Problem 3

$$\min_{x} f(x) = 4.3x_1 + 31.8x_2 + 63.3x_3 + 15.8x_4 + 68.5x_5 + 4.7x_6,$$

Subject to:

$$g_1(x) = 17.1x_1 + 38.2x_2 + 204.2x_3 + 212.3x_4 + 623.4x_5 + 1495.5x_6 - 169x_1x_3 - 3580x_3x_5 - 3810x_4x_5 - 18500x_4x_6 - 24300x_5x_6 - 4.97 \ge 0,$$
$$g_2(x) = 1.88 + 17.9x_1 + 36.8x_2 + 113.9x_3 + 169.7x_4 + 337.8x_5 + 1385.2x_6 - 139x_1x_3 - 2450x_4x_5 - 600x_4x_6 - 17200x_5x_6 \ge 0,$$
$$g_3(x) = 429.08 - 273x_2 - 70x_4 - 819x_5 + 26000x_4x_5 \ge 0,$$
$$g_4(x) = 159.9x_1 - 311x_2 + 587x_4 + 391x_5 + 2198x_6 - 14000x_1x_6 + 78.02 \ge 0.$$

This problem has global minima at (0, 0, 0, 0, 0, 0.00333) with f_{min} = 0.0156. The bounds on the variables are $0 \le x_1 \le 0.31$, $0 \le x_2 \le 0.046$, $0 \le x_3 \le 0.068$, $0 \le x_4 \le 0.042$, $0 \le x_5 \le 0.028$, $0 \le x_6 \le 0.0134$.

Problem 4

$$\max_{x} f(x) = 25(x_1 - 2)^2 + (x_2 - 2)^2 + (x_3 - 1)^2 + (x_4 - 4)^2 + (x_5 - 1)^2 + (x_6 - 4)^2,$$

Subject to:

$$g_1(x) = x_1 + x_2 - 2 \ge 0,$$
$$g_2(x) = -x_1 + x_2 + 6 \ge 0,$$
$$g_3(x) = x_1 - x_2 + 2 \ge 0,$$
$$g_4(x) = -x_1 + 3x_2 + 2 \ge 0,$$
$$g_5(x) = (x_3 - 3)^2 + x_4 - 4 \ge 0,$$
$$g_6(x) = (x_5 - 3)^2 + x_6 - 4 \ge 0,$$

This problem has 18 global maxima and one global maxima at (5, 1, 5, 0, 5, 10) with f_{max} = 310. The bounds on the variables are $0 \le x_1 \le 5$, $0 \le x_2 \le 1$, $1 \le x_3 \le 5$, $0 \le x_4 \le 6$, $0 \le x_5 \le 5$, $0 \le x_6 \le 10$.

Problem 5

$$\min_x f(x) = \left(x_1^2 + x_2 - 11\right)^2 + \left(x_1 + x_2^2 - 7\right)^2,$$

Subject to:

$$g_1(x) = 4.84 - (x_1 - 0.05)^2 - (x_2 - 2.5)^2 \geq 0,$$

$$g_2(x) = x_1^2 + (x_2 - 2.5)^2 - 4.84 \geq 0.$$

This problem has two decision variables. It has global minima at $x^* = (2.246826, 2.381865)$ with f_{min} = 13.59085. The bounds on the variables are $0 \leq x_i \leq 6, \ for \ i = 1, 2$.

Problem 6

$$\min_x f(x) = 5\sum_{i=1}^{4} x_i - 5\sum_{i=1}^{4} x_i^2 - \sum_{i=5}^{13} x_i,$$

$$g_1(x) = 10 - (2x_1 + 2x_2 + x_{10} + x_{11}) \geq 0,$$

$$g_2(x) = 10 - (2x_1 + 2x_3 + x_{10} + x_{12}) \geq 0,$$

$$g_3(x) = 10 - (2x_2 + 2x_3 + x_{11} + x_{12}) \geq 0,$$

$$g_4(x) = 8x_1 - x_{10} \geq 0,$$

$$g_5(x) = 8x_2 - x_{11} \geq 0,$$

$$g_6(x) = 8x_3 - x_{12} \geq 0,$$

$$g_7(x) = 2x_4 + x_5 - x_{10} \geq 0,$$

$$g_8(x) = 2x_6 + x_7 - x_{11} \geq 0,$$

$$g_9(x) = 2x_8 + x_9 - x_{12} \geq 0,$$

$$x_i \geq 0, \quad i = 1, \ldots, 13,$$

$$x_i \leq 1, \quad i = 1, \ldots 9, 13.$$

This problem has global minima at $x^* = (1,\ 1,\ \ldots 1,\ 3,\ 3,\ 3,\ 1)$ with $f_{min} = -15$. The bounds on the variables are $0 \leq x_i \leq u_i,\ i = 1, 2, \ldots, n,$ where $u = (1,\ 1, \ldots, 1,\ 100,\ 100,\ 100,\ 1)$.

Problem 7

$$\max_x f(x) = \frac{\sin^3(2\pi x_1)\sin(2\pi x_2)}{x_1^3(x_1 + x_2)},$$

Subject to:

$$g_1(x) = -x_1^2 + x_2 - 1 \geq 0,$$

$$g_2(x) = -1 + x_1 - (x_2 - 4)^2 \geq 0.$$

This problem has global maxima at $x^* = (1.2279713,\ 4.2453733)$, $f(x^*) =$ 0.095 with f_{max} = 0.095. The bounds on the variables are $0 \leq x_i \leq 10,\ i = 1, 2$.

Problem 8

$$\min_x f(x) = x_1^2 + (x_2 - 1)^2$$

Subject to:

$$h_1(x) = x_2 - x_1^2 = 0.$$

This problem has global minima at $x^* = \pm\left({}^1\!/_{2^{0.5}}, {}^1\!/_2\right)$, $f(x^*) = 0.75$ with f_{min} = 0.75. The bounds on the variables are: $-1 \le x_i \le 1,\ i = 1, 2.$

Problem 9

$$\max_x f(x) = \left(\sqrt{n}\right)^n \prod_{i=1}^{n} x_i,$$

Subject to: <!endaligned >

$$h_1(x) = \sum_{i=1}^{n} x_i^2 - 1 = 0.$$

This problem has global maxima at $x^* = \left({}^1\!/_{n^{0.5}}, \ldots {}^1\!/_{n^{0.5}}\right)$ with f_{max} = 1. The bounds on the variables are: $0 \le x_i \le 1,\ i = 1, 2, \ldots, n.$

Problem 10

$$\max_x f(x) = \left| \frac{\sum_{i=1}^{n} \cos^4(x_i) - 2\prod_{i=1}^{n} \cos^2(x_i)}{\sqrt{\sum_{i=1}^{n} i x_i^2}} \right|,$$

Subject to:

$$g_1(x) = \prod_{i=1}^{n} x_i - 0.75 \ge 0,$$

$$g_2(x) = 7.5n - \sum_{i=1}^{n} x_i \ge 0,$$

This problem has global maxima at f_{max} = 0.803619 for n = 20. The bounds on the variables are: $0 \le x_i \le 10,\ i = 1, 2, \ldots, n.$

References

1. Kim, J.H., Myung, H.: A two phase evolutionary programming for general constrained optimization problem. In: Proceedings of the Fifth Annual Conference on Evolutionary Programming, San Diego (1996)

2. Michalewicz, Z.: Genetic algorithms, numerical optimization and constraints. In: Echelman L. J. (ed.) Proceedings of the Sixth International Conference on Genetic Algorithms, pp. 151–158 (1995)
3. Myung, H., Kim, J.H.: Hybrid evolutionary programming for heavily constrained problems. Bio-Systems **38**, 29–43 (1996)
4. Orvosh, D., Davis, L.: Using a genetic algorithm to optimize problems with feasibility constraints. In: Echelman L.J. (ed.) Proceedings of the Sixth International Conference on Genetic Algorithms, pp. 548–552 (1995)
5. Michalewicz, Z., Attia, N.: Evolutionary optimization of constrained problems. In: Proceedings of Third Annual Conference on Evolutionary Programming, pp. 998–1008. World Scientific, River Edge (1994)
6. Joines, J., Houck, C.: On the use of non-stationary penalty functions to solve nonlinear constrained optimization problems with GAs. In: Proceedings of the First IEEE Conference on Evolutionary Computation, pp. 587–602. IEEE Press, Orlando (1994)
7. Homaifar, A.A., Lai, S.H.Y., Qi, X.: Constrained optimization via genetic algorithms. Simulation **62**, 242–254 (1994)
8. Smith, A.E., Coit, D.W.: Constraint Handling Techniques-Penalty Functions, Handbook of Evolutionary Computation. Oxford University Press and Institute of Physics Publishing, Oxford (Chapter C 5.2) (1997)
9. Coello, C.A.: Theoretical and numerical constraint handling techniques used with evolutionary algorithms: a survey of the state of the art. Comput. Methods Appl. Mech. Eng. **191**, 1245–1287 (2002)
10. Deb, K.: An efficient constraint handling method for genetic algorithms. Comput. Methods Appl. Mech. Eng. **186**, 311–338 (2000)
11. Coello, C.A., Mezura-Montes, E.: Constraint-handling in genetic algorithms through the use of dominance-based tournamen selection. Adv. Eng. Inf. **16**, 193–203 (2002)
12. Deb, K., Agarwal, S.: A niched-penalty approach for constraint handling in genetic algorithms. In: Proceedings of the ICANNGA, Portoroz, Slovenia (1999)
13. Akhtar, S., Tai, K., Ray, T.: A Socio-behavioural simulation model for engineering design optimization. Eng. Optim. **34**, 341–354 (2002)
14. Eskandar, A., Sadollah, A., Bahreininejad, A., Hamdi, M.: Water cycle algorithm—a novel metaheuristic optimization method for solving constrained engineering optimization problems. Comput. Struct. **110–111**, 151–166 (2012)
15. Eskandar, A., Sadollah, A., Bahreininejad, A., Hamdi, M.: Mine blast algorithms: a new population based algorithm for solving constrained engineering optimization problems. Appl. Soft Comput. (in press) (2012)
16. Pant, M., Thangaraj, R., Abraham, A.: A new PSO algorithm with crossover operator for global optimization problems. In: Corchado E. et al. (eds.) Second International Symposium on Hybrid Artificial Intelligent Systems (HAIS'07), Soft computing Series, Innovations in Hybrid Intelligent Systems, vol. 44, pp. 215–222. Springer, Germany (2007)
17. Deep, K., Dipti, S.: A self organizing migrating genetic algorithm for constrained optimization. Appl. Math. Comput. **198**, 237–250 (2008)
18. Pant, M., Thangaraj, R., Abraham, A.: New mutation schemes for differential evolution algorithm and their application to the optimization of directional over-current relay settings. Appl. Math. Comput. **216**, 532–544 (2010)
19. Deep, K., Bansal, J.C.: Quadratic approximation PSO for economic dispatch problems with valve-point effects. In: International Conference on Swarm, Evolutionary and Memetic computing, SRM University, Chennai, pp. 460–467, Proceedings Springer (2010)
20. Deep, K., Das, K. N.: Hybrid Binary Coded Genetic Algorithm for Constrained Optimization. In: ICGST AIML-11 Conference, Dubai, UAE (2011)
21. Singh, D., Agrawal, S., Singh, N.: A novel variant of self organizing migrating algorithm for function optimization. In Proceedings of the 3rd International Conference on Soft Computing for Problem Solving. Advances in Intelligent and Soft Computing, vol. 258, pp. 225–234. Springer, India (2014)

22. Zelinka, I., Lampinen, J.: SOMA—Self Organizing Migrating Algorithm. In: Proceedings of the 6th International Mendel Conference on Soft Computing, pp. 177–187. Brno, Czech, Republic (2000)
23. Zelinka, I.: SOMA—Self Organizing Migrating Algorithm. In: Onwubolu G.C., Babu B.V. (eds.) New Optimization Techniques in Engineering. Springer, Berlin (2004)
24. Nolle, L., Zelinka, I.: SOMA applied to optimum work roll profile selection in the hot rolling of wide steel. In: Proceedings of the 17th European Simulation Multiconference ESM 2003, pp. 53–58. Nottingham, UK, ISBN 3-936150-25-7, 9–11 June 2003 (2003)
25. Nolle, L., Zelinka, I., Hopgood, A.A., Goodyear, A.: Comparision of an self organizing migration algorithm with simulated annealing and differential evolution for automated waveform tuning. Adv. Eng. Softw. **36**, 645–653 (2005)
26. Oplatkova, Z., Zelinka, I.: Investigation on Shannon-Kotelnik theorem impact on soma algorithm performance. In: Proceedings 19th European Conference on Modelling and Simulation Yuri Merkuryev, Richard Zobel (2005)

Optimization of Directional Overcurrent Relay Times Using C-SOMGA

Kusum Deep and Dipti Singh

Abstract An important problem in electrical engineering is to determine the optimal directional overcurrent relay times. The problem is modeled as a constrained nonlinear optimization problem in which the decision variables are the devices that control the act of isolation of faulty lines from the system without disturbing the healthy lines. Three models are considered namely IEEE-3 bus system, IEEE-4 bus system and IEEE-6 bus system. The problem is solved using self organizing migrating genetic algorithm for constrained optimization (C-SOMGA) which is a genetic algorithm hybridized with self organizing migrating algorithm. The results obtained by C-SOMGA are compared with the results obtained by C-GA, C-SOMA, RST2 and MATLAB TOOL BOX. It is shown that C-SOMGA is able to provide superior results in terms of optimality and feasibility in comparison to other methods considered. The main purpose of this chapter is to show the efficiency and robustness of the algorithm C-SOMGA to solve real life problem with very small population size.

1 Introduction

The problem considered in this chapter has its origin in electrical power systems. It requires finding the optimal values of decision variables subject to intricately interconnected non-linear inequality constraints. The problem is about computing the values of the decision variables of the devices called "Relays", which control the act of isolation of faulty lines from the system without disturbing the healthy lines.

K. Deep
Indian Institute of Technology Roorkee, Roorkee, India
e-mail: kusumfma@iitr.ac.in

D. Singh (✉)
Department of Applied Sciences, Gautam Buddha University, Greater Noida, India
e-mail: diptipma@rediffmail.com

D. Davendra and I. Zelinka (eds.), *Self-Organizing Migrating Algorithm*,
Studies in Computational Intelligence 626, DOI 10.1007/978-3-319-28161-2_8

Directional Overcurrent Relays (DOCRs) are provided in electrical power systems to isolate only the faulty lines, in the event of the faults in the system. These relays are placed at both ends of each line. Thus, number of directional overcurrent relays in an electrical power system is twice the number of the lines. To maintain the continuity of supply to healthy sections and to isolate the faulty section only, relays are coordinated. This ensures that minimum lines are disrupted when fault occurs. This is done in DOCRs by properly fixing the two adjustable parameters of each relay called "settings". The two settings of each relay are plug setting (*PS*) and time dial setting (*TDS*). There can be many relays in the system depending on the size of the system. Thus, each relay introduces two decision variables (one *TDS* and one *PS*) in the problem. The above stated problem of coordinating each DOCR with one another in electrical power systems can be modeled as a non-linear constrained optimization problem. Objective function for this problem is the sum of the operating times of all the primary relays, which are expected to operate in order to clear the faults of their corresponding zones. The constraints of this problem are bounds on all decision variables, complexly interrelated times of the various relays (called selectivity constraints) and restrictions on each term of the objective function to be within the specified limits.

In Sect. 1 the introduction to the problem is given. In Sect. 2 review of the literature for this problem is given. In Sect. 3, the methodology of the technique used to solve this problem is given. The general formulation of the problem is stated in Sect. 4. The formulation of the optimization problem is given in Sect. 5. In Sects. 6–8 the IEEE 3-bus, IEEE 4-bus and IEEE 6-bus models are given, respectively. The method of solution and discussion of results are given in Sect. 9. Finally the conclusions are given in Sect. 10.

2 Previous Work

As the dimension of the problem increases for the modern interconnected power systems, the complexity of the problem increases. Also, due to the complexities of non-linear programming techniques, most of the researchers have solved the problem in linear environment by presuming the values of decision variables (all plug settings), which make the problem non-linear. This presumption is made based on the basis of engineering experience (e.g. Irving and Elrafie [13], Chattopadhyay et al. [5], Urdaneta et al. [23, 26], etc.).

Not many non-linear approaches have been applied in this area due to the complexities of large dimension of the problem. However, linear approaches cannot ensure correct settings of the relays [15]. These approaches cannot consider all possible operating conditions of the system. The results obtained may be trapped in local optimum relay settings [23]. Non-linear methods can produce optimal results by optimizing all settings of relays and thus, avoid undesired tripping of those relays, which are not supposed to operate for a fault under consideration. First optimization attempt in this area used simplex-based linear approach for optimizing

TDS settings for presumed *PS* settings and Generalized Reduced Gradient non-linear technique to optimize the *PS* settings for already optimized *TDS* [25]. This procedure was iterated till convergence was achieved.

The use of optimization techniques in relay coordination was first suggested by Urdaneta et al. [25]. Irving and Elrafie [13] used Sparse Dual Revised Simplex method of linear programming suggested by Irving and Sterling [14] to optimize *TDS* settings for preassumed non-linear *PS* settings. Laway and Gupta [15] applied Simplex and Rosenbrock-Hillclimb methods to optimize *TDS* and *PS* settings respectively, in a similar way, as used by Urdaneta et al. [25]. These approaches were further followed by simplex-based approaches with more and more sophistications about finer aspects of the relays Urdaneta et al. [26], Chattopadhyay et al. [5], Urdaneta and Perez [24], Abdelaziz et al. [1]. So and Li [22] used evolutionary programming. A survey of all coordination philosophies used by various researchers in the past has been presented recently by Birla et al. [4].

Recently, Birla et al. [2] made an attempt to use "MATLAB Toolbox" and "Numeric Algorithm Group" Sequential Quadratic Programming routines Birla et al. [3]. Deep et al. [10] used RST2 of Shanker and Mohan (now Deep) [21] to solve the relay coordination problem for a IEEE 3 bus and IEEE 4 bus models. The results obtained by RST2 are compared with the results obtained by MATLAB Toolbox. (MATLAB Toolbox uses Sequential Quadratic Programming (SQP) method to solve the constrained non-linear optimization problems.) It is observed that although MATLAB Toolbox gives a lower value of the objective function its quality is inferior to the one obtained by RST2 because the solution obtained by MATLAB Toolbox violates some constraints whereas the solution obtained by RST2 does not violate any constraint at all.

3 Methodology

The technique used in this paper to solve this problem is a hybridized genetic algorithm for constrained optimization problems (C-SOMGA) which is extended version of Self Organizing Migrating Genetic Algorithm (SOMGA). SOMGA is a hybridized variant of GA for solving unconstrained nonlinear optimization problems, Deep and Dipti [6], which is inspired by the features of Self Organizing Migrating Algorithm (SOMA). SOMA is a population based stochastic search technique which is based on the social behavior of group of individuals. This algorithm is presented by Zelinka and Lampinen [27]. The more detail about this technique can be found in many research papers and books Oplatkova and Zelinka [19], Zelinka [28], Nolle and Zelinka [17], Nolle et al. [18], Nolle [16], Zelinka et al. [29, 30], Godfrey and Babu [11], etc. The main features that motivate us to incorporate this technique in GA are that it works with very low population size and it has more exploration capabilities than other low population based approaches. The selection operator used in C-SOMGA approach is well known tournament selection method developed by Osyczka and Krenich [20] for constrained optimization. This method is very effective while solving highly constrained single criterion optimization problems as well as the problems with computationally expensive objective function. In this selection the tournament between two chromosomes is carried out in the following way:

1. If both chromosomes are not in the feasible region the one which is closer to the feasible region is taken to the next generation. The values of the objective function are not calculated for either of chromosomes.
2. If one chromosome is in the feasible region and the other one is out of the feasible region the one which is in the feasible region is taken to the next generation. The values of the objective function are not calculated for either chromosome.
3. If both chromosomes are in the feasible region, the values of the objective function are calculated for both chromosomes and the one, which has a better value of the objective function, is taken to the next generation.

3.1 Methodology of C-SOMGA

As discussed earlier C-SOMGA presented by Deep and Dipti [7], is an extended version of SOMGA. Earlier C-SOMGA has been used to solve many real life problems Deep and Dipti [8, 9] and the success of this approach over these problems motivates us to use this approach to solve this problem. The methodology of this approach is almost similar to the methodology of SOMGA only difference is in the constraints handling for which tournament selection method as discussed above is used. The working steps of C-SOMGA is as follows:

First the individuals are generated randomly. These individuals compete with each other through well known constraint tournament selection method. Create new individuals via single point crossover and bitwise mutation. Then the best individual among them is considered as leader and all others are considered as active. For each active individual a new population of size N is created. Where N = path length/step size. This population is nothing but the new positions of the active individual, proceeds in the direction of the leader in n steps of the defined length. This path is perturbed randomly by a parameter called as PRT parameter. It is defined in the range $\langle 0, 1 \rangle$. A PRT vector is created using PRT parameter value, before an individual proceeds towards leader. The movement of an individual is given as follows:

$$\begin{aligned} &x_{i,j}^{MLnew} = x_{i,j,start}^{ML} + \left(x_{L,j}^{ML} - x_{i,j,start}^{ML}\right) tPRTVector_j \\ &where \quad t \in \langle 0,\ by\, Step\, to,\ PathLength \rangle, \\ &ML\ is\ actual\ migration\ loop. \end{aligned} \tag{1}$$

$x_{i,j}^{MLnew}$ is the new positions of an individual.
$x_{i,j,start}^{ML}$ is the positions of active individual.
$x_{L,j}^{ML}$ is the positions of leader.

Then sort this population according to the fitness value in decreasing order. Starting from the best one of the new population, evaluates the constraint violation function described below:

$$\psi(x) = \sum_{m=1}^{M} [h_m(x)]^2 + \sum_{k=1}^{K} G_k[g_k(x)]^2 \tag{2}$$

If $\psi(x) = 0$, replace the active individual with the current position and move to the next active individual. And If $\psi(x) > 0$ then move to the next best position of the sorted new population. In this way, all the active individuals are replaced by the new updated feasible position. If no feasible solution is available then active individual remains the same. At last the best individuals (number equal to population size) from the previous and current generations are selected for the next generation. The computational steps of this approach are given below:

Step 1 Generate the initial population.
Step 2 Evaluate all individuals.
Step 3 Apply tournament selection for constrained optimization on all individuals to select the better individuals for the next generation.
Step 4 Apply crossover operator on all individuals with crossover probability P_c to produce new individuals.
Step 5 Evaluate the new individuals.
Step 6 Apply mutation operator on every bit of every individual of the population with mutation probability P_m.
Step 7 Evaluate the mutated individuals.
Step 8 Find leader (best fitted individual) of the population and consider all others as active individuals of the population.
Step 9 For each active individual a new population of size N is created. This population is nothing but the new positions of the active individual towards the leader in n steps of the defined length. The movement of this individual is given in Eq. (1).
Step 9.1 Sort new population with respect to fitness in decreasing order.
Step 9.2 For each individual in the sorted population, check feasibility criterion.
Step 9.3 If feasibility criterion is satisfied replace the active individual with the new position, else move to next position in sort order and go to step 9.2.

Step 10 Select the best individuals (in fitness) of previous and current generation for the next generation via tournament selection.
Step 11 If termination criterion is satisfied go to step 12 else go to step 3.
Step 12 Report the best chromosome as the final optimal solution.

4 General form of the Problem

The operatings time (T) of a DOCR is non-linear function of the relay settings (Time Dial Settings (TDS) and Plug Settings (PS) and the fault current (I) seen by the relay). Therefore, Relay operating-time equation for a directional overcurrent relay is given by a non-linear equation as given below:

$$T = \frac{\alpha * TDS}{\left(\frac{I}{PS * CT_{pri-rating}}\right)^{\beta} - \gamma} \tag{3}$$

* denotes the multiplication. Only TDS and PS are unknown variables in the above equation. These are the "decision variables" of the problem. α, β and γ are the constants representing the behavior of characteristic in a mathematical way, in which operating time of the DOCR varies and are given as 0.14, 0.02 and 1.0 respectively as per IEEE std. [12]. Value of CTpri_rating depends upon the number of turns in the equipment CT (Current Transformer). CT is used to reduce the level of the current so that relay can withstand it. With each relay one "Current Transformer" is used and thus, CTpri_rating is known in the problem. Value of I (Fault current passing through the relay) is also known, as it is a system dependent parameter and continuously measured by measuring instruments.

Number of constraints for systems of bigger sizes will be dependent upon the number of lines in the system (see Table 1). In practice, electrical engineering power systems may be of even bigger sizes and there are other types of relays also besides DOCRs. Coordinating DOCRs with other types of relays generates even larger number of constraints than shown in Table 1. It is evident from Table 1 that simultaneous optimization of both the settings (TDS and PS) of each DOCR of the system is a complex problem.

Table 1 The complexity of the DOCR problem as the bus size increases

	IEEE 3-bus	IEEE 4-bus	IEEE 6-bus
No. of lines	3	4	7
No. of DOCRs (relays)	6	8	14
No. of decision variables	12	16	28
No. of selectivity constraints	8	9	38

5 The Optimization Problem

The relay, which is supposed to operate first to clear the fault, is called primary relay. A fault close to relay is known as the close-in fault for the relay and a fault at the other end of the line is known as a far-bus fault for this relay. Conventionally, objective function in coordination studies is constituted as the summation of operating-times of all primary relays, responding to clear all close-in and far-bus faults. The objective function is as follows:

$$\text{Minimize } OBJ = \sum_{i=1}^{N_{cl}} T^{i}_{pri\_cl\_in} + \sum_{j=1}^{N_{far}} T^{j}_{pri\_far\_bus} \tag{4}$$

where,

N_{cl} is number of relays responding for close-in fault.
N_{far} is number of relays responding for far-bus fault.
$T_{pri\text{-}cl\text{-}in}$ is primary relay operating-time for close-in fault.
$T_{pri\text{-}far\text{-}bus}$ is primary relay operating-time for far-bus fault.

The constraints are:

(a) Bounds on variables TDSs

$$TDS^{i}_{min} \leq TDS^{i} \leq TDS^{i}_{max}$$

where, i varies from 1 to N_{cl}.
TDS^{i}_{min} is lower limit and TDS^{i}_{max} is upper limit of TDS^{i}. These limits are 0.05 and 1.1, respectively.

(b) Bounds on variables PSs

$$PS^{j}_{min} \leq PS^{j} \leq PS^{j}_{max}$$

where, j varies from 1 to N_{cl}.
PS^{j}_{min} is lower limit and PS^{j}_{max} is upper limit of PS^{j}. These are 1.25 and 1.50, respectively.

(c) Limits on primary operation times: This constraint imposes constraint on each term of objective function to lie between 0.05 and 1.0.
(d) Selectivity constraints for all relay pairs:

$$T_{backup} - T_{primary} - CTI \geq 0$$

T_{backup} is operating time of backup relay and $T_{primary}$ is operating time of primary relay. Value of CTI is known.

6 Model I (The IEEE 3-Bus Model)

For the coordination problem of IEEE 3 bus model, value of each of N_{cl} and N_{far} is 6 (equal to number of relays or twice the lines). Accordingly, there are 12 decision variables (two for each relay) in this problem i.e. TDS^1 to TDS^6 and PS^1 to PS^6. The 3 bus system can be visualized as shown in Fig. 1.

Objective function (OBJ) to be minimized as given by Eq. (4) is:

$$OBJ = \sum_{i=1}^{N_{cl}} T^i_{pri\_cl\_in} + \sum_{j=1}^{N_{far}} T^j_{pri\_far\_bus}$$

Here,

$$T^i_{pri\_cl\_in} = \frac{0.14 * TDS^i}{\left(\frac{a^i}{PS^i * b^i}\right)^{0.02} - 1}; \quad T^i_{pri\_far\_bus} = \frac{0.14 * TDS^j}{\left(\frac{c^i}{PS^i * d^i}\right)^{0.02} - 1};$$

The values of constants a^i, b^i, c^i and d^i are given in Table 2.

Constraints

Constraints for the modelI will be as under:

Bounds on variables TDSs (Time dial setting of each relay)

$$TDS^i_{\min} \leq TDS^i \leq TDS^i_{\max} \; where, \quad i = 1, 2, \ldots, N_{cl}$$

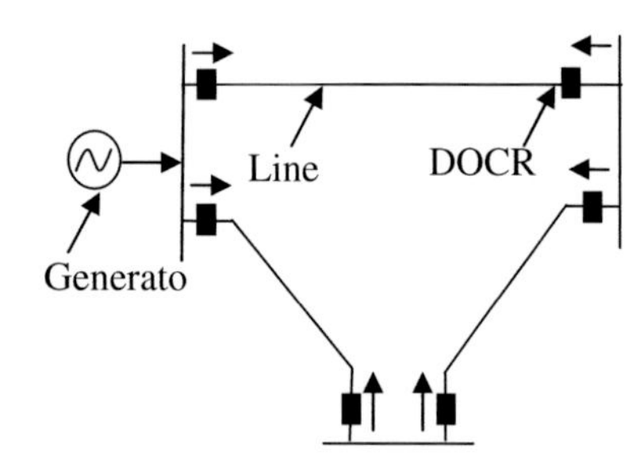

Fig. 1 A typical IEEE 3-bus DOCR coordination problem model

Table 2 Values of constants a^i, b^i, c^i and d^i for model-I

$T^i_{pri\_cl\_in}$			$T^i_{pri\_far\_bus}$		
TDS^i	a^i	b^i	TDS^j	c^i	d^i
TDS^1	9.4600	2.0600	TDS^2	100.6300	2.0600
TDS^2	26.9100	2.0600	TDS^1	14.0800	2.0600
TDS^3	8.8100	2.2300	TDS^4	136.2300	2.2300
TDS^4	37.6800	2.2300	TDS^3	12.0700	2.2300
TDS^5	17.9300	0.8000	TDS^6	19.2000	0.8000
TDS^6	14.3500	0.8000	TDS^5	25.9000	0.8000

Table 3 Values of constants e^i, f^i, g^i and h^i for model-I

T^i_{backup}			$T^i_{primary}$		
p	e^i	f^i	q	g^i	h^i
5	14.0800	0.8000	1	14.0800	2.0600
6	12.0700	0.8000	3	12.0700	2.2300
4	25.9000	2.2300	5	25.9000	0.8000
2	14.3500	0.8000	6	14.3500	2.0600
5	9.4600	0.8000	1	9.4600	2.0600
6	8.8100	0.8000	3	8.8100	2.2300
2	19.2000	2.0600	6	19.2000	0.8000
4	17.9300	2.2300	5	17.9300	0.8000

$TDS^i_{\min}$ is lower limit and $TDS^i_{\max}$ is upper limit of TDS^i. These limits are 0.05 and 1.1, respectively.

Bounds on variables PSs (Plug setting of each relay)

$$\mathrm{PS}^j_{\min} \leq \mathrm{PS}^j \leq \mathrm{PS}^j_{\max}, \ \ Where, j = 1, 2, \ldots, N_{cl}$$

$\mathrm{PS}^j_{\min}$ is lower limit and $\mathrm{PS}^j_{\max}$ is upper limit of PS^j. These are 1.25 and 1.50, respectively.

Limits on primary operation times: This constraint imposes constraint on each term of objective function to lie between 0.05 and 1.0.

Selectivity constraints for all relay pairs:

$$T_{backup} - T_{primary} - CTI \geq 0$$

T_{backup} is operating time of backup relay and $T_{primary}$ is operating time of primary relay. Value of CTI is 0.3. Here,

$$T^i_{backup} = \frac{0.14 * TDS^p}{\left(\frac{e^i}{PS^p * f^i}\right)^{0.02} - 1} \text{ and } T^i_{primary} = \frac{0.14 * TDS^q}{\left(\frac{g^i}{PS^q * h^i}\right)^{0.02} - 1}$$

The values of constants e^i, f^i, g^i and h^i are given in Table 3.

7 Model II (The IEEE 4-Bus Model)

The next coordination problem is of IEEE 4 bus model, value of each of N_{cl} and N_{far} is 8 (equal to number of relays or twice the lines). Accordingly, there are 16 decision variables (two for each relay) in this problem i.e. TDS^1 to TDS^8 and PS^1 to PS^8. The value of CTI in 4 bus model is 0.3. The 4 bus system can be visualized as shown in Fig. 2.

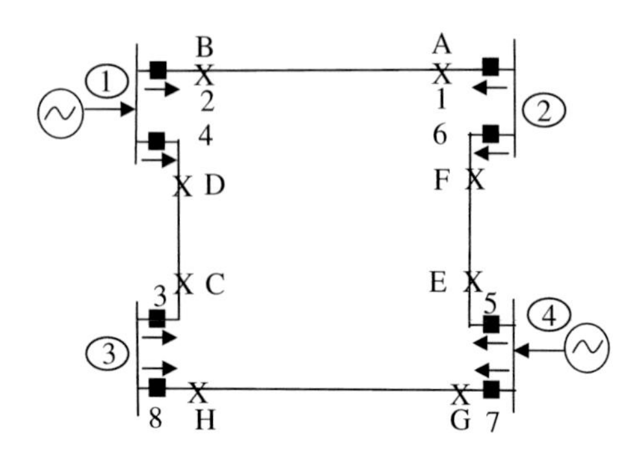

Fig. 2 A typical IEEE 4-bus DOCR coordination problem model

The objective function and constraints for the model will be of same form as in the case of Model-I problem (with $N_{cl} = 8$) described in Sect. 5. The values of constants a^i, b^i, c^i, d^i and e^i, f^i, g^i, h^i for Model-II are given in Tables 4 and 5 respectively.

Table 4 Values of constants a^i, b^i, c^i and d^i for model II

$T^i_{pri\_cl\_in}$			$T^i_{pri\_far\_bus}$		
TDS^i	a^i	b^i	TDS^j	c^i	d^i
TDS^1	20.3200	0.4800	TDS^2	23.7500	0.4800
TDS^2	88.8500	0.4800	TDS^1	12.4800	0.4800
TDS^3	13.6100	1.1789	TDS^4	31.9200	1.1789
TDS^4	116.8100	1.1789	TDS^3	10.3800	1.1789
TDS^5	116.7000	1.5259	TDS^6	12.0700	1.5259
TDS^6	16.6700	1.5259	TDS^5	31.9200	1.5259
TDS^7	71.7000	1.2018	TDS^8	11.0000	1.2018
TDS^8	19.2700	1.2018	TDS^7	18.9100	1.2018

Table 5 Values of constants e^i, f^i, g^i and h^i for model II

T^i_{backup}			$T^i_{primary}$		
p	e^i	f^i	q	g^i	h^i
5	20.3200	1.5259	1	20.3200	0.4800
5	12.4800	1.5259	1	12.4800	0.4800
7	13.6100	1.2018	3	13.6100	1.1789
7	10.3800	1.2018	3	10.3800	1.1789
1	1.1600	0.4800	4	116.8100	1.1789
2	12.0700	0.4800	6	12.0700	1.1789
2	16.6700	0.4800	6	16.6700	1.5259
4	11.0000	1.1789	8	11.0000	1.2018
4	19.2700	1.1789	8	19.2700	1.2018

8 Model III (The IEEE 6-Bus Model)

The third coordination problem is of IEEE 6 bus model, value of each of N_{cl} and N_{far} is 14 (equal to number of relays or twice the lines). Accordingly, there are 28 decision variables (two for each relay) in this problem i.e. TDS^1 to TDS^{14} and PS^1 to PS^{14}. The 6 bus system can be visualized as shown in Fig. 3.

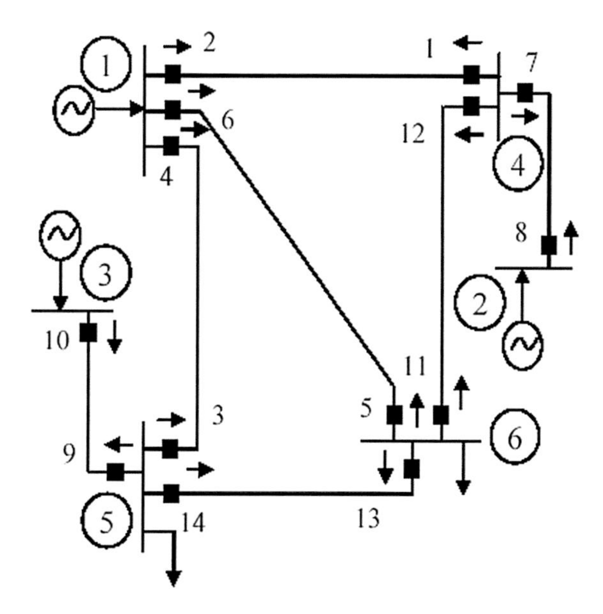

Fig. 3 A typical IEEE 6-bus DOCR coordination problem model

Table 6 Values of constants a^i, b^i, c^i and d^i for model III

$T^i_{pri\_cl\_in}$			$T^i_{pri\_far\_bus}$		
TDS^i	a^i	b^i	TDS^j	c^i	d^i
TDS^1	2.5311	0.2585	TDS^2	5.9495	0.2585
TDS^2	2.7376	0.2585	TDS^1	5.3752	0.2585
TDS^3	2.9723	0.4863	TDS^4	6.6641	0.4863
TDS^4	4.1477	0.4863	TDS^3	4.5897	0.4863
TDS^5	1.9545	0.7138	TDS^6	6.2345	0.7138
TDS^6	2.7678	0.7138	TDS^5	4.2573	0.7138
TDS^7	3.8423	1.746	TDS^8	6.3694	1.746
TDS^8	5.618	1.746	TDS^7	4.1783	1.746
TDS^9	4.6538	1.0424	TDS^{10}	3.87	1.0424
TDS^{10}	3.5261	1.0424	TDS^9	5.2696	1.0424
TDS^{11}	2.584	0.7729	TDS^{12}	6.1144	0.7729
TDS^{12}	3.8006	0.7729	TDS^{11}	3.9005	0.7729
TDS^{13}	2.4143	0.5879	TDS^{14}	2.9011	0.5879
TDS^{14}	2.9011	0.5879	TDS^{13}	4.335	0.5879

The objective function and constraints for the model will be of same form as in the case of Model-I problem (with $N_{cl} = 14$) described in Sect. 5. The values of constants a^i, b^i, c^i, d^i and e^i, f^i, g^i, h^i for Model-III are given in Tables 6 and 7 respectively.

Table 7 Values of constants e^i, f^i, g^i and h^i for model III

T^i_{backup}			$T^i_{primary}$		
p	e^i	f^i	q	g^i	h^i
8	4.0909	1.746	**1**	5.3752	0.2585
8	2.9323	1.746	**1**	2.5311	0.2585
3	0.6213	0.4863	**2**	2.7376	0.2585
3	1.6658	0.4863	**2**	5.9495	0.2585
10	3.0923	1.0424	**3**	4.5897	0.4863
10	2.561	1.0424	**3**	2.9723	0.4863
1	0.8869	0.2585	**4**	4.1477	0.4863
1	1.5243	0.2585	**4**	6.6641	0.4863
12	2.5444	0.7729	**5**	4.2573	0.7138
12	1.4549	0.7729	**5**	1.9545	0.7138
13	1.8321	0.5879	**9**	5.2696	1.0424
13	1.618	0.5879	**9**	4.6538	1.0424
11	2.1436	0.7729	**7**	4.1783	1.746
11	1.9712	0.7729	**7**	3.8423	1.746
10	2.7784	1.0424	**14**	5.3541	0.5879
10	2.026	1.0424	**14**	2.9011	0.5879
2	1.8718	0.2585	**7**	3.8423	1.746
2	2.0355	0.2585	**7**	4.1783	1.746
14	2.0871	0.5879	**11**	3.9005	0.7729
14	1.4744	0.5879	**11**	2.584	0.7729
6	1.8138	0.7138	**11**	3.9005	0.7729
6	1.1099	0.7138	**11**	2.584	0.7729
4	3.4386	0.4863	**9**	5.2696	1.0424
4	3.0368	0.4863	**9**	4.6538	1.0424
2	0.4734	0.2585	**12**	3.8006	0.7729
2	1.5432	0.2585	**12**	6.1144	0.7729
8	4.5736	1.746	**12**	6.1144	0.7729
8	3.3286	1.746	**12**	3.8006	0.7729
4	0.8757	0.4863	**14**	2.9011	0.5879
4	2.5823	0.4863	**14**	5.3541	0.5879
12	2.7269	0.7729	**13**	4.335	0.5879
12	1.836	0.7729	**13**	2.4143	0.5879
1	1.1231	0.2585	**6**	6.2345	0.7138
6	1.6085	0.7138	**13**	4.335	0.5879
14	1.7142	0.5879	**5**	4.2573	0.7138
11	1.2886	0.7729	**1**	5.3752	0.2585
13	1.4995	0.5879	**3**	4.5897	0.4863
3	1.4658	0.4863	**6**	6.2345	0.7138

9 Method of Solution and Discussion of Results

The technique used to solve these IEEE bus model problems is C-SOMGA. The experimental set up for solving these three problems is given in Table 8. Each problem has been run 5 times and the best solution obtained out of 5 is reported as the global optimal solution. For comparison, previously quoted results by MATLAB Tool Box and RST2 of Shanker and Mohan [21] are used. The results are also compared with C-GA and C-SOMA. For fair comparison in C-GA same selection, crossover and mutation operators and in C-SOMA same tournament selection method are used as in C-SOMGA. Other parameter values are also kept same in all the three algorithms.

In Table 9, the results obtained by C-SOMGA for IEEE 3-Bus model are reported. These results are compared with C-GA, C-SOMA, MATLAB Tool Box and RST2. It can be observed in Table 9 that MATLAB Tool Box achieved the best minimum value, C-SOMGA second best, RST2 third best, C-GA is at fourth place and C-SOMA is at last position. Since it is a constrained optimization problem, it is necessary that the solution must be in the feasible domain. In Table 10, it is observed that the solution obtained by MATLAB Tool Box and C-SOMA are infeasible. Hence the solution obtained by these two algorithms cannot be accepted for this problem. Therefore C-SOMGA provides best feasible solution of this problem which is better than C-GA and RST2.

This problem has been run five times and the best and worst results of each run are plotted in Fig. 4 with the results of MATLAB Tool Box and RST2. It can be seen that in all the five runs the best and worst results obtained by C-SOMGA are better than that obtained by RST2. It is also better than MATLAB Tool Box since the result by MATLAB Tool Box is infeasible.

In Table 11, the results for IEEE 4-Bus problem obtained by all the five techniques are given. Again the results obtained by MATLAB Tool Box is the best minimum solution but infeasible also. For infeasibility see Table 12. The second best minimum is obtained by C-SOMGA which is feasible. RST2, SOMA and C-GA also provide feasible solution at third, fourth and fifth rank. On the ranking basis of providing feasible solutions C-SOMGA is the best. In Fig. 5 the best and

Table 8 Experimental setup

	IEEE 3-Bus	IEEE 4-Bus	IEEE 6-Bus
Population size	20	20	30
P_c	0.95	0.95	0.95
P_m	0.005	0.005	0.005
Step size	0.21	0.21	0.21
Path length	3	3	3
String length	30	30	30
Total function evaluations allowed	500,000	500,000	500,000

Table 9 Optimal decision variables and objective function optimal values for IEEE 3-bus model

Decision variables	Value by C-GA algorithm	Value by C-SOMA algorithm	Value by MATLAB Tool Box	Value by RST2 Algorithm	Value by C-SOMGA algorithm
TDS^1	0.0500000000	0.1296280000	0.0500000000	0.0500620000	0.0500000000
TDS^2	0.2468750000	0.5477830000	0.1976466710	0.2107300000	0.2002870000
TDS^3	0.0500000000	0.0999874000	0.0500000000	0.0500210000	0.0500000000
TDS^4	0.2468750000	0.2058190000	0.2090317120	0.2188270000	0.2092140000
TDS^5	0.1935550000	0.0835336000	0.1812052361	0.1881400000	0.1813980000
TDS^6	0.1976560000	0.2777830000	0.1806755223	0.1953780000	0.1860030000
PS^1	1.2500000000	1.3921900000	1.2500000000	1.2512340000	1.2500100000
PS^2	1.2500000000	1.4880500000	1.5000000000	1.3534360000	1.4719700000
PS^3	1.2500000000	1.4465400000	1.2500000000	1.2500000000	1.2500000000
PS^4	1.2500000000	1.3838900000	1.5000000000	1.3817690000	1.4978300000
PS^5	1.2739300000	1.3544100000	1.5000000000	1.3743430000	1.4962000000
PS^6	1.2500000000	1.3489200000	1.5000000000	1.2501860000	1.4035300000
Optimal value of objective function	5.0761600000	8.0101600000	4.7806507047	4.8354270193	4.7898900000

Table 10 Value of each selectivity constraint for IEEE 3-bus model

Constraint No.	Constraint values in second				
	Value by C-GA algorithm	Value by C-SOMA algorithm	MATLAB tool box allows tolerance 1.0e−08 in constraints	RST2 algorithm	C-SOMGA algorithm
1	0.0291460000	−0.7872000	0.00000000000000	0.00051468116110	0.0361232000
2	0.0000259000	−0.6390300	0.00000000000001	0.00013627506057	0.0000030909
3	0.0214740000	−0.0829000	−0.00000000000002	0.00050754672332	0.0184409000
4	0.0064460000	−0.0374200	0.09008111400397	0.08576410572325	0.0000000851
5	0.0884650000	1.09562000	0.04221324579101	0.03879991683310	0.0000021494
6	0.1521860000	0.30189700	0.02117707464442	0.01422080195843	0.1003800000
7	0.0527230000	0.18450000	−0.0000000000000	0.00050469099876	0.0000201169
8	0.1830670000	1.41382000	0.10042972713526	0.09584778802172	0.0898825000
	Feasible	Infeasible	Infeasible	Feasible	Feasible

worst solutions of the five runs of C-SOMGA are plotted. In all the runs, results obtained by C-SOMGA are better than RST2 and MATLAB Tool Box.

Table 13 contains the best results obtained by C-SOMGA out of five runs and compared with other four techniques for 6-bus model. The result obtained by MATLAB Tool Box is the best but reported infeasible in Birla et al. [2]. C-SOMGA provides second best minimum which is feasible also. RST2 and C-GA

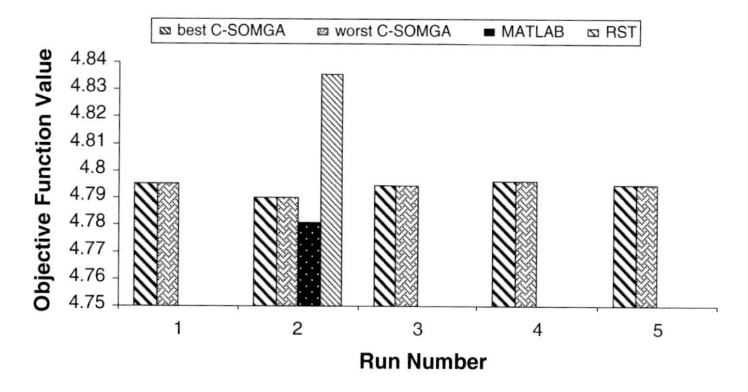

Fig. 4 Comparative results of C-SOMGA and previously quoted results for 3-Bus system

Table 11 Optimal decision variables and objective function optimal values for IEEE 4-bus model

Decision variables	Value by C-GA algorithm	Value by C-SOMA algorithm	Value by MATLAB Tool Box	Value by RST2 Algorithm	Value by C-SOMGA algorithm
TDS^1	0.0500000000	0.0500000000	0.0500000000	0.0500260000	0.0500570000
TDS^2	0.2281250000	0.2166920000	0.2121687898	0.2242050000	0.2155220000
TDS^3	0.0500000000	0.0500000000	0.0500000000	0.0500070000	0.0500000000
TDS^4	0.1687500000	0.1641810000	0.1515761615	0.1586850000	0.1516590000
TDS^5	0.1390630000	0.1377150000	0.1264004560	0.1366540000	0.1284240000
TDS^6	0.0500000000	0.0500000000	0.0500000000	0.0500170000	0.0500000000
TDS^7	0.1687500000	0.1563940000	0.1337862054	0.1387680000	0.1340360000
TDS^8	0.0500000000	0.0500004000	0.0500000000	0.0500380000	0.0500000000
PS^1	1.3750000000	1.3464000000	1.2733272654	1.2910100000	1.2724900000
PS^2	1.2500000000	1.4057700000	1.5000000000	1.2645400000	1.4295500000
PS^3	1.2500000000	1.2500000000	1.2500000000	1.2500000000	1.2500000000
PS^4	1.2500000000	1.2500300000	1.5000000000	1.3460040000	1.4980900000
PS^5	1.2500000000	1.2510700000	1.5000000000	1.2669120000	1.4503100000
PS^6	1.2500000000	1.2500000000	1.2500000000	1.2512300000	1.2500000000
PS^7	1.2500000000	1.3769800000	1.5000000000	1.3937230000	1.4944700000
PS^8	1.2500000000	1.2500000000	1.2500000000	1.2507740000	1.2500000000
Optimal value of objective function	3.8587400000	3.7892200000	3.6697457859	3.7050183128	3.6745300000

are at the third and fourth place respectively. The result obtained by C-SOMA is infeasible. In Fig. 6, best and worst solution of the five runs of C-SOMGA is plotted. In this case the best solution in all five runs is better that RST2 and MATLAB Tool Box. But the worst solution is inferior to RST2.

Table 12 Value of each selectivity constraint for IEEE 4-bus model (Constraint values in second)

Constraint No.	Value by C-GA	Value by C-SOMA	MATLAB Tool Box Allows tolerance 1.0e−08 in constraints	RST2 algorithm	C-SOMGA algorithm
1	0.0031260000	0.0000010400	−0.0000000000	0.0002684130	0.0000010955
2	0.0929590000	0.0891080000	0.1003129217	0.0902582981	0.0980991000
3	0.0703050000	0.0547230000	0.0000000000	0.0000299943	0.0000012618
4	0.1235750000	0.1095270000	0.0498406752	0.0470376507	0.0496938000
5	0.0586510000	0.0433470000	-0.0000000000	0.0075951410	0.0000022420
6	0.0299840000	0.0246850000	0.0258683209	0.0229711095	0.0249862000
7	0.0065800000	0.0000004550	-0.0000000000	0.0001173834	0.0000003936
8	0.1035830000	0.0879960000	0.0976159913	0.0901641137	0.0975229000
9	0.0140720000	0.0019520000	0.0000000000	0.0000327579	0.0000000932
	Feasible	Feasible	Infeasible	Feasible	Feasible

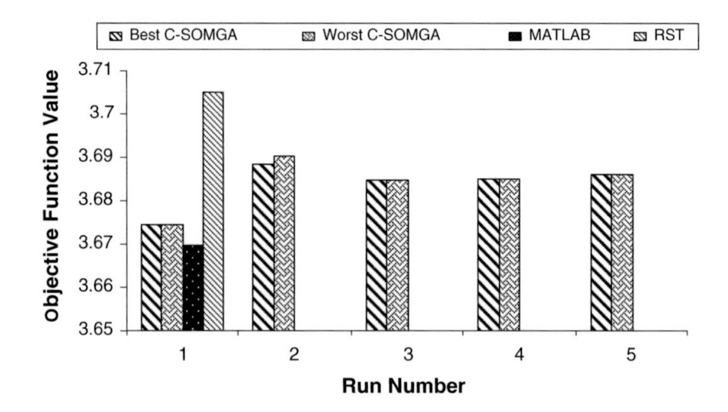

Fig. 5 Comparative results of C-SOMGA and previously quoted results for 4-Bus system

Main advantage of using C-SOMGA is that it requires very less population size for solving these kinds of problems. Another advantage of using C-SOMGA is that for solving such a highly constrained non linear optimization problem, it takes only 2–4 min (1 run for 1 model). On the basis of the results for IEEE 3-Bus, IEEE 4-Bus and IEEE 6-Bus, it is concluded that C-SOMGA provides better solutions than other techniques.

Table 13 Optimal decision variables and objective function optimal values for IEEE 6-bus model

Relay number	Optimal values of settings									
	C-GA		C-SOMA		MATLAB Tool Box		RST2		C-SOMGA	
	TDS	PS	TDS	PS	TDS	PS	TDS	PS	TDS	PS
1	0.16875	1.25	0.265438	1.25239	0.1014	1.500	0.1170	1.2567	0.1109290	1.32407
2	0.2875	1.25	0.25962	1.25012	0.1863	1.500	0.2080	1.2531	0.1865800	1.49635
3	0.16875	1.25	0.355199	1.25	0.0791	1.500	0.0857	1.4389	0.0950484	1.27194
4	0.16875	1.25	0.246928	1.27744	0.1006	1.500	0.1126	1.2507	0.1015420	1.47800
5	0.05	1.25	0.0501097	1.33236	0.0500	1.2500	0.0500	1.2502	0.0500009	1.25006
6	0.0537109	1.37524	0.341553	1.3864	0.0500	1.3808	0.0525	1.4006	0.0500067	1.39429
7	0.05	1.25	0.0500049	1.29868	0.0500	1.2500	0.0500	1.2501	0.0500001	1.25000
8	0.0796875	1.25	0.262765	1.25786	0.0500	1.2500	0.0500	1.2502	0.0500017	1.25001
9	0.05	1.25	0.0500349	1.27688	0.0500	1.2500	0.0500	1.2504	0.0500017	1.25000
10	0.109375	1.25	0.177264	1.25	0.0500	1.4708	0.0639	1.2501	0.0594232	1.35235
11	0.0796875	1.3125	0.0882636	1.2501	0.0650	1.500	0.0827	1.2700	0.0684055	1.45239
12	0.0796875	1.25	0.219663	1.25	0.0505	1.500	0.0604	1.3270	0.0588452	1.30544
13	0.0796875	1.25	0.128998	1.34693	0.0500	1.4329	0.0588	1.2507	0.0500016	1.43296
14	0.109375	1.25	0.278954	1.33107	0.0708	1.500	0.0924	1.2521	0.0757100	1.43487
Optimal value of objective function	13.7996		26.1495		10.1386		10.6192		10.357800000	
Feasibility	Feasible		Infeasible		Infeasible		Feasible		Feasible	

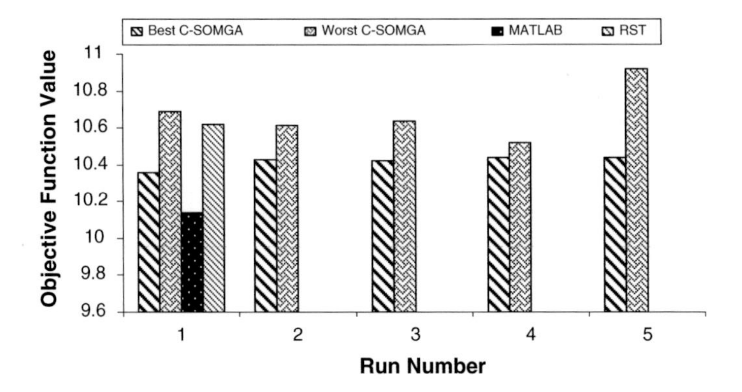

Fig. 6 Comparative results of C-SOMGA and previously quoted results for 6-Bus system

10 Conclusions

In this paper, electrical engineering power system DOCR coordination problem is solved by modeling it as constrained non-linear optimization problem using C-SOMGA. The problem is to determine the optimal value of Time dial setting and Plug setting so that the relay time can be minimized. Three models of this problem namely IEEE 3-Bus, IEEE 4-Bus and IEEE 6-Bus are solved using C-SOMGA. The complexities of all the three models are different due to different decision variables and constraints.

The results obtained by C-SOMGA are compared with C-GA, C-SOMA and MATLAB Tool Box and RST2 algorithm. In all the three models, C-SOMGA outperforms C-GA, C-SOMA, MATLAB Tool Box and RST2. The advantage of using C-SOMGA is that for such a complex problem it requires very less population size and computation time. C-SOMGA is found to be a robust technique for such type of constrained nonlinear optimization problems.

References

1. Abdelaziz, A.Y., Talaat, H.E.A., Nosseir, A.I., Hajjar, A.A.: An adaptive protection scheme for optimal coordination of overcurrent relays. Int. J. Electr. Power Syst. Res. **61**(1), 1–9 (2002)
2. Birla, D., Maheshwari, R.P., Gupta, H.O.: A new non-linear directional overcurrent relay coordination technique, and banes and boons of near-end faults based approach. IEEE Trans. Power Deliv. **21**(3), 1176–1182 (2006)
3. Birla, D., Maheshwari, R.P., Gupta, H.O.: Optimal DOCR coordination in non-linear environment by simultaneously optimizing settings. In: International Conference on Computer Application in Electrical Engineering Recent Advances, September 28–October 01, IIT, Roorkee, (India) (2005)

4. Birla, D., Maheshwari, R.P., Gupta, H.O.: Time-overcurrent relay coordination: a review. Int. J. Emerg. Electr. Power Syst. **2**(2), 1–13 (2005)
5. Chattopadhyay, B., Sachdev, M.S., Sidhu, T.S.: An on-line relay coordination algorithm for adaptive protection using linear programming technique. IEEE Trans. Power Deliv. **11**, 165–173 (1996)
6. Deep, K., Dipti, S.: A new hybrid self organizing migrating genetic algorithm for function optimization. In: IEEE Congress on Evolutionary Computation, pp. 2796–2803 (2007)
7. Deep, K., Dipti: A self organizing migrating genetic algorithm for constrained optimization. Appl. Math. Comput. **198**(1), 237–250 (2008)
8. Deep, K., Dipti: A self organizing migrating genetic algorithm for reliability optimization. J. Inf. Comput. Sci. **4**(3), 163–172 (2009)
9. Deep, K., Dipti: Engineering optimization using SOMGA. Advances in intelligent systems and computing, vol. 236, pp. 323–336. Springer, Berlin (2012)
10. Deep, K., Birla, D., Maheshwari, R.P., Gupta, H.O., Takur, M.: A population based heuristic algorithm for optimal relay operating time. World J. Modell. Simul. **3**(3), 167–176 (2006)
11. Godfrey, C.O., Babu, B.V.: New optimization techniques in engineering. Springer, Heidelberg, Germany (2004). ISBN 3-540-20167-X
12. IEEE Std. C37.112-1996, IEEE standard inverse-time characteristic equations for overcurrent relays. Approved by IEEE standards Board, September (1996)
13. Irving, M.R., Elrafie, H.B.: Linear programming for directional overcurrent relay coordination in interconnected power systems with constraint relaxation. Electric Power Syst. Res. **27**(3), pp. 209–216 (1993)
14. Irving, M.R., Sterling, M.J.H.: Economic dispatch of active power with constraint relaxation. Proceeding IEEE, Part C, pp. 172–177 (1983)
15. Laway, N.A., Gupta, H.O.: A method for adaptive coordination of directional relays in an interconnected power system. Developments in Power System Protection, IEEE conference pub. No. 368, pp. 240–243, London (1993)
16. Nolle, L.: SASS applied to optimum work roll profile selection in the hot rolling of wide steel. Knowl.-Based Syst. **20**(2), 203–208 (2007)
17. Nolle, L., Zelinka, I.: SOMA applied to optimum work roll profile selection in the hot rolling of wide steel. In: Proceedings of the 17th European Simulation Multiconference ESM 2003, Nottingham, UK, pp. 53–58, ISBN 3-936150-25-7, 9–11 June 2003
18. Nolle, L., Zelinka, I., Hopgood, A.A., Goodyear, A.: Comparision of an self organizing migration algorithm with simulated annealing and differential evolution for automated waveform tuning. Adv. Eng. Soft. **36**, 645–653 (2005)
19. Oplatkova, Z., Zelinka, I.: Investigation on Shannon-Kotelnik theorem impact on SOMA algorithm performance. In: Proceedings 19th European Conference on Modelling and Simulation Yuri Merkuryev, Richard Zobel (2005)
20. Osyczka, A., Krenich, S.: A new method of solution of nonlinear programming problems using genetic algorithms. In: 3rd polish conference on evolutionary algorithms and global optimization, Potok Zloty, 25–28 May 1999 (in Polish)
21. Shanker, K., Mohan, C.: A random search technique for the global minima of constrained nonlinear optimization problems. In: Proceedings of the International Conference on Optimization Techniques and Applications, Singapore, pp. 905–918 (1987)
22. So, C.W., Li, K.K.: Overcurrent relay coordination by evolutionary programming. Electr. Power Syst. Res. **53**, 83–90 (2000)
23. Urdaneta A. J., Perez L.G., Gomez J.F., Feijoo, B., Gonzalez, M.: Presolve analysis and interior point solutions of the linear programming coordination problem of directional overcurrent relays. Int. J. Electr. Power Energy Syst. **23**(8), 819–825 (2001)
24. Urdaneta, A.J., Perez, L.G.: Optimal coordination of directional overcurrent relays considering definite time back-up relaying. IEEE Trans. Power Deliv. **14**(4), 1276–12184 (1999)
25. Urdaneta, A.J., Nadira, R., Perez, L.: Optimal coordination of directional overcurrent relay in interconnected power systems. IEEE Trans. Power Deliv. **3**, 903–911 (1988)

26. Urdaneta, A.J., Resterbo, H., Sanchez, J., Fajardo, J.: Coordination of directional overcurrent relays timing using linear programming. IEEE Trans. Power Deliv. **11**, 122–129 (1996)
27. Zelinka I., Lampinen, J.: SOMA-self organizing migrating algorithm, Mendal. In: 6th International Conference on Soft Computing, vol. 80(2), Brno, Czech Republic, pp. 214 (2000)
28. Zelinka, I.: Analytic programming by means of soma algorithm. In: Proceeding of 8th International Conference on Soft Computing Mendel'02, Brno, Czech Republic, pp. 93–101. ISBN 80-214-2135-5 (2002)
29. Zelinka, I., Lampinen, J., Nolle, L.: On the theoretical proof of convergence for a class of SOMA search algorithms. In: Proceedings of the 7th International MENDEL Conference on Soft Computing, Brno, CZ, pp. 103–110, ISBN 80-214-1894-X, 6-8 June 2001
30. Zelinka, I., Oplatkova, Z., Nolle, L.: Boolean symmetry function synthesis by means of arbitrary evolutionary algorithms—comparative study. In: Proceedings of the 18th European simulation multiconference ESM 2004, Magdeburg, Germany, pp. 143–148, ISBN 3-936150-35-4, 13–14 June 2004

SOMGA for Large Scale Function Optimization and Its Application

Dipti Singh and Kusum Deep

Abstract Self Organizing Migrating Genetic Algorithm (SOMGA) is a hybridized variant of Genetic Algorithm (GA) inspired by the features of Self Organizing Migrating Algorithm, presented by Deep and Dipti (IEEE Congr Evol Comput, pp 2796–2803, 2007) [1]. SOMGA extracts the features of binary coded GA and real coded SOMA in such a way that diversity of the solution space can be maintained and thoroughly exploited keeping function evaluation low. It works with very less population size and tries to achieve global optimal solution faster in less number of function evaluations. Earlier SOMGA has been used to solve problems up to 10 dimensions with population size 10 only. This chapter is brake into three sections. In first section a possibility of using SOMGA to solve large scale problem (dimension up to 200) has been analyzed with the help of 13 test problems. The reason behind extension is that SOMGA works with very small population size and to solve large scale problems (dimension 200) only 20 population size is required. On the basis of results it has been concluded that SOMGA is efficient to solve large scale global optimization problems with small population size and hence required lesser function evaluations. In second section, two real life problems from the field of engineering as an application have been solved using SOMGA. In third section, a comparison between two ways of hybridization has been analyzed. There can be two approaches to hybridize a population based technique. Either by incorporating a deterministic local search in it or by merging it with other population based technique. To see the effect of both the approaches on GA, the results of SOMGA on five test problems are compared with the results of MA (GA+ deterministic local search). Results clearly indicates that SOMGA is less expensive and effective to solve these problems.

D. Singh (✉)
Department of Applied Sciences, Gautam Buddha University, Greater Noida, India
e-mail: diptipma@rediffmail.com

K. Deep
Indian Institute of Technology Roorkee, Roorkee, India
e-mail: kusumfma@iitr.ac.in

D. Davendra and I. Zelinka (eds.), *Self-Organizing Migrating Algorithm*,
Studies in Computational Intelligence 626, DOI 10.1007/978-3-319-28161-2_9

Keywords Self organizing migrating algorithm · Genetic algorithm · Large scale global optimization · Real life problems

1 Introduction

A variety of computational techniques have appeared in literature for solving nonlinear optimization problems. However, there is no single technique that can claim to efficiently solve each and every nonlinear optimization problem. In fact, a technique, which is efficient for one type of nonlinear optimization problem, may be inefficient for solving other types of nonlinear optimization problems. Moreover, the computational time and memory space requirements of most of these techniques are quite large. Keeping this in view, it is desirable to develop more efficient and reliable computational techniques for solving a large variety of nonlinear real life optimization problems of practical interest which require less memory space so that these can be easily implemented on personal computers. Some of the general requirements of a computational algorithm for solving global optimization problems are: (i) wide applicability i.e., it should be applicable to a wide class of real life problems (ii) simplicity in structure, which would allow it to be easily implemented on a computer system (iii) minimum mathematical complexities, so that it can be used conveniently, even by non expert users.

Optimization problems arise in several fields such as system engineering, telecommunication and manufacturing systems etc. In fact the newly developed optimization techniques are now being extensively used in various spheres of human activity where decisions have to be taken in some complex situations that can be represented by mathematical models. A real life optimization problem may have a number of local as well as global optimal solutions. It is desired by most of the users to design optimization techniques which determine the global optimal solution rather than the local optimal solution of nonlinear optimization problems. With the advent of computers, population based heuristics are becoming popular day by day, not only because of their ease of implementation, but also due to their wide applicability. Population based stochastic search methods have been frequently used in the literature to solve real life global optimization problems. Although these probabilistic search algorithms do not give absolute guarantee to determine the global optimum solution, these methods are preferred over traditional methods. Evolutionary Algorithms, particularly Genetic Algorithms (GAs) are the most commonly used heuristics for solving real life problems. Though GAs are efficient to solve global optimization problem but usually converges very slow and generally required large population size also. Many attempts have been made in literature to improve the efficiency of Genetic algorithms either by designing new operators or by incorporating the features of other techniques. Grefensette [2] introduced a hybrid variant of GA which uses a traditional hill climbing routine for improving the fitness of newly generated points known as Fawwin and Lamarckian evolution approach and then the new offspring compete with their parents for selection as a member of the next population. Kasprzyk and Jasku [3] developed a variant of GA which is hybridization of GA and simplex method known as genetic—simplex algorithm, in this approach first the solution area is explored by GA operators and then simplex method uses starting points provided by the GA to determine an optimum solution. Chelouah and Siarry [4] also proposed a hybrid method by combining the features of continuous Tabu search and Nelder-Mead simplex algorithm. This approach is used to find the global minima of nonlinear optimization problems. Wang et al. [5] apply quantum computing to GAs to develop a class of quantum-inspired GAs.

Javadi et al. [6] presented a neural-network based genetic algorithm which uses neural network to improve solution quality and convergence speed of GAs. Fan et al. [7] integrate the Nelder–Mead Simplex search method with genetic algorithm and particle swarm optimization in an attempt to locate the global optimal solution of nonlinear continuous variable functions focusing mainly on response surface methodology. Comparative performance on ten test problems is demonstrated. Hwang and Song [8] present a novel adaptive real-parameter simulated annealing genetic algorithm which maintain the merits of GAs and simulated annealing. Zhang and Lu [9] define a new real valued mutation operator and use it to design a hybrid real coded GA with quasi-simplex technique. A nitche hybrid genetic algorithm is proposed by Wei and Zhao [10] and results are reported on 3 benchmark functions. Premalatha and Nataranjan [11] established hybrid PSO which proposes the modification strategies in PSO using GA to solve the optimization problems. Khosravi et al. [12] proposes a novel hybrid algorithm that uses the abilities of evolutionary and conventional algorithm simultaneously. Ghatei et al. [13] designed a new variant of particle swarm optimization by including Great Deluge Algorithm (GDA) as local search factor. Esmin and Matwin [14] presented a hybrid approach of using the features of PSO and GA known as HPSOM algorithm. The main idea of this approach is to integrate the PSO with genetic algorithm mutation method.

It is clear from the literature that several variants of GA are available in literature to improve the efficiency of these algorithms. Deep and Dipti [1] presented a variant of GA named as SOMGA in which GA has been hybridized with a new emerging population based technique, Self Organizing Migrating Algorithm (SOMA). SOMA is an emergent search technique in the field of population based techniques, developed by Zelinka and Lampinen [15]. This algorithm is based on the self organizing behavior of group of individuals looking for food. For this all individuals follow the path of one individual known as Leader selected among them based on the best fitness value. In the whole process of this algorithm, no new solutions are created during the search. Instead, only the positions of the solutions are changed during a generation, called a migration loop and it works with small population size. The details of this algorithm can be found in many research papers and books Oplatkova and Zelinka [16], Zelinka [17], Nolle and Zelinka [18], Nolle et al. [19], Nolle [20], Zelinka et al. [21, 22], Onwubolu and Babu [23], etc. The common feature between GA and SOMA is that both are population based stochastic search heuristics. Mutation and crossover is done (but the way in which they are applied is different). Some differences of the two algorithms are as follows:

- In GA new points are generated, whereas in SOMA no new points are generated, but instead their positions are updated.
- Individuals can proceed in any direction in GA, whereas in SOMA individuals proceed only in the direction of the leader.
- GA has a competitive behavior, but SOMA has competitive-cooperative behavior.
- GA works with a large population size, whereas SOMA works with a small population size.

2 Previous Work Done

The algorithm SOMGA has been designed to solve the unconstrained non-linear optimization problems of the type:

$$\left.\begin{array}{r} \operatorname{Min} f(X), X=(x_1, x_2, \ldots x_n) \\ a_i \leq x_i \leq b_i, i=1,2, \ldots n. \end{array}\right\} \quad (1)$$

where a_i and b_i are the lower and upper bounds on the variables.

As discussed above several variants of population based techniques are available in literature, for improving the convergence of these algorithms. The main reason of slow convergence and premature convergence of these algorithms is considered as diversity mechanism. If one algorithm fails to maintain the diversity during the search then there are more chances to be converging premature. SOMGA is an effort to improve the efficiency of both the algorithms GA and SOMA by extracting the best features of these algorithms. In the hybridization of SOMGA, binary coded GA and real coded SOMA has been used. It derives the features of selection, crossover and mutation from binary coded GA and derives the features of small population size, organization and migration from real coded SOMA. The features of GA and SOMA are combined in such a way that solution search space can be thoroughly exploited and diversity of the search domain can be preserved by generating new points in solution space. The methodology of this algorithm is:

Methodology

First the individuals are generated randomly. These individuals compete with each other through tournament selection; create new individuals via single point crossover and bitwise mutation. Then the best individual among them is considered as leader and the worst individual is considered as active. The active individual proceeds in the direction of the leader in n steps of the defined length. This path is perturbed randomly by a parameter known as PRT parameter. It is defined in the range $\langle 0, 1\rangle$. Using this PRT parameter value, PRT vector is created before an individual proceeds towards leader. This parameter has the same effect as mutation in GA. The movement of an individual is given as follows:

$$x_{i,j}^{MLnew} = x_{i,j,start}^{ML} + \left(x_{L,j}^{ML} - x_{i,j,start}^{ML}\right) tPRTVector_j \quad (2)$$

where $t \in \langle 0, by\ Step\ to, PathLength \rangle$,

ML	is actual migration loop.
$x_{i,j}^{MLnew}$	is the new positions of an individual.
$x_{i,j,start}^{ML}$	is the positions of active individual.
$x_{L,j}^{ML}$	is the positions of leader.

At last the best individuals (number equal to population size) from the previous and current generations are selected for the next generation. The computational steps of this approach are given below:

Step 1: Generate the initial population.
Step 2: Evaluate all individuals
Step 3: Apply tournament selection on all individuals to select the better individuals for the next generation.
Step 4: Apply crossover operator on all individuals with crossover probability P_c to produce new individuals.
Step 5: Evaluate the new individuals.
Step 6: Apply mutation operator on every bit of every individual of the population with mutation probability P_m.
Step 7: Evaluate the mutated individuals.
Step 8: Find leader (best fitted individual) and active (worst fitted individual) of the population.
Step 9: For active individual a new population of size N is created. where N = (Path Length/step size). This population is nothing but the new positions of the active individual towards the leader in n steps of the defined length. The movement of this individual is given in Eq. (2).
Step 9.1: Select the best individual of the new population and replace the active individual with this best individual.
Step 10: Select the best individuals (in fitness) of previous and current generation for the next generation via tournament selection.
Step 11: If termination criterion is satisfied stop else go to Step 3.
Step 12: Report the best chromosome as the final optimal solution.

Salient Features of SOMGA

The salient features of the SOMGA are:

1. SOMGA attempts to determine the global optimal solution of nonlinear unconstrained optimization problems.
2. SOMGA does not require the continuity and/or differentiability conditions of the objective function.

3. SOMGA works on purely function evaluations and hence can be used in situations where the objective function is discontinuous in nature.
4. SOMGA does not require an initial guess value to start, but instead SOMGA requires a lower and upper bound for the unknown variables.

3 Section 1

3.1 Solution of Large Scale Problems Using SOMGA

Now a days, to solve large scale optimization problems using evolutionary algorithms has become the new field of research. In large scale problems, where the number of unknown variables vary up to 200 not only the computational time but also the memory space requirements becomes very large. Besides this the complexity of the problem also increases significantly that many algorithms fails to reach the optimal solution. One algorithm able to reach optimal solution requires large population size as well as functional evaluations. The possibility of using SOMGA for solving large scale problems is considered in this section. One possible way to reduce the computational time and also to reduce the memory space is to reduce the population size. Since SOMGA already works with less population size hence there is no need of reducing population size. Reduction in memory space is also not required because the population generated for the active individuals takes memory space only for short time. After choosing the best one of the generated population, active individual gets replaced by this best individual and other individuals release the memory space. So here the chances for using it to solve large scale problems are very strong. The problems up to 200 variables can be solved on Pentium IV normally configured system.

3.2 Results and Discussion

In order to observe the performance of SOMGA on large scale problems a set of thirteen test problems have been selected given in appendix. These problems are scalable in nature that is their size can be increased or decreased as per the user's choice. In general, the complexity of the problem increases as the dimension of the problem increases. Here in Griewank function the complexity at dimension 10 is maximum. The complexity of this function decreases as the dimension increases above 10.

The parameters of SOMGA used to solve 100 and 200 dimension are same and are given in Table 1. These parameters are population size, crossover rate, mutation rate, string length, PRT, step size, path length and total number of function calls

Table 1 Parameters of SOMGA for problem size 100 and 200

Parameter	Value
Population size	20
Crossover rate	0.95
Mutation rate	0.0001
String length	30
PRT	1
Step	0.091
Path length	3
Total number of function calls allowed	350,000

Table 2 Results of large scale problems for problem size 10

Problem	Mean	Standard deviation	% Success	Function evaluations
Cosine mixture function	0.996	0.003	100	4336
Exponential function	0.994	0.003	100	2028
Ackley function	0.007	0.002	100	9644
Sphere function	0.007	0.003	100	3790
Griewank function	0.068	0.036	20	13,639
Axis parallel hyper ellipsoid function	0.006	0.003	100	4137
Schwefel's double sum function	0.007	0.003	100	7324
Restrigin's function	0.003	0.003	100	16,895
Rosenbrock function	4.699	1.889	0	NA
Schwefel's function	0.004	0.003	100	11,660
Zakhrov's function	0.005	0.003	100	4939
Ellipsoidal function	0.006	0.003	100	3468
Schwefel's problem 4 function	0.005	0.003	100	310

allowed. Main thing to notice in Table 1 is that for solving 100 and 200 dimensional problems, only 20 population size is required. The results for problem size 10 are given in Table 2 taken from Deep and Dipti [1].

Here, first we apply SOMGA for problem size 100. The percentage of success, average function evaluations, mean and standard deviation of the objective function values of 30 runs are recorded in Table 3. The same information for problem size 200 is shown in Table 4. These values are also shown graphically in Figs. 1 and 2. It can be observed in Table 3 that SOMGA gives 100 % success in 11 problems out of thirteen. In Griewank problem the success rate is 40 % and Rosenbrock could not be solved at all.

In Table 4, the results for a problem size of 200 are presented. It is notable that SOMGA gives 100 % success in 10 problems out of thirteen. In Griewank problem,

Table 3 Results of large scale problems for problem size 100

Problem	Mean	Standard deviation	% Success	Function evaluations
Cosine mixture function	−9.99544	0.003721	100	66,865
Exponential function	−0.99245	0.002478	100	35,523
Ackley function	0.008684	0.00129	100	126,121
Sphere function	0.0077	0.002573	100	75,449
Griewank function	0.050896	0.071455	40	259,559
Axis parallel hyper ellipsoid function	0.008083	0.001724	100	85,889
Schwefel's double sum function	0.006285	0.002476	100	126,078
Restrigin's function	0.004288	0.004288	100	185,626
Rosenbrock function	95.24293	1.177368	0	NA
Schwefel's function	0.009224	0.000703	100	248,930
Zakhrov's function	0.006024	0.002964	100	156,021
Ellipsoidal function	0.008618	0.002317	100	118,629
Schwefel's problem 4 function	0.004494	0.002769	100	521

Table 4 Results of large scale problems for problem size 200

Problem	Mean	Standard deviation	% Success	Function evaluations
Cosine mixture function	−19.9926	0.003517	100	114,165
Exponential function	−0.99016	0.00015	100	61,248
Ackley function	0.008969	0.001514	100	203,392
Sphere function	0.009564	0.000559	100	115,071
Griewank function	0.05995	0.109189	60	257,840
Axis parallel hyper ellipsoid function	0.00962	0.000466	100	180,245
Schwefel's double sum function	0.008717	0.002257	100	246,508
Restrigin's function	0.005262	0.004334	100	272,246
Rosenbrock function	221.4295	35.95709	0	NA
Schwefel's function	4159.555	166.9408	0	NA
Zakhrov's function	0.009576	0.000521	100	255,085
Ellipsoidal function	0.008911	0.001696	100	311,626
Schwefel's problem 4 function	0.004653	0.003564	100	527

the success rate is 60 % for problem size 200 where as for problem size 100 it is 40 %. The success rate in Griewank problem increases as the dimension of the problem increases. This shows the behavior of Griewank problem on increasing the

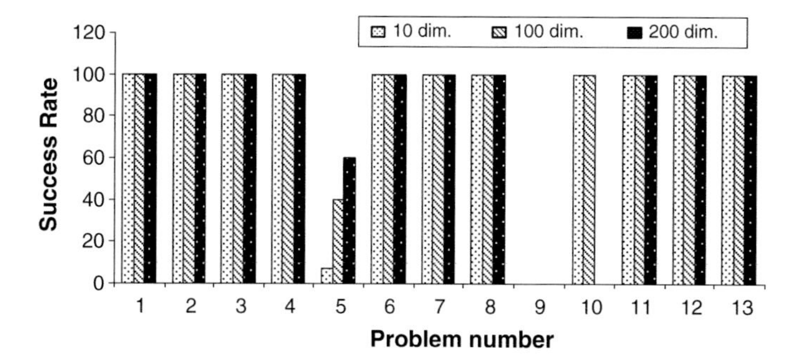

Fig. 1 Performance of SOMGA in terms of success for 100 dimension and 200 dimension problems

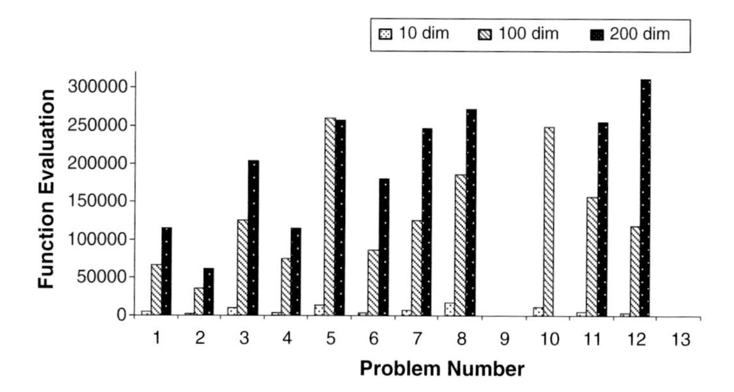

Fig. 2 Performance of SOMGA in terms of function evaluations for 100 dimension and 200 dimension problems

dimension that is, the complexity of the Griewank problem decreases as the dimension increases. In Rosenbrock problem SOMGA fails. Schwefel problem gives 100 % success for problem size 100, but could not be solved at all for problem size 200. This shows that the complexity of the schewefel problem increases as the dimension increases.

In Fig. 1, the success rate obtained by SOMGA in 10 dimension, 100 dimension and 200 dimension problems is plotted. Since the complexity of the Griewank function decreases as the dimension of the problem increases. Same kind of behavior can be seen in Fig. 2 problem number 5. In Fig. 2, the function evaluations required by SOMGA in 10 dimension, 100 dimension and 200 dimension problems are plotted. The behavior is obvious. Function evaluations are increasing as the dimension of the problem is increasing. In Griewank problem the function evaluations in 100 dimesion and 200 dimensions are almost same.

4 Section 2

Solution of Two Real Life Problems

To show the efficiency of algorithm SOMGA, it has been tested to solve two real life problems from the field of engineering. The problems and their solutions are described as follows.

4.1 Optimal Thermohydraulic Performance of an Artificially Roughened Air Heater

This problem is taken from Prasad and Saini [24]. In this problem the optimal thermo hydraulic performance of an artificially roughened solar air heater is considered. Optimization of the roughness and flow parameters (p/e, e/D, R_e) is considered to maximize the heat transfer while keeping the friction losses to be minimum. This is an unconstrained optimization problem. It has three decision variables. The mathematical model of the problem, as given in Prasad and Saini [24] is:

$$\text{Maximize } L = 2.5 \log e^{+} + 5.5 - 0.1 R_M - G_H$$

where

$$R_M = 0.95 x_2^{0.53}$$
$$G_H = 4.5 (e^{+})^{0.28} (0.7)^{0.57}$$
$$e^{+} = x_1 x_3 (\bar{f}/2)^{1/2}$$
$$\bar{f} = (f_s + f_r)/2$$
$$f_s = 0.079 x_3^{-0.25}$$
$$f_r = 2\left[0.95 x_3^{0.53} + 2.5 \log(1/2x_1)^2 - 3.75\right]^{-2}$$

The notations used are as follows:

e^+	roughness height.
p	pitch of the roughness element.
D	the hydraulic diameter of solar heater.
$x_1 = e/D$	relative roughness height.
$x_2 = p/e$	relative roughness pitch.
$x_3 = R_e$	Reynolds number.

Table 5 Optimal thermo hydraulic performance of an artificially roughened air heater

	Value of objective	Values of variables
Solution obtained by SOMGA	4.18241	$x_1 = 0.10558676$, $x_2 = 10.000276$, $x_3 = 4567.99$
Solution given in Pant [25]	4.182	$x_1 = 0.052$, $x_2 = 10.00$, $x_3 = 10258.46$
Solution given in Prasad and Saini [24]	≈4.18	$x_1 \approx 0.0205$, $x_2 \approx 10.00$

The bounds on the variables are:

$$0.02 \leq x_1 \leq 0.8, \quad 10 \leq x_2 \leq 40, \quad 3000 \leq x_3 \leq 20{,}000$$

This is an unconstrained problem and has been solved by SOMGA. It is earlier solved by Prasad and Saini [24] and Pant [25]. The results obtained by SOMGA are compared with the available results and are presented in Table 5. It is clear from the Table 5 that the solution obtained by SOMGA is better than previously quoted results.

4.2 *Frequency Modulation Sounds Parameter Identification Problem*

This problem has been taken from Tsutsui and Fujimoto [26]. It is an unconstrained optimization problem. This problem is to determine the six parameters a_1, ω_1, a_2, ω_2, a_3, ω_3 of the Frequency Modulation Sound model represented by

$$y(t) = a_1 \cdot \sin(\omega_1 \cdot t \cdot \theta + a_2 \cdot \sin(\omega_2 \cdot t \cdot \theta + a_3 \cdot \sin(\omega_3 \cdot t \cdot \theta))),$$

with $\theta = \frac{2 \cdot \pi}{100}$. An evaluation function P_{fms} is defined as the summation of 101 square errors between the evolved data and the model data as follows:

$$P_{fms}(a_1, \omega_1, a_2, \omega_2, a_3, \omega_3) = \sum_{t=0}^{100} (y(t) - y_0(t))^2,$$

where the model data are given by the following equation:

$$y_0(t) = 1.0 \cdot \sin(5.0 \cdot t \cdot \theta - 1.5 \cdot \sin(4.8 \cdot t \cdot \theta + 2.0 \cdot \sin(4.9 \cdot t \cdot \theta))).$$

Each parameter is in the range −6.40 to 6.35. In this problem, a generated sound wave and its evaluation function P_{fms} are extremely sensitive to some of these parameters a_1, w_1, a_2, w_2, a_3, w_3. This makes the problem difficult to reach the optimal point. This problem is a highly complex multimodal one having strong epistasis, with minimum value $P_{fms}(x^*) = 0$. Empirical results are shown in Table 6.

Table 6 Solution of Frequency modulation sounds parameter identification problem

Method	Result obtained	Total runs required	# Evaluation
SOMGA	0.00941009	156	7207
p-fGA	–	22621	–

Tsutsui and Fujimoto [26] have used this problem for the testing of their technique that is Phenotypic Forking Genetic Algorithm (p-fGA). This technique takes 22621 generations on average to solve this problem. On the other hand SOMGA solves this problem in 156 generation or 7207 function evaluation only.

5 Section 3

5.1 Comparison with the Memetic Algorithm

Generally two kinds of local search methods can be used in GAs to improve the efficiency of these algorithms. One is the deterministic method and the other is to use another population based stochastic search strategies as local searches. Although deterministic methods have less exploration capabilities but they converges very fast in small neighborhood. In this section a comparison has been made between the two approaches of hybridization. One approach is SOMGA described in Sect. 2, Previous work done, in which GA is hybridized with another population based technique SOMA. Another approach is hybridization of GA with a deterministic local search method. This approach is based on the idea that first let the population evolve using GA and when the search domain become narrow and convergence slow down after certain number of generations apply local search on selected individual to faster the convergence of it. The methodology of this approach is as follows:

Step 1: Generate the initial population.
Step 2: Evaluate all individuals.
Step 3: Apply selection operator on all individuals to select the better individuals for the next generation.
Step 4: Apply crossover operator on all individuals with crossover probability P_c to produce new individuals.
Step 5: Evaluate the new individuals.
Step 6: Apply mutation operator on every bit of every individual of the population with mutation probability P_m.
Step 7: Evaluate the mutated individuals.
Step 8: If generation is less than the specified generations go to Step 3 else go to Step 9.
Step 9: Select best q% individuals and apply LS.
Step 10: If termination criterion is satisfied stop else go to Step 3.

Table 7 Parameters of MA

Dimension	n = 10
Population size	100
Crossover rate p_c	0.95
Mutation rate p_m	0.001
Specified number of generations	100
Total number of function calls allowed	40,000

Table 8 Comparative results of SOMGA and MA

Problem	SOMGA				MA			
	Mean	S.D.	# of success	# of evaluation	Mean	S.D.	# of success	# of evaluation
Ackley	0.007	0.002	30	9644	0.06023	0.30033	28	22,521
Schwefel	0.004	0.003	30	11,660	0.00013	0	30	21,634
Griewank	0.068	0.036	6	13,639	0.04943	0.04123	5	33,143
Restrigin	0.003	0.003	30	16,895	0.09532	0.26137	26	32,465
Rosenbrock	4.699	1.889	0	–	0.00103	0.00167	30	21,012

The algorithm terminates as soon as the best fit solution is obtained within 1 % accuracy of the known global optimum solution or it crosses the allowed number of function evaluations. Any derivative free local search method can be used in this approach. We use the well known Hooke and Jeeves direct search method described in Bazaraa et al. [27]. It is multidimensional search method with discrete steps and works without using derivatives. Hence it is able to solve a wide range of problems. It is also very quick and robust at finding the local optimal solution of a problem. It is tested on five well-known benchmark test problems, namely, Ackley function, Schwefel's function, Griewank's function, Restrigin's function, Rosenbrock's function taken from Ali et al. [28]. These five test problems are most commonly used for evaluating the performance of evolutionary algorithms. The parameters used in this approach namely population size, probability of crossover (P_c), probability of mutation (P_m), specified number of generations (taken in our experiments after fine-tuning) after which LS has to be activated and total number of function calls allowed are given in Table 7 and the results obtained by this approach are presented in Table 8.

5.2 *Results and Discussion*

For this approach, it is suggested that for solving an n dimensional problem, population size 5 * n should be taken and LS should be applied after 10 * n generations. But minimum population size 100 is required for 10 or more than 10 dimensional problem. All the results have been taken using these specifications. Figure 3 shows the success attained by MA and SOMGA. The greater the portion covered by an algorithm in a column, the better the method is, in that problem. It is clear from the graph that the portion covered by both the algorithms GA and SOMGA is almost equal, except in one problem. The main difference between the two is that MA is providing success at 100 population size for 10 variable problem and SOMGA is providing the same success at 10 population size for 10 variable problem. Hence on the basis of population size SOMGA is better than MA.

Figure 4 shows the function evaluations taken by MA and SOMGA in successful runs. In this graph, the lesser the portion covered by an algorithm in a column, the better the method is in that problem. It is evident from the graph that the portion covered by SOMGA is much lesser than MA. This means SOMGA required less number of function evaluations than MA for obtaining the global optimal solution of unconstrained optimization problems. Hence SOMGA is more efficient than MA.

On the basis of these results, it is concluded that SOMGA is far more superior to MA in terms of function evaluations. Hence SOMGA is recommended for solving the unconstrained nonlinear optimization problems.

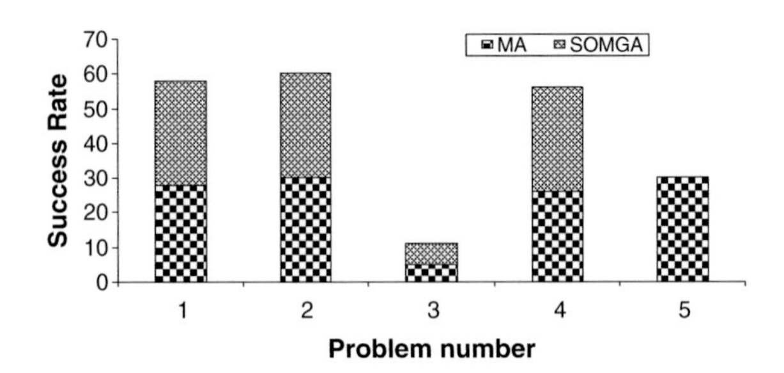

Fig. 3 Graph showing success attained by MA and SOMGA. Greater portion ⇒ better method

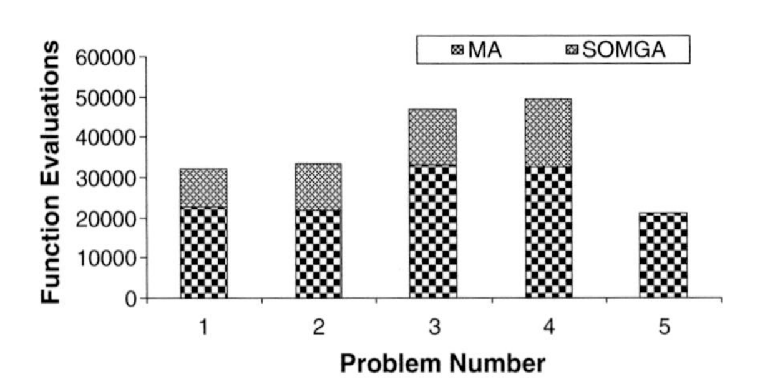

Fig. 4 Graph showing the function evaluations taken by MA and SOMGA. Lesser portion ⇒ better method

6 Conclusions

In this chapter a variant of GA and SOMA known as SOMGA has been used to solve large scale optimization problems. From the discussion part of results for large scale optimization it is clear that for solving up to 200 dimension problem SOMGA requires very less population size as well as function evaluations. Later two real life problems from the field of engineering have been solved using the same technique and SOMGA proves its efficiency to solve these problems. At last a comparison between two variants of GA i.e. SOMGA and MA has been made. Though MA is able to solve all the test problems with good success rate but it is expensive in terms of function evaluations. One drawback of using deterministic methods is that they are usually problem specific hence restricted to solve some specific classes of problems. Another drawback of using these traditional local searches is that if LS is not implemented properly, the chances of getting trapped in local minima is more because these methods have less exploration qualities. It means that these hybridized algorithms demand careful fine tuning of local search parameters. On the other hand population based stochastic search techniques are not problem specific and are applicable to solve a wide range of problems. Another advantage of using these techniques is that they use multiple guesses to improve a solution. Hence have more exploration qualities.

Appendix

This Appendix contains the list of 13 benchmark test problems taken from literature, which are used to evaluate the performance of the algorithm. These problems are unconstrained nonlinear optimization problems having a number of local as well as global optimal solutions. All the problems have varying difficulty level and contain unimodal as well as multi modal problems.

Problem 1: (Cosine Mixture Problem)
This problem is Cosine Mixture Function. The global optimum of this function is at (0, 0,…, 0) with $f_{min} = -0.1n$. where n is the dimension of the problem. The functional form is as follows:

$$\min_x f(x) = \sum_{i=1}^{n} x_i^2 - 0.1 \sum_{i=1}^{n} \cos(5\pi x_i), \quad \textit{for } x_i \in [-1, 1].$$

Problem 2: (Exponential Problem)
This problem is Exponential Function. The global optimum of this function is at (0, 0,…, 0) with $f_{min} = -1$. The functional form is as follows:

$$\min_x f(x) = \exp\left(-0.5 \sum_{i=1}^{n} x_i^2\right), \quad for\ x_i \in [-1, 1].$$

Problem 3: (Ackley Function)
This problem is the Ackley function. The surface of the Ackley function has numerous local minima due to its exponential terms. Any search algorithm based on the gradient information will be trapped in local optima, but any search strategy that analyzes a wider region will be able to cross the valley among the optima and achieve better results. Its global minimum is at (0, 0,…, 0) with $f_{min} = 0$. The functional form is as follows:

$$f(x) = 20 + e - 20e^{-\left(\frac{1}{5}\sqrt{\frac{1}{n}\sum_{i=1}^{n} x_i^2}\right)} - e^{-\left(\frac{i}{n}\sum_{i=1}^{n} \cos(2\pi x_i)\right)}, \quad for\ x_i \in [-15, 30].$$

Problem 4: (Sphere Function Problem)
The next problem is Sphere Function. This problem is continuous convex and unimodal. Its global minimum is at (0, 0,…, 0) with $f_{min} = 0$. The functional form is as follows:

$$\min_x f(x) = \sum_{i=1}^{n} x_i^2, \quad for\ x_i \in [-5.12, 5.12].$$

Problem 5: (Griewank Function)
This problem is a widely employed test function for global optimization, the Griewank function. While this function has an exponentially increasing number of local minima as its dimension increases, it turns out that a simple Multistart algorithm is able to detect its global minimum more and more easily as the dimension increases. The optima of this function are regularly distributed. Number of local minima for arbitrary n is unknown, but in two dimensional case there are some 500 local minima. Its global minimum is at (0, 0,…, 0) with $f_{min} = 0$. The functional form is as follows:

$$f(x) = \sum_{i=1}^{n} \frac{x_i^2}{4000} - \prod_{i=1}^{n} \cos\left(\frac{x_i}{\sqrt{i}}\right) - 1, \quad for\ x_i \in [-5.12, 5.12].$$

Problem 6: (Axis Parallel Hyper Ellipsoid)
This problem is Axis Parallel Hyper Ellipsoid Function. This test problem is similar to sphere problem function. It is also known as the weighted sphere model. It is continuous convex and unimodal. Its global minimum is at (0, 0,…, 0) with $f_{min} = 0$. The functional form is as follows:

$$\min_x f(x) = \sum_{i=1}^{n} i x_i^2, \quad for\ x_i \in [-5.12, 5.12].$$

Problem 7: (Schwefel's Double Sum)

This problem is Schwefel's double sum Function. This function is an extension of axis parallel hyper ellipsoid function. It produces a rotated hype-ellipsoid. It is continuous convex and unimodal. Its global minimum is at (0, 0,…, 0) with $f_{min} = 0$. The functional form is as follows:

$$\min_x f(x) = \sum_{i=1}^{n} \left(\sum_{j=1}^{i} x_j \right)^2, \quad for\ x_i \in [-65.536,\ 65.536].$$

Problem 8: (Rastrigin Function)

This problem is the Rastrigin Function. It is the extended form of the sphere function with a modulator term $\alpha \cdot \cos(2\pi x_i)$. This function consists of a large number of local minima (not exactly known) whose value increases with the distance to the global minimum. Its global minimum is at (0, 0,…, 0) with $f_{min} = 0$. The functional form is as follows:

$$f(x) = 10n + \sum_{i=1}^{n} \left(x_i^2 - 10\cos(2\pi x_i) \right), \quad for\ x_i \in [-5.12, 5.12].$$

Problem 9: (Rosenbrock Function)

This problem is the Rosenbrock function, also known as the banana function. It is a continuous, differentiable, unimodal and non separable function. Its difficulty arises due to nonlinear interaction between parameters. The global optimum is inside a long narrow parabolic shaped flat valley. Its global minimum is at (1, 1,…, 1) with $f_{min} = 0$. The functional form is as follows:

$$f(x) = \sum_{i=1}^{n-1} \left(100(x_{i+1} - x_i^2)^2 + \left(x_i - 1\right)^2\right), \quad for\ x_i \in [-2.048, 2.048].$$

Problem 10: (Schwefel Function)

This problem is Schwefel Function. The contour of this function is made up of a great number of peaks and valleys. This function has a second best minimum far from the global minimum, so it is difficult for many algorithms to locate the global optimum of this function. Its global minimum is at (1, 1,…, 1) with $f_{min} = 0$. The functional form is as follows:

$$f(x) = 418.9829n - \sum_{i=1}^{n} \left(x_i \sin \sqrt{|x_i|} \right), \quad for\ x_i \in [-500, 500].$$

Problem 11: (Zakharov's Problem)

This problem is Zakharov Function. Its global minimum is at (0, 0,…, 0) with $f_{min} = 0$. The functional form is as follows:

$$\min_x f(x) = \sum_{i=1}^{n} x_i^2 + \left(\sum_{i=1}^{n} \frac{i}{2} x_i\right)^2 + \left(\sum_{i=1}^{n} \frac{i}{2} x_i\right)^4, \quad for\ x_i \in [-5.12, 5.12].$$

Problem 12: (Ellipsoidal Function)

This problem is Ellipsoidal Function. Its global minimum is at (1, 2,…, n) with $f_{min} = 0$. The functional form is as follows:

$$\min_x f(x) = \sum_{i=1}^{n} (x_i - i)^2, \quad for\ x_i \in [-n, n].$$

Problem 13: (Schwefel Problem 4)

This problem is Schwefel Problem 4 Function. Its global minimum is at (0, 0,…, 0) with $f_{min} = 0$. The functional form is as follows:

$$\min_x f(x) = \max_i \{|x_i|, 1 \le i \le n\}, \quad for\ x_i \in [-100, 100].$$

References

1. Deep, K., Singh, D.: A new hybrid self organizing migrating genetic algorithm for function optimization. In: IEEE Congress on Evolutionary Computation, pp. 2796–2803 (2007)
2. Grefensette, J.: Lamarckian learning in multi-agent environment In: Proceedings of the Fourth International Conference on Genetic Algorithms, San Mateo, CA, Morgan Kauffman (1994)
3. Kasprzyk, G.P., Jasku, M.: Application of hybrid genetic algorithms for deconvulation of electrochemical responses in SSLSV method. J. Electroanal. Chem. **567**, 39–66 (2004)
4. Chelouah, R., Siarry, P.: A hybrid method combining continuous Tabu search and Nelder–Mead simplex algorithm for global optimization of multiminima functions. Eur. J. Oper. Res. **161**, 636–654 (2005)
5. Wang, L., Tang, F., Wu, H.: Hybrid genetic algorithm based on quantum computing for numerical optimization and parameter estimation. Appl. Math. Comput. **171**, 1141–1156 (2005)
6. Javadi, A., Farmani, A.R., Tan, T.P.: A hybrid intelligent genetic algorithm. Adv. Eng. Inf. **19**, 255–262 (2005)
7. Fan, S.K.S., Liang, Y.C., Zahara, E.: a genetic algorithm and a particle swarm optimizer hybridized with Nelder–Mead simplex search. Comput. Ind. Eng. **50**, 401–425 (2006)
8. Hwang, F.S., Song, H.R.: A hybrid real parameter genetic algorithm for function optimization. Adv. Eng. Inf. **20**, 7–21 (2006)
9. Zhang, G., Lu, H.: Hybrid real coded genetic algorithm with quasi-simplex technique. Int. J. Comput. Sci. Netw. Secur. **6**(10), 246–255 (2006)
10. Wei, L., Zhao, M.: A Nitche hybrid genetic algorithm for global optimization of continuous multi modal functions. Appl. Math. Comput. **160**, 649–661 (2005)
11. Premalatha, K., Nataranjan, A.M.: Hybrid PSO and GA for global optimization. Int. J. Open Probl. Comput. Math. **2** (2009)

12. Khosravi, A., Lari, A., Addeh, J.: A new hybrid of evolutionary and conventional optimization algorithm. Appl. Math. Sci. **6**, 815–825 (2012)
13. Ghatei, S., et al.: A new hybrid algorithm for optimization using PSO and GDA. J. Basic Appl. Sci. Res. **2**, 2336–2341 (2012)
14. Esmin, A., Matwin, S.: A hybrid particle swarm optimization algorithm with genetic mutation. Int. J. Innovative Comput. Inf. Control **9**, 1919–1934 (2013)
15. Zelinka I., Lampinen, J.: SOMA-self organizing migrating algorithm. In: Mendal, 6th International Conference on Soft Computing, Brno, Czech Republic, vol. 80, issue-2, p. 214 (2000)
16. Oplatkova, Z., Zelinka, I.: Investigation on Shannon-Kotelnik theorem impact on SOMA algorithm performance. In: Proceedings 19th European Conference on Modelling and Simulation Yuri Merkuryev, Richard Zobel (2005)
17. Zelinka, I.: Analytic programming by means of soma algorithm. In: Proceeding of 8th International Conference on Soft Computing Mendel '02, Brno, Czech Republic, pp. 93–101 (2002). ISBN 80-214-2135-5
18. Nolle, L., Zelinka, I.: SOMA applied to optimum work roll profile selection in the hot rolling of wide steel. In: Proceedings of the 17th European Simulation Multiconference ESM 2003, Nottingham, UK, pp. 53–58 (2003). ISBN 3-936150-25-7, 9-11
19. Nolle, L., Zelinka, I., Hopgood, A.A., Goodyear, A.: Comparision of an self organizing migration algorithm with simulated annealing and differential evolution for automated waveform tuning. Adv. Eng. Softw. **36**, 645–653 (2005)
20. Nolle, L.: SASS applied to optimum work roll profile selection in the hot rolling of wide steel. Knowl. Based Syst. **20**(2), 203–208 (2007)
21. Zelinka, I., Lampinen, J., Nolle, L.: On the theoretical proof of convergence for a class of SOMA search algorithms. In: Proceedings of the 7th International MENDEL Conference on Soft Computing, Brno, CZ, pp. 103–110, 6–8 June 2001. ISBN 80-214-1894-X
22. Zelinka, I., Oplatkova, Z., Nolle, L.: Boolean symmetry function synthesis by means of arbitrary evolutionary algorithms—comparative study. In: Proceedings of the 18th European Simulation Multiconference ESM 2004, Magdeburg, Germany, pp. 143–148, June 2004. ISBN 3-936150-35-4, 13-14
23. Onwubolu, C.G., Babu, B.V.: New Optimization Techniques in Engineering. Springer, Heidelberg (2004). ISBN 3-540-20167-X
24. Prasad, B.N., Saini, J.S.: Optimal thermo hydraulic performance of artificially roughened solar air heaters. J. Solar Energy **47**, 91–96 (1991)
25. Pant, M.: Genetic Algorithms for Global Optimization and their Applications. Ph.D. thesis, Department of Mathematics, IIT Roorkee, Formerly University of Roorkee (2003)
26. Tsutsui, S., Fujimoto, Y.: Phenotypic forking genetic algorithm (p-fGA). In: IEEE International Conference on Evolutionary Computing (ICEC '95), Vol. 2, pp. 556–572 (1995)
27. Bazaraa, M.S., Sherali, H.D., Shetty, C.M.: Nonlinear Programming Theory and Algorithms. Wiley, New York (1993)
28. Ali, M.M., Khompatraporn, C., Zabinasky, Z.: A numerical evaluation of several global optimization algorithms on selected benchmark test problems. J. Global Optim. **31**, 635–672 (2005)

Solving the Routing Problems with Time Windows

Zuzana Čičková, Ivan Brezina and Juraj Pekár

Abstract The routing problems are considered to be a very important part in the field of transportation that can be applied to solve a wide variety of real problems. The economic impact of optimization of these problems can lead to considerable savings in logistics costs and in that way to increase the competitive advantage of many companies. In general, the importance of routing problems follows not only from many practical applications but also from its computational complexity (importance for theoretical research). The corresponding problems have long been solved by classical mathematical apparatus and standard methods, but such a solution is sometimes impossible or complicated and lengthy, therefore the increasing interest in "non-traditional" methods of solution is evident in the past decade. This chapter deals with self-organizing migrating algorithm (SOMA) for solving the routing problems, focused on the traveling salesman problem, capacitated vehicle routing problem and its modifications based on time restrictions. Our interest in time restricted routing problems arises from a real-life distribution problem in one of the regions of Slovakia (individual customers' commodities delivering times were restricted by their available service time), where the previous distribution was realized on the base of solution derived with heuristic Clarke & Wright's savings algorithm with time windows.

Z. Čičková (✉) · I. Brezina · J. Pekár
Department of Operations Research and Econometrics, University of Economics in Bratislava, Dolnozemská cesta 1/b, 852 35, Bratislava, Slovakia
e-mail: cickova@euba.sk

I. Brezina
e-mail: brezina@euba.sk

J. Pekár
e-mail: juraj.pekar@euba.sk

D. Davendra and I. Zelinka (eds.), *Self-Organizing Migrating Algorithm*,
Studies in Computational Intelligence 626, DOI 10.1007/978-3-319-28161-2_10

1 Introduction

Nowadays, the threat of depletion of non-renewable resources which are necessary for car propulsion is the reason for development and utilization of instruments that take advantage of the optimization. The efficiency may be increased by force of quantitative approaches that are aimed at optimization of physical distribution of the commodities. Related optimization problems are known as routing problems.

The traveling salesman problem (TSP) is one of the most popular routing problems. In its simplest form it asks the following: given a list of cities and the known distances between each pair of cities. The aim is to find the shortest circular route, which visits each city exactly once. Despite its relatively simple verbal formulation, this is one of the most famous problems of discrete mathematics, which is fascinating in that there is generally no known polynomial algorithm that would allow an optimal solution of large instances in real time. Thus, this issue remains at the heart of further research. That argument is supported by lots of algorithms developed, created computer programs and published scientific papers. And if it is not realistic to obtain the optimal solution of specific instance, it still can be satisfied with the techniques that provide "good" solution "almost always" and thus prioritize "good" and relatively fast heuristic solution before slowly algorithm that ensures returning of optimal solution.

The basic formulation of the traveling salesman problem considers only one vehicle with unlimited capacity. This problem may be modified by various additional conditions and thus take into account the real constraints of practical issues. The routing problems generally involve the assignment of vehicle (fleet of vehicles) to trips such that corresponding costs are as low as possible. The capacitated vehicle routing problem (CVRP) is one of the most intensively studied problems in optimization. The standard CVRP is the generalization of travelling salesman problem, where we consider that the capacity of vehicle (fleet of vehicles) is limited and that the non-negative demands of customers (cities) are known. The vehicle (vehicles) is (are) located in the central depot. This problem consists in designing the optimal set of routes for a vehicle in order to serve a given set of customers (the vehicle (vehicles) needs (need) to be returned to a depot). The practical problems of physical distribution often include the need to respect the time restriction. Frequently used time restrictions are considered as the earliest possible time of service, the possible time of service or the need to serve during the given time interval. The above mentioned terms are known as time windows and the corresponding problems are known as routing problems with time windows.

The theoretical attractiveness of the traveling salesman problem is evident from a number of published works. The first successful solutions are associated with the names of Dantzig, Fulkerson and Johnson [23]. Their results were followed by Robacker [45], who described the cheapest-insertion heuristic. Croes in [22] used a heuristic local search algorithm called 2-opt. method. The use of dynamic programming to solve the traveling salesman problem is presented in [3]. Formulation of TSP as an integer programming problem demonstrated Miller et al. [35].

The frequently used methods for solving TSP include branch and bound method described in [32]. Popular heuristic algorithm for solving a traveling salesman problem is presented in [31] and another well-known heuristic in Christofides [10]. Improved Lin-Kernighan heuristics is presented in [29] as Lin-Kernighan-Helsgaun algorithm. Despite the large number of published works until the 90s of the 20th century, presented experiments included only small instances. To date, currently reported isolated cases of optimal solutions of big instances [19, 27, 38] required using large computer networks (calculations often require tens of years of CPU time). It is obvious that using one computer to solve big instances will likely continue the necessity to use heuristic methods to find sub-optimal solutions. The popular heuristics include, e.g. Clarke & Wright's savings heuristics, which is also used to solve the vehicle routing problem instances. Improving it using Holmer and Parker's perturbation scheme can be found in [39]. Interesting traveling salesman problem solution using a graph pyramid algorithm is described in [40]. Various traditional heuristics include e.g. savings-insertion methods, improvement-exchange heuristics etc. (e.g. [39]).

Classical mathematical apparatus and standard methods have long been used to solve TSP or other routing problems. Such a solution is sometimes impossible or complicated and lengthy; therefore the increasing interest in "non-traditional" methods of solution is evident in the past decade. For all it can be mentioned: the application of principle of differential evolution [37] or [41], using the self-organizing migration algorithm [5, 16], the possibility of using neural network [12], ants colony algorithm [30].

Different approaches of constrained optimization can be found, e.g. in [21, 28, 33, 34, 49, 53] etc. The TSP Challenge Centre of Rutgers University[1] page provides summary the results obtained from the classical solution heuristics, as well as some metaheuristic approaches.

As was mentioned before, the routing problems often include the need to respect time restrictions. Here are many ways to define time windows. In general, the problem is known as the problem with soft delivery time windows consider only the earliest possible service time or the last possible service time and problems with hard delivery time windows consider the time restriction given by time interval with lower and upper limit. Some models include waiting [20], or the violation of time restriction is allowed, although incurring some cost [47]. In [24] is a time window understood only as the time interval during which the vehicle must arrive at their service (while not intending to limit service time or completion of service). The most popular corresponding routing problems are named traveling salesman problem with time windows (TSPTW) and capacitated vehicle routing problems with time windows (CVRPTW) regardless of the type of the time window. Furthermore, some other works dealing with the solution of the corresponding problems with time windows in classical and alternative ways will be presented.

---

[1]http://www2.research.att.com/~dsj/chtsp/ (1.1.2014).

In [48] is proposed heuristic method for addressing related problems based on the 2 and 3-exchange procedures. Example of exact solution based on dynamic programming and branch and bound method can be found in [2] or [11], which both solved instances with dimensions up to 50 nodes. Solomon's heuristics [50] and Potvin and Rousseau's heuristics [44] can be mentioned as representatives of classical methods. Other applications can be found in [6, 42, 54] or in [52]. A representative of classical heuristics includes Solomon's heuristics [50] and Potvin and Rousseau's heuristics [44]. Widely used metaheurics include e.g. Tabu search [7, 46], simulated annealing [51], particle swarm optimization [1], ants colony [47], genetic algorithms [9, 26, 36], self-organizing migrating algorithm [4, 14, 18], neural networks [57].

This chapter is focused on self-organizing migrating algorithm (SOMA) for solving the capacitated vehicle routing problem with time windows (CVRPTW). Our interest in CVRPTW arises from a real-life distribution problem in one of the regions of Slovakia (individual customers' commodities delivering times were restricted by their available service time).

The chapter consists of two main parts. First, the brief view on basic types of routing problems (traveling salesman problem, vehicle routing problem and their modification considering time restriction), as well as literature overview aimed on solving possibilities, is given in the introduction. Generally, that problem can be easy formulated in the terms of binary programming. The corresponding models are introduced in the second part of the chapter. Such formulation allows the use of standard software to solve specific instances. The possibility to obtain optimal solution is rather limited taking into account the size of instance. Thus, the other way is the use of evolutionary techniques. Following this idea, the next part of the chapter is focused on application of self-organizing migrating algorithm (SOMA) to routing problems and it is divided as following: some modification of basic version of SOMA that takes into account the specificity of routing problems is presented and its efficiency is validated on the basis of publicly available instances.

Second, the real-life vehicle routing problem with time windows that has appeared in Slovakia is solved. Solution obtained was compared with solution that was derived with heuristic Clarke & Wright's savings algorithm with time windows that belong to the group of classical heuristics.

2 Mathematical Models of Selected Routing Problems

The classical versions of routing problems can be described in terms of graph theory. Further on, the following notation is implemented: Let $N = \{1, 2, \ldots n\}$ be the set of served nodes (customers) and let $N_0 = N \cup \{0\}$ be a set of nodes that represents the customers together with the initial node (origin). The routing problems can be formulated on finite connected directed and weighted graph $G = (V, H)$, where $V = V_S \cup V_Z$ represents set of all nodes ($n + 1$ elements) and the set $V_S = \{v_0\}$ represents the origin and $V_Z = \{v_i,\ i \in N\}$ represents the set of customers. Let the set $H \subset V \times V$ be the arc set $h_{ij} = (v_i, v_j)$ considered to be directed from the node v_i to node v_j, $i, j \in N_0$. A number or weight $o(h_{ij})$ is assigned to each edge. Let $\mathbf{C} = \{c_{ij}\}$ be a matrix of size $n + 1$, while its elements are defined as follows:

$$c_{ij} = \begin{cases} o(h_{ij}), & \text{if there exists the edge } h_{ij} \\ 0, & \text{if } i = j \\ M, & \text{if there is no edge } h_{ij} \end{cases} \tag{1}$$

where M represents a big positive number.

The routing problems are often described on complete weighted directed graph $\overline{G} = (V, \overline{H})$ in which every pair of distinct vertices is connected by a pair of unique edges (one in each direction), so that $\overline{H}$ represents the set of edges. The weight of each edge is equal to minimal distance between nodes v_i a v_j in the original graph. Let d_{ij} be a shortest distance between the nodes v_i and v_j, $i, j \in N_0$, than a matrix $\mathbf{D} = \{d_{ij}\}$ of size $n + 1$ can be named the matrix of shortest distances.

Mathematical models of selected routing problems are easy to formulate in the terms of binary programming. Further on, the models involve binary x_{ij} $(i, j \in N_0,$ $i \neq j)$ with the following notation:

$$x_{ij} = \begin{cases} 1, & \text{if the final route goes from } v_i \text{ to } v_j \\ 0, & \text{otherwise} \end{cases} \tag{2}$$

Next, the mathematical models of traveling salesman problem, vehicle routing problem, traveling salesman problem with time windows and vehicle routing problem with time windows are introduced.

2.1 Traveling Salesman Problem (TSP)

The mathematical formulation of TSP is based on formulation of the assignment problem with addition of sub-tour elimination constraints. So that formulation is known as Miller-Tucker-Zemlin's formulation [35].

$$\min f(\mathbf{X}, \mathbf{u}) = \sum_{i \in N_0} \sum_{\substack{j \in N_0 \\ i \neq j}} d_{ij} x_{ij} \tag{3}$$

$$\sum_{j \in N_0} x_{ij} = 1, \quad j \in N_0, \; i \neq j \tag{4}$$

$$\sum_{j \in N_0} x_{ij} = 1, \quad i \in N_0, \; i \neq j \tag{5}$$

$$u_i - u_j + (n+1)x_{ij} \leq n, \quad i \in N_0, j \in N, i \neq j \tag{6}$$

$$1 \leq u_i \leq n, \quad i \in N \tag{7}$$

$$u_0 = 0 \tag{8}$$

$$x_{ij} \in \{0,1\}, \quad i,j \in N_0, i \neq j \tag{9}$$

The objective function (3) minimizes the total travelled distance. Equations (4) and (5) ensure that a vehicle leaves each node and vehicle enters each node exactly ones. The model considers new set of variables u_i, $i \in N_0$ representing the sequence in which nodes are being visited (6) and (7). For convenience one may add (8) ensuring the node indexed 0 is required to be the origin.

2.2 Traveling Salesman Problem with Time Windows (TSPTW)

Next, it will be discussed the possibility of taking into account the time restrictions in the formulation of the traveling salesman problem. The time window of each customer v_i, $i \in N$ is considered as the earliest possible start of service in different nodes (e_i, $i \in N$) and the last acceptable time of service in different nodes (l_i, $i \in N$). Further on, there is the known service duration o_i, $i \in N$, while the parameter o_0 is set to 0. Let the variables τ_i, $i \in N$ represent the real starting service time of corresponding customer, so that $e_i \leq \tau_i$ and $\tau_i + o_i \leq l_i$, $(i \in N)$. Let the elements of the matrix $\mathbf{D}$ represent shortest time distance between all nodes d_{ij}, $i,j \in N_0$. The model takes into account the waiting time so that the vehicle is allowed to wait for service if it arrives before the earliest possible start of service. Denote the w_j be the waiting time at the j-served customer, so it is considered that after operating the ith customer it is immediately followed by service of jth customer, which will start in the time e_j.

Thus, mathematical model of TSPTW could be formulated as follows:

$$\min f(\mathbf{X}, \mathbf{w}, \boldsymbol{\tau}) = \sum_{i \in N_0} \sum_{\substack{j \in N_0 \\ i \neq j}} x_{ij} + \sum_{i \in N} o_i + \sum_{j \in N} w_j \tag{10}$$

$$\sum_{i \in N_0} x_{ij} = 1, \quad j \in N_0, i \neq j \tag{11}$$

$$\sum_{j \in N_0} x_{ij} = 1, \quad i \in N_0, i \neq j \tag{12}$$

$$\tau_i + o_i + d_{ij} - M(1 - x_{ij}) \leq \tau_j, \quad i \in N_0, j \in N \tag{13}$$

$$w_j \geq \tau_j - \tau_i - o_i - d_{ij} - M(1 - x_{ij}), \quad i \in N_0, j \in N, i \neq j \tag{14}$$

$$e_i \leq \tau_i, \quad i \in N \tag{15}$$

$$\tau_i + o_i \leq l_i, \quad i \in N \tag{16}$$

$$\tau_0 = 0 \tag{17}$$

$$x_{ij} \in \{0, 1\}, \quad i, j \in N_0, i \neq j \tag{18}$$

$$w_i \geq 0, \quad i \in N \tag{19}$$

Objective function (10) ensures the minimization of the total time required for transfer, the total service time and the total waiting time. Equations (11) and (12) ensure that each final route goes through every node exactly ones. However, the anti-cyclical conditions (6) are replaced by conditions (13), whose meaning is as follows: if the edge (i, j) is used, then the real beginning time of the service of jth customer is greater than or equal to the starting time of the previous service increased by its operating and transit time between customers. In the case of waiting it is also increased by the value of the waiting time at the jth customer [calculated by Eq. (14)]. Equations (15) and (16) ensure that the time windows of all nodes on the route are met.

2.3 *Capacitated Vehicle Routing Problem (CVRP)*

As it was mentioned before, the CVRP is one of the famous generalizations of traveling salesman problem, which importance is undoubtedly conditioned also by practical utility. Main difference follows from a presumption that each customer located at the node v_i $i \in N$ has a certain demand g_i, which have to be met from the initial node ($i = 0$)—origin. The operation is performed using a vehicle with a certain capacity (g). The goal is to identify those routes of vehicle so that the total travelled distance is as low as possible (we suppose that shortest distances between all nodes d_{ij}, $i,j \in N_0$ are known) with respect to the following restrictions: the origin represents initial node and also the final node of every route, from the origin the demands q_i, $i \in N$ of all the other nodes are met (in full), each node (except central node) is visited exactly once and total demand on route must not exceed the capacity of the vehicle (g). The model implicitly assumes that $q_i \leq g$ for all $i \in N$, i.e. the demand of each customer does not exceed the capacity of the vehicle. Based on this assumption, the model can be stated as follows:

$$\min f(\mathbf{X}, \mathbf{u}) = \sum_{i \in N_0} \sum_{\substack{j \in N_0 \\ i \neq j}} d_{ij} x_{ij} \tag{20}$$

$$\sum_{i \in N_0} x_{ij} = 1, \quad j \in N, \, i \neq j \tag{21}$$

$$\sum_{j \in N_0} x_{ij} = 1, \quad i \in N, \, i \neq j \tag{22}$$

$$u_i + q_j - g(1 - x_{ij}) \leq u_j, \quad i \in N_0, \, j \in N, \, i \neq j \tag{23}$$

$$q_i \leq u_i \leq g, \quad i \in N \tag{24}$$

$$u_0 = 0 \tag{25}$$

$$x_{ij} \in \{0, 1\}, \quad i, j \in N_0, \; i \neq j \tag{26}$$

Objective function (20) determines the total distance traveled. Meaning of the Eqs. (21) and (22) is analogous to the previous models, namely ensure that each customer (except the origin) is visited exactly ones. Equation (23) are anti-cyclical conditions that prevent the formation of such sub-cycles which do not contain a starting node v_0. The set of variables u_i, $i \in N$ ensures the calculation of current load of vehicles in its route to ith customer (including) (23). Equation (25) ensures that load of the vehicle is set to zero in the origin. Equation (24) ensure that all demands on the route must not exceed the capacity of the vehicle.

2.4 Capacitated Vehicle Routing Problem with Time Windows (CVRPTW)

Formulation of vehicle routing problem with time windows takes into account the same assumption as mentioned in the part 2.3 as well as the time restrictions (part 2.2). Mathematical formulation combining these assumptions is given below:

$$\min f(\mathbf{X}, \mathbf{u}, \mathbf{w}, \boldsymbol{\tau}) = \sum_{i \in N_0} \sum_{\substack{j \in N_0 \\ i \neq j}} d_{ij} x_{ij} + \sum_{i \in N} o_i + \sum_{j \in N} w_j \tag{27}$$

$$\sum_{i \in N_0} x_{ij} = 1, \quad j \in N, \, i \neq j \tag{28}$$

$$\sum_{j \in N_0} x_{ij} = 1, \quad i \in N, \, i \neq j \tag{29}$$

$$u_i + q_j - g(1 - x_{ij}) \leq u_j, \quad i \in N_0, j \in N, i \neq j \tag{30}$$

$$q_i \leq u_i \leq g, \quad i \in N \tag{31}$$

$$u_0 = 0 \tag{32}$$

$$\tau_i + o_i + d_{ij} - M\left(1 - x_{ij}\right) \leq \tau_j, \quad i \in N_0, j \in N \tag{33}$$

$$w_j \geq \tau_j - \tau_i - o_i - d_{ij} - M\left(1 - x_{ij}\right), \quad i \in N_0, j \in N, i \neq j \tag{34}$$

$$e_i \leq \tau_i, \quad i \in N \tag{35}$$

$$\tau_i + o_i \leq l_i, \quad i \in N \tag{36}$$

$$\tau_0 = 0 \tag{37}$$

$$x_{ij} \in \{0, 1\}, \quad i, j \in N_0, i \neq j \tag{38}$$

$$w_i \geq 0, \quad i \in N \tag{39}$$

Objective function (27) minimizes the sum of the total time needed to move the vehicle, the total service time of customers and the total waiting time. Meaning of the Eqs. (28) and (29) is analogous to the previous models. Each vertex are simultaneously assigned to two variables u_i, which represent the load of the vehicle, and u_i, which return the real starting service time of corresponding customer. There are also two types of anti-cycling Eqs. (30) and (33). Equations (30), (31) and (32) reflect the load the vehicle and also ensure that the capacity restriction of the vehicle is satisfied. Equations (33)–(37) provide the calculation of the real starting time of service of the corresponding customer within a given time window.

3 Self-organizing Migrating Algorithm for the Routing Problems

Self-organizing migrating algorithm (SOMA) was originally designed to solve non-constrained problems with continuous variables [56], so if one wants to apply it for solving routing problems, it is necessary to consider the following factors:

- selection of an appropriate representation of individual,
- formulation of objective function,
- transformation of parameters of individual to the real numbers,
- transformation of unfeasible solutions,
- setting of the control parameters.

Selection of an appropriate representation of individual. One way of representing solutions (individuals) is the use of natural representation of individual. In response to the vehicle routing problem, each node (city) except initial node (origin) is assigned with integer from 1 to n (n represents the number of nodes except origin), which clearly represents corresponding node in an individual. Each individual is then represented by n-dimensional vector of integers, representing the sequence of nodes visits. Parameters on individual in the initial population can be established as a random permutation of n integers, so it is the random permutation of the sequence of nodes. Each individual in the population is also assigned with its fitness that represents total cost of the route.

Formulation of objective function. Formulation of objective function is illustrated for these routing problems: TSP, TSPTW, CVRP, and CVRPTW.

The objective for traveling salesman problem (TSP) takes into account only the fact that different nodes are connected in one circular route in order given by parameters of an individual. The fitness can be presented in the following pseudocode:

Procedure Fitness TSP
Input:
D matrix of shortest distances (of size n+1) between all the nodes (first row and first column are associated with the origin)
Output:
fitness (total route traveled)
s total distance needed to serve corresponding supply nodes
s= d(x[0], x[1])
For i=1 to n-1
    s=s+d(x[i], x[i+1])
EndFor
s=s+ d(x[n], x[0])
Fitness = s
EndProcedure

Formulation of objective function for traveling salesman with time windows (TSPTW) requires to ensure the limits given by time windows so that all nodes except the origin must be served within the time window (earliest possible start of service and no later than the end of the permissible service). In doing so, each node (except the origin) is associated with its service time at the same time. The service is provided including waiting, so in the event of early arrival of the vehicle it is possible to wait until the opening of the time window. The fitness calculation can be presented by the pseudocode:

Procedure Fitness TSPTW
Input:
D matrix of shortest time distances (of size n+1) between all the nodes (first row and first column are associated with the origin)
o vector of service time of each customer (of size n)
e vector of earliest possible time of service of each customer (of size n)

l vector of last possible time of service of each customer (of size n)
p penalty constant
Output:
fitness (total duration of the route)
s total time needed to serve corresponding customer
$s = d(\mathrm{x}[0], \mathrm{x}[1])$
If $s < e[\mathrm{x}[1]]$, then $s = e[\mathrm{x}[1]]$
EndIf
$s = s + o[\mathrm{x}[1]]$
For $i = 1$ to n-1
    $s = s + d(\mathrm{x}[i], \mathrm{x}[i+1])$
        If $s \geq e[\mathrm{x}[i+1]] \wedge s + o[\mathrm{x}[i+1]] \leq l[\mathrm{x}[i+1]]$, then $s = s + o[\mathrm{x}[i+1]]$
        ElseIf $s < e[\mathrm{x}[i+1]]$, then $s = e[\mathrm{x}[i+1]] + o[\mathrm{x}[i+1]]$
        ElseIf $s + o[\mathrm{x}[i+1]] > l[\mathrm{x}[i+1]]$, then $s = s+p$
        EndIf
EndFor
s=s+ $d(\mathrm{x}[n], \mathrm{x}[0])$
Fitness = s
EndProcedure

Formulation of objective function for capacitated vehicle routing problem (CVRP), requires to take into account the demand of the nodes as well as the capacity of the vehicle in that way, the total demand on route must not exceed the capacity of the vehicle. Considering the goal to find the shortest route, the nodes are connected to one route only if non-zero savings (calculated using the Clarke & Wright's algorithm, it is thus possible to avoid the route preference associated solely on the basis of the vehicle's capacity compared with the shortest routes).

Procedure Fitness CVRP
Input:
D matrix of shortest distances (of size n+1) between all the nodes (first row and first column are associated with the origin)
q vector of customers demand (of size n)
g capacity of the vehicle
Output:
fitness (total route traveled)
k current load of vehicle (sum of the demands that are served up to the moment)
s total distance needed to serve corresponding customer
U matrix (of size n) of savings based on Clarke & Wright's algorithm
Calculate **U**
$s = d(\mathrm{x}[0], \mathrm{x}[1])$
$k = q[1]$
For i=1 to n-1
    If $k + q[i+1] \leq g \wedge u(\mathrm{x}[i], \mathrm{x}[i+1]) \neq 0$, then $s = s + d(\mathrm{x}[i], \mathrm{x}[i+1])$, $k = k + g[i+1]$

Else s=s+d(x[i], x[0])+ d(x[0],x[i+1],), k= q[i+1]
EndIf
EndFor
s=s+ d(x[n], x[0])
Fitness = s
EndProcedure

Formulation of objective function for capacitated vehicle routing problem with time windows (CVRPTW) must respect previous capacity and time constraints.

Procedure Fitness CVRPTW
Input:
D matrix of shortest time distances (of size n+1) between all the nodes (first row and first column are associated with the origin)
q vector of customers demand (of size n)
o vector of service time of each customer (of size n)
e vector of earliest possible time of service of each customer (of size n)
l vector of last possible time of service of each customer (of size n)
g capacity of the vehicle
Output:
fitness (total duration of the route)
k current load of vehicle (sum of the demands that are served up to the moment)
s total time needed to serve corresponding customer
j total time needed to serve corresponding customer in corresponding route
U matrix of savings (of size n) based on Clarke & Wright's heuristics
Calculate **U**
j = d(x[0], x[1])
k = q[1]
If j < e[x[1]], then j = e[x[1]]
EndIf
$j = j + o[x[1]]$
For i = 1 to n -1
j = j + d(x[i], x[i+1])
k = k + q[i+1]
If $j \geq e$[x[i+1]] $\wedge$ j + o[x[i+1]] $\leq$ l[x[i+1]] $\wedge$ $k \leq g$ $\wedge$ u[x[i],x[i+1]] $\neq$ 0 then j = j + o[x[i+1]]
ElseIf j < e[x[i+1]] $\wedge$ $k \leq g$ $\wedge$ u[x[i],x[i+1]] $\neq$ 0 then j = e[x[i+1]] + o[x[i+1]]
ElseIf k >g $\vee$ u[x[i],x[i+1]]=0 $\vee$ j + o[x[i]]>l[x[i]], then j = j- d(x[i], x[i+1])+d(x[i], x[*0*]), s=s+j, j = d(x[0], x[i+1])
If j > e[x[i+1]], then j = e[x[i+1]]
Endif
j = j + o[x[i+1]]
k = q[i+1],
EndIf

j=j+ d(x[n], x[0])
s=s+j
Fitness = s
EndProcedure

Transformation of parameters of individual to the real numbers. Because SOMA was originally designed to solve problems with continuous variables and the used natural representation consists of integer variables, it is desirable to transform integers to real numbers. The used method for transformation was presented in [37] for solving travelling salesman problem.

Transformation of unfeasible solutions. The use of SOMA for routing problems does not require the formation of feasible solution in case of natural representation of individual; therefore it is necessary to choose an appropriate method of transformation of the unfeasible solutions [17]. The problem of infeasibility occurs in two cases:

(a) Parameter of individual after the transformation from real numbers to integers is less than 1 or greater then n, in this case the relevant parameter is replaced by new randomly generated parameters in range $\langle 1, n \rangle$.
(b) Created individual does not comprise a permutation of n integers. In this case, the correction approach that was presented in [4] was used.

  (1) Let **m** be the vector of parameters of the individual (of size n) containing k different elements. If $n - k = 0$, go to step (4). Otherwise, go to step (2).
  (2) Create the vector **p** (dimension $n - k$) of random permutation of such elements, which are not included in the vector **m**. If the number of non-zero components of the vector $p = 0$, go to step (4). Otherwise, find the first repeated element of vector **m**. Let this element be m_c and let the first nonzero element of vector **p** be p_k. Set $m_c = p_k$ and go to step (3).
  (3) Set $p_k = 0$ and return to the step (2).
  (4) Return **m**.

Setting of the control parameters of SOMA. The control parameters can be set on the base of the article [13], which describes the possibility of setting the parameters with the help some statistical methods e.g. Kruskal-Wallis test, Bartlett's test, Cochran-Hartley's test.

The success of using SOMA and other evolutionary techniques generally can be greatly affected by selecting an appropriate method which ensures the feasibility of solutions. Generally applicable methods usually tolerate the presence of infeasible individuals in the population, which can radically affect the processing time of algorithm. In general, the use of evolutionary techniques also requires a priori parameters settings, for which there is no a deeper theoretical base. Use specialized methods may limit their complex proposal (normally applicable only for specific type of problem), and is also associated with the detailed knowledge of research problems. The difference in efficiency approaches is visible from e.g. [16], which compares penalization approach for solving the traveling salesman problem with algorithm of differential evolution as well as with self-organizing migration algorithm. The compared approaches are: penalty approach and an approach based on the transformation of infeasible solutions.

4 Computational Experiments

The computational experiments were conducted to analyze the possibility of solving traveling salesman problem, as well as vehicle routing problem compared to their modifications containing time restrictions. The experiments were carried out in exact way and also by self-organizing migration algorithm (SOMA). The experiments were performed on PC with Intel® Core™ i7-3770 CPU with a frequency of 3.40 GHz and 8 GB of RAM under MS Windows 8. Since it was desirable that the calculations were made in a sufficiently short time, the selected instances contains up to 26 nodes.

Exact solution [using mathematical formulations (3)–(41)] have been implemented using software GAMS (solver Cplex 12.2.0.0). The self-organizing migration algorithm (based on the methodology described in Sect. 3) was implemented in MatLab 8.3.

The primary aim of the experiments was to compare the quality of the obtained solution and the time needed to solve the instances for traveling salesman problem with time windows (TSPTW) from [43]. Data were derived from Solomon's RC2 VRPTW instances [50].

SOMA was run with the following settings of control parameters: PopSize = 500n (where n represents the set of nodes excluding depot), Migrations = 300, Step = 0.91 and PRT = 0.7. Parameters were set on the basis of the procedure described in [16]. Due to the computational limits of software GAMS were instances with time (TSPTW, CVRPTW) with size 19 and 20 solved only by SOMA and the quality of the results was compared with known published optimal solution.[2] Instances of traveling salesman problem with time windows (TSPTW) were based on [25], in which individual instances difference is mainly based on the "length" of time windows.

The instances for capacitated vehicle routing problem (CVRP) and for capacitated vehicle routing problem with time windows (CVRPTW) were also set according to [43]. SOMA was also applied to VRPTW instances (of size 26) from [50]. In this case, solving was realized for a time window of type "soft", i.e. in the formulation is only considered starting of service within a given time window (because of the known optimal solution for this type of time window).

The quality of solution obtained by SOMA compared to optimal solution is evident from Tables 1, 2, 3 and 4 Solution obtained using GAMS (Cplex) is

[2]http://myweb.uiowa.edu/bthoa/TSPTWBenchmarkDataSets.htm (1.1.2014).

Table 1 Traveling salesman problem (TSP)

TSP		GAMS	SOMA			
Instance	Size	Time (s)	Avg. time (s)	% deviation	Best solution (s)	% deviation
rc206_1	4	0.01	0.04	0	0.7	0
rc207_4	6	0.02	0.03	0	0.03	0
rc205_1	14	0.09	12.51	0.9	38.13	0
rc202_2	14	0.94	7.48	0	3.9	0
rc203_4	15	0.11	4.58	0	2.03	0
rc203_1	19	38.72	226.04	0.27	182	0
rc201_1	20	4.14	635.75	4.31	643.76	0
n20w120_1	20	0.14	195.08	0	30.64	0
n20w140_1	20	0.13	129.08	0.61	65.74	0
n20w160_1	20	2.17	1394.12	4.37	1729.32	2.25

Table 2 Traveling salesman problem with time windows (TSPTW)

TSPTW		GAMS	SOMA			
Instance	Size	Time (s)	Avg. time (s)	% deviation	Best solution (s)	% deviation
rc206_1	4	0.01	1.85	0	1	0
rc207_4	6	0.02	0.02	0	0.01	0
rc205_1	14	1.34	1279.71	0.9	208.13	0
rc202_2	14	1.53	94.98	0	50.29	0
rc203_4	15	1.09	12.29	0	5.42	0
rc203_1	19	–	2070.21	3.97	515.3	3.11
rc201_1	20	–	528.41	6.84	365.98	0
n20w120_1	20	659.19	3427.94	0.14	3414.51	0
n20w140_1	20	54.41	1333.61	6.59	187.51	0
n20w160_1	20	–	423.33	0	388.33	0

evaluated in terms of time that was needed to achieve optimal solution, solution obtained by SOMA is analyzed in terms of average performance (10 simulations were conducted for each instance)—avg. time and also it was evaluated the best result achieved from the realized simulations in terms of time—best solution, respectively percentage deviation from the optimal solution provided by GAMS system or known from other sources—% deviation.

The experiments on instances for TSP are listed on Table 1. From the presented results it is clear that for all the instances the optimal solution (by GAMS) was achieved relatively quickly (up to about 39 s.). The percent accuracy by SOMA was, on average, from 0 to 4.37 %. In 9 cases, the value of the best obtained solution was identical to the optimal solution, in one case, the best solution differs from the optimum by 2.25 %. In terms of time consumption better achievement of GAMS before **SOMA** is apparent.

Table 3 Capacitated vehicle routing problem (CVRP)

CVRP		GAMS	SOMA			
Instance	Size	Time (s)	Avg. time (s)	% deviation	Best solution (s)	% deviation
rc206_1	4	0.01	0.5	0	0.8	0
rc207_4	6	0.13	1	0	0.03	0
rc205_1	14	208.59	21.62	0	12.31	0
rc202_2	14	92.25	25.02	0.7	7.2	0.7
rc203_4	15	36.73	400.67	0	73.05	0
r101_25	26	1365.81	1029.74	4.64	711	0
r201_25	26	1526.87	1714.36	5.2	2124.91	0.8
c101_25	26	292,115.52	1023.41	6.35	1532.43	3.84
c201_25	26	229,634.87	9533.96	4.57	15,149.12	0
rc101_25	26	200,681.34	6585.2	0.26	5976.97	0
rc201_25	26	127,563.22	5829.53	4.15	4169.53	0

Extension of TSP by time windows (Table 2) caused a slight increase in the time required to identify the optimal solution by GAMS. Even, the computational limits were insufficient in some case of instances with size 19 and 20. The solutions obtained by SOMA showed an average deviation from 0 to 6.6 %, the best solution deviation from the optimum showed 0 in 9 cases, in one case, the deviation showed 3.11 %.

The results for CVRP instances are presented in Table 3 Based on the results presented, it can be stated that although GAMS identifies the optimal solution in all cases, time consumed by calculations was significantly increased in case of instances of classes C and RC (of dimension 26). The quality of solutions identified by SOMA was in range to 6.35 % in the average case and to 3.84 % in the best case (such solutions are achieved in a reasonable time).

Table 4 Capacitated vehicle routing problem with time windows (CVRPTW)

CVRPTW		GAMS	SOMA			
Instance	Size	Time (s)	Avg. time (s)	% deviation	Best solution (s)	% deviation
rc206_1	4	0.08	1	0	0.5	0
rc207_4	6	0.05	0.92	0	0.8	0
rc205_1	14	104.98	172.62	0	73.12	0
rc202_2	14	3316.27	102.18	0	41.51	0
rc203_4	15	45,572.67	41.12	0.01	22.28	0.01
r101_25	26	–	5032.89	1.08	5157.19	0.3
r201_25	26	–	3055.44	5.12	3255.08	3.27
c101_25	26	–	3315.36	7.8	6251.18	0.27
c201_25	26	–	4646.82	6.21	5261.12	0.8
rc101_25	26	–	1270.35	7.83	1725.33	4.41
rc201_25	26	–	7409.74	9.19	1826.33	5.11

Extension of CVRP by delivery time windows (CVRPTW) caused not only an increase in the time required for identification of the optimal solution by GAMS, also pointed out the impossibility of obtaining the optimal solution for instances with dimension 26 (Table 4). A mean deviation of the average solution from optimal solution by SOMA was from 0 to 9.19 % and the difference between optimal solution and the best identified solution was to 5.11 %.

5 Solving the Real-Life Vehicle Routing Problem with Time Windows

The aforementioned models provide the base for other modifications that allow implicitly take into account many possible limitations encountered in solving practical problems.

Next, the real-life vehicle routing problem with time windows is presented. The problem deals with the real distribution scheduling in the region of Banská Bystrica in Slovakia. Distribution center was situated in the city of Banská Bystrica and distribution was made to 29 municipalities, in which the stores were situated with a daily calling for certain number of crates of merchandise. Also, there are known earliest possible start and latest acceptable end of service (between 6:00 and 9:00) for each customer, i.e. start and end time when it was possible to realize the supply. Further on, service time (unloading time of vehicle) in certain municipalities was also estimated. The input data for each customer i, $i \in N$, with the demand g_i, earliest possible start of service e_i, last acceptable end of service l_i, and estimated service duration o_i were set as follows [in structure (i name—q_i, e_i, l_i, o_i)]: (1 Kyncečová—5, 0, 120, 3); (2 Nemce—4, 60, 180, 3); (3 Malachov—6, 0, 120, 3); (4 Tajov—10, 0, 180, 5); (5 Riečka—4, 0, 120, 3); (6 Selce—12, 30, 180, 5); (7 Š. Dolina—4, 60, 120, 3); (8 H. Mičiná—10, 0, 180, 5); (9 Vlkanová—6, 60, 120, 3); (10 Horné Pršany—4, 30, 180, 3); (11 S. Ľupča—15, 0, 180, 8); (12 Badín—11, 0, 180, 5); (13 Králiky—3, 60, 180, 3); (14 Harmanec—17, 0, 180, 8); (15 Môlča—4, 120, 180, 3); (16 Kordíky—6, 60, 180, 3); (17 Hronsek—5, 30, 90, 3); (18 Sielnica—5, 120, 180, 3); (19 Priechod—4, 60, 150, 3); (20 Staré Hory—9, 0, 180, 5); (21 Lučatín—7, 30, 150, 3); (22 D. Mičiná—10, 0, 180, 5); (23 D. Harmanec—12, 0, 180, 5); (24 Sliač—18, 0, 180, 8); (25 Kováčová—16, 0, 180, 8); (26 Medzibrod—4, 120, 180, 3); (27 Čerín—5, 60, 180, 3); (28 Podkonice—7, 0, 180, 3); (29 Brusno—11, 0, 180, 5).

The capacity of the available vehicles was set to 80 crates and service duration in the distribution center 0 Banská Bystrica is estimated as 30 min. Further on, time distances in minutes between the center and individual customers themselves were known (matrix **D** with elements d_{ij}, $i, j \in N_0$—Table 5).

The objective is to minimize the total service time (transfer time, waiting time, service time), as well as the minimization of the number of vehicles, i.e. the determination of minimal number of vehicles which must be used daily (the limit on driving time of a vehicle is also considered −180 min). At the same time it is

Table 5 Input data: time distances in minutes

	0	1	2	3	4	5	6	7	8	9	10	11	12	13	14
0	0.0	4.7	5.4	6.0	6.3	7.1	7.7	7.9	8.2	8.3	8.4	8.8	9.1	9.3	9.9
1	4.7	0.0	0.8	10.7	10.9	11.7	3.0	12.6	12.8	13.0	13.0	8.7	13.7	13.9	14.6
2	5.4	0.8	0.0	11.4	11.7	12.5	3.8	13.3	13.6	13.7	13.8	9.4	14.5	14.7	15.3
3	6.0	10.7	11.4	0.0	12.3	13.1	13.7	13.9	14.2	14.3	14.4	14.8	15.1	15.3	15.9
4	6.3	10.9	11.7	12.3	0.0	5.2	13.9	12.6	14.4	14.6	14.6	15.1	15.3	3.0	14.6
5	7.1	11.7	12.5	13.1	5.2	0.0	14.7	13.4	15.3	15.4	15.4	15.9	16.1	8.2	15.4
6	7.7	3.0	3.8	13.7	13.9	14.7	0.0	15.6	15.8	16.0	16.0	11.7	16.7	16.9	17.6
7	7.9	12.6	13.3	13.9	12.6	13.4	15.6	0.0	16.1	16.2	16.3	16.7	17.0	15.6	5.0
8	8.2	12.8	13.6	14.2	14.4	15.3	15.8	16.1	0.0	16.5	16.5	15.7	17.3	17.4	18.1
9	8.3	13.0	13.7	14.3	14.6	15.4	16.0	16.2	16.5	0.0	13.5	17.1	2.3	17.6	18.2
10	8.4	13.0	13.8	14.4	14.6	15.4	16.0	16.3	16.5	13.5	0.0	17.2	14.2	17.6	18.3
11	8.8	8.7	9.4	14.8	15.1	15.9	11.7	16.7	15.7	17.1	17.2	0.0	17.9	18.1	18.7
12	9.1	13.7	14.5	15.1	15.3	16.1	16.7	17.0	17.3	2.3	14.2	17.9	0.0	18.3	19.0
13	9.3	13.9	14.7	15.3	3.0	8.2	16.9	15.6	17.4	17.6	17.6	18.1	18.3	0.0	17.6
14	9.9	14.6	15.3	15.9	14.6	15.4	17.6	5.0	18.1	18.2	18.3	18.7	19.0	17.6	0.0
15	10.4	10.3	11.0	16.4	16.7	17.5	13.3	18.3	5.3	18.7	18.8	10.4	19.5	19.7	20.3
16	11.2	15.8	16.6	17.2	4.9	10.1	18.8	17.5	19.4	19.5	19.5	20.0	20.2	7.9	19.5
17	11.3	16.0	16.7	17.3	17.6	18.4	19.0	19.2	18.6	3.0	16.5	20.1	5.3	20.6	21.2
18	11.3	16.0	16.7	17.3	17.6	18.4	19.0	19.2	19.5	10.7	16.5	20.1	11.5	20.6	21.2
19	12.2	7.5	8.3	18.2	18.4	19.2	4.5	20.1	20.3	20.5	20.5	16.2	21.2	21.4	22.1
20	12.4	17.1	17.8	18.4	17.1	17.9	20.1	7.5	20.6	20.7	20.8	21.2	21.5	20.1	7.9
21	12.4	12.3	13.0	18.4	18.7	19.5	15.3	20.3	19.3	20.7	20.8	3.6	21.5	21.7	22.3
22	12.6	17.2	18.0	18.6	18.8	19.6	20.2	20.5	4.4	17.3	20.9	20.0	19.5	21.8	22.5
23	12.7	17.3	18.1	18.7	17.3	18.1	20.3	7.8	20.8	21.0	21.0	21.5	21.7	20.3	2.7
24	14.1	18.8	19.5	20.1	20.4	21.2	21.8	22.0	18.6	7.3	19.3	22.9	8.1	23.4	24.0
25	14.1	18.8	19.5	20.1	20.4	21.2	21.8	22.0	21.3	10.0	19.3	22.9	10.7	23.4	24.1
26	14.4	14.3	15.0	20.4	20.7	21.5	17.3	22.3	21.3	22.7	22.8	5.6	23.5	23.7	24.3
27	14.7	19.4	20.1	20.7	21.0	21.8	22.4	22.6	6.6	19.4	23.1	22.2	21.7	24.0	24.7
28	14.8	14.7	15.4	20.8	21.1	21.9	17.7	22.7	21.7	23.1	23.2	6.0	23.9	24.1	24.7
29	15.6	15.5	16.2	21.6	21.9	22.7	18.5	23.5	22.5	23.9	24.0	6.8	24.7	24.9	25.5

1	15	16	17	18	19	20	21	22	23	24	25	26	27	28	29
0	10.4	11.2	11.3	11.3	12.2	12.4	12.4	12.6	12.7	14.1	14.1	14.4	14.7	14.8	15.6
1	10.3	15.8	16.0	16.0	7.5	17.1	12.3	17.2	17.3	18.8	18.8	14.3	19.4	14.7	15.5
2	11.0	16.6	16.7	16.7	8.3	17.8	13.0	18.0	18.1	19.5	19.5	15.0	20.1	15.4	16.2
3	16.4	17.2	17.3	17.3	18.2	18.4	18.4	18.6	18.7	20.1	20.1	20.4	20.7	20.8	21.6
4	16.7	4.9	17.6	17.6	18.4	17.1	18.7	18.8	17.3	20.4	20.4	20.7	21.0	21.1	21.9
5	17.5	10.1	18.4	18.4	19.2	17.9	19.5	19.6	18.1	21.2	21.2	21.5	21.8	21.9	22.7
6	13.3	18.8	19.0	19.0	4.5	20.1	15.3	20.2	20.3	21.8	21.8	17.3	22.4	17.7	18.5
7	18.3	17.5	19.2	19.2	20.1	7.5	20.3	20.5	7.8	22.0	22.0	22.3	22.6	22.7	23.5
8	5.3	19.4	18.6	19.5	20.3	20.6	19.3	4.4	20.8	18.6	21.3	21.3	6.6	21.7	22.5
9	18.7	19.5	3.0	10.7	20.5	20.7	20.7	17.3	21.0	7.3	10.0	22.7	19.4	23.1	23.9
10	18.8	19.5	16.5	16.5	20.5	20.8	20.8	20.9	21.0	19.3	19.3	22.8	23.1	23.2	24.0
11	10.4	20.0	20.1	20.1	16.2	21.2	3.6	20.0	21.5	22.9	22.9	5.6	22.2	6.0	6.8
12	19.5	20.2	5.3	11.5	21.2	21.5	21.5	19.5	21.7	8.1	10.7	23.5	21.7	23.9	24.7

(continued)

Table 5 (continued)

1	15	16	17	18	19	20	21	22	23	24	25	26	27	28	29
13	19.7	7.9	20.6	20.6	21.4	20.1	21.7	21.8	20.3	23.4	23.4	23.7	24.0	24.1	24.9
14	20.3	19.5	21.2	21.2	22.1	7.9	22.3	22.5	2.7	24.0	24.1	24.3	24.7	24.7	25.5
15	0.0	21.6	21.7	21.7	17.8	22.8	14.0	9.6	23.1	23.9	24.5	16.0	11.8	16.4	17.2
16	21.6	0.0	22.5	22.5	23.3	22.0	23.6	23.7	22.2	25.3	25.3	25.6	25.9	26.0	26.8
17	21.7	22.5	0.0	12.4	23.5	23.7	23.7	14.3	24.0	9.0	11.7	25.7	16.4	26.1	26.9
18	21.7	22.5	12.4	0.0	23.5	23.7	23.7	17.7	24.0	3.4	2.8	25.7	19.8	26.1	26.9
19	17.8	23.3	23.5	23.5	0.0	24.6	19.8	24.7	24.8	26.3	26.3	21.8	26.9	22.2	23.0
20	22.8	22.0	23.7	23.7	24.6	0.0	24.8	25.0	10.7	26.5	26.5	26.8	27.1	27.2	28.0
21	14.0	23.6	23.7	23.7	19.8	24.8	0.0	23.6	25.1	26.5	26.5	2.0	25.8	9.6	3.2
22	9.6	23.7	14.3	17.7	24.7	25.0	23.6	0.0	25.2	14.3	17.0	25.6	2.2	26.0	26.8
23	23.1	22.2	24.0	24.0	24.8	10.7	25.1	25.2	0.0	26.8	26.8	27.1	27.4	27.5	28.3
24	23.9	25.3	9.0	3.4	26.3	26.5	26.5	14.3	26.8	0.0	2.7	28.5	16.4	28.9	29.7
25	24.5	25.3	11.7	2.8	26.3	26.5	26.5	17.0	26.8	2.7	0.0	28.5	19.1	28.9	29.7
26	16.0	25.6	25.7	25.7	21.8	26.8	2.0	25.6	27.1	28.5	28.5	0.0	27.8	11.6	1.2
27	11.8	25.9	16.4	19.8	26.9	27.1	25.8	2.2	27.4	16.4	19.1	27.8	0.0	28.2	29.0
28	16.4	26.0	26.1	26.1	22.2	27.2	9.6	26.0	27.5	28.9	28.9	11.6	28.2	0.0	12.8
29	17.2	26.8	26.9	26.9	23.0	28.0	3.2	26.8	28.3	29.7	29.7	1.2	29.0	12.8	0.0

necessary to distinguish two facts: (1) if the vehicle returns to the origin when the capacity of a vehicle is exceeded, then the service time in the origin (time to re-loaded) is added to the total time of the corresponding route and the next route will be realized with the same vehicle, (2) if the vehicle returns to the origin because of the last possible service time violation, the service time in the origin is not added to the total time of the route and the next route is realized by another vehicle.

Modeling such a situation is only possible with considerable modification of mentioned models of routing problems. Further on, following notation is used: Let $N = \{1, 2, \ldots n\}$ be the set representing customers and let $N_0 = N \cup \{0\}$ be the set of all nodes (including origin). Let $H = \{1, 2, \ldots\}$ be the set of vehicles, where r represents number of vehicles, while each vehicle $h \in H$ is the same capacity g and let the set $K = \{1, 2, \ldots k\}$ represents order of arc in sequence of hth vehicle, where $2n$ represents maximal number of arcs in a sequence. Then variables x_{ijkh}, $i, j \in N_0,\ i \neq j, h \in H, k \in K$ can be defined as follows:

$$x_{ijkh} = \begin{cases} 1, & \text{if the edge between nodes } i \text{ and } j \text{ is used} \\ & \text{by vehicle } h \text{ as } k\text{th in sequence} \\ 0, & \text{otherwise} \end{cases} \qquad (40)$$

Each customer has certain demand (g_i, $i \in N$) and a service duration (o_i, $i \in N$). Further on, there is the known time window of each customer: as the earliest possible start of service in different nodes (e_i, $i \in N$) and the last acceptable time of service in different nodes (l_i, $i \in N$). The demand is fulfilled from initial node ($i = 0$)—origin. We suppose that the service time at the origin is set to o_0 (this time is added to the total time only when the vehicle returns to the origin due to violation of capacity limit, and it is not able to serve the nodes on the next route).

The goal is to determine the minimal number of vehicles so that the total travelled time or distance is as low as possible (we suppose that there is known the shortest time distance between all nodes d_{ij}, $i,j \in N_0$) with respect to the following restrictions: the origin represents initial node and also the final node of every route, from the origin the demand g_i, $i \in N$ of all the other nodes is met within their time windows (the earliest possible start of service e_i, $i \in N$, the last acceptable time of service l_i, $i \in N$, each node (except central node) is visited exactly once and total demand on route must not exceed the capacity of the vehicle (g). The total time of the route of a vehicle could not exceed the given time (l_0).

The model takes into account the waiting time so that the vehicle is allowed to wait for service if it arrives before the earliest possible start of service.

The following decision variables will be used (besides the binary variables x_{ijkh}):

w_j—nonnegative variables that indicates waiting time at node j, $j \in N$,
τ_j—variables that indicates real starting time of service at node j, $j \in N_0$,
u_j—variables that represents remaining capacity of vehicle at the node j, $j \in N$.
Mathematical formulation of the model is given below [18]:

$$\begin{aligned} \min f(\mathbf{X}, \mathbf{u}, \mathbf{w}, \boldsymbol{\tau}) = & \sum_{i \in N_0} \sum_{\substack{j \in N_0 \\ i \neq j}} \sum_{h \in H} \sum_{k \in K} d_{ij} x_{ijkh} + \sum_{i \in N} w_i + \sum_{i \in N} o_i \\ & + \sum_{j \in N} \sum_{k \in K - \{1\}} \sum_{h \in H} o_0 x_{0jkh} + p \sum_{i \in N} \sum_{h \in H} x_{0j1h} \end{aligned} \tag{41}$$

$$\sum_{j \in N_0} \sum_{k \in K} \sum_{h \in H} x_{ijkh} = 1, \quad i \in N, i \neq j \tag{42}$$

$$\sum_{i \in N_0} \sum_{k \in K} \sum_{h \in H} x_{ijkh} = 1, \quad j \in N, i \neq j \tag{43}$$

$$\sum_{i \in N_0} x_{ij(k-1)h} = \sum_{i \in N_0} x_{jikh}, \quad j \in N, h \in H, k \in K - \{1\}, i \neq j \tag{44}$$

$$\tau_i + o_i + d_{ij} \leq \tau_j + M\left(1 - x_{ijkh}\right), \quad i,j \in N, h \in H, k \in K, i \neq j \tag{45}$$

$$w_j \geq \tau_j - \tau_i - o_i - M\left(1 - x_{ijkh}\right), \quad i,j \in N, h \in H, k \in K, i \neq j \tag{46}$$

$$\begin{aligned} & \tau_i + o_i + d_{i0} + o_0 + d_{0j} \leq \tau_j + M\left(\left(1 - x_{i0(k-1)h}\right) + \left(1 - x_{0jkh}\right)\right), \\ & i,j \in N, h \in H, k \in K - \{1\}, i \neq j, \end{aligned} \tag{47}$$

$$\begin{aligned} & w_j \geq \tau_j - \tau_i - o_i - d_{i0} - o_0 - d_{0j} - M\left(\left(1 - x_{i0(k-1)h}\right) - \left(1 - x_{0jkh}\right)\right), \\ & i,j \in N, h \in H, k \in K - \{1\}, i \neq j, \end{aligned} \tag{48}$$

$$d_{0j} \le \tau_j + M\left(1 - x_{0j1h}\right), \quad j \in N, h \in H \tag{49}$$

$$\tau_j + o_j + d_{0j} \le l_0 + \left(1 - x_{j0kh}\right), \quad j \in N, h \in H, k \in K \tag{50}$$

$$u_i + q_j - u_j \le M\left(1 - x_{ijkh}\right), \quad i \in N_0, j \in N, h \in H, k \in K, i \ne j \tag{51}$$

$$\sum_{i \in N} x_{0jkh} \le \sum_{i \in N} x_{j0(k-1)h}, \quad j \in N,\ k \in K - \{1\} \tag{52}$$

$$\sum_{j \in N} x_{0j1h} \le 1, \quad h \in H \tag{53}$$

$$x_{0j1h} \ge x_{0jkh} \quad j \in N, h \in H, k \in H \tag{54}$$

$$u_0 = 0 \tag{55}$$

$$u_i \le g, \quad i \in N \tag{56}$$

$$\tau_0 = 0 \tag{57}$$

$$e_i \le \tau_i, \quad i \in N \tag{58}$$

$$\tau_i + o_i \le l_i, \quad i \in N \tag{59}$$

$$x_{ijh} \in \{0, 1\}, \quad i, j \in N_0, i \ne j, h \in H, k \in K \tag{60}$$

$$w_i \ge 0,\ u_i \ge 0, \quad \tau_i \ge 0,\ i \in N \tag{61}$$

The objective function (41) minimizes the total duration travelled and also the number of vehicles (p represents the penalty constant). Equations (42) and (43) ensure that a vehicle leaves each node and vehicle enters each node except the origin exactly ones. Equations (44) and (52) ensure the connectivity of the route. Equations (45) and (46) calculate the real starting time of service for the next node on the route (except the origin) on the base of previous node. Equations (47) and (48) ensure the calculation of starting time of service for the next node on the route, in case that the route goes through origin. Equation (49) calculate the real starting time of service of the first node on the route of the vehicle. Equation (50) ensures that the total vehicle time travelled must not exceed the given time (T). Equations (51), (55) and (56) ensure that all demands on the route must not exceed the capacity of the vehicle. Equations (53) and (54) ensure that each route starts at the origin exactly once. Equations (57), (58) and (59) ensure that the time windows of all nodes on the route are met.

The use of SOMA requires the formulation of objective function (fitness) with respect to the following facts:

(1) if the vehicle returns to the origin when the capacity of a vehicle is exceeded, then the service time in the origin (time to re-loaded) is added to the total time of the corresponding route (in practice, the next route will be realized with the same vehicle), (2) if the vehicle returns to the origin when last possible service time is violated, the objective function is penalized by penalty constant, which represents the fact that the next route is realized by another vehicle and the real duration of distribution is calculated by subtracting the total of these penalties.
Goal: to minimize number of vehicles, so that the total distance travelled by vehicles is as low as possible.

The fitness calculation can be presented by the pseudocode:

Procedure Fitness VRPTW2
Input:
D matrix of shortest time distances (of size $n+1$) between all the nodes (first row and first column are associated with the origin)
q vector of customers demand (of size n)
o vector of service time of each customer including depot (of size $n+1$)
e vector of earliest possible time of service of each customer (of size n)
l vector of last possible time of service of each customer including depot (of size $n+1$)
g capacity of the vehicles
p penalty constant
Output:
fitness (total duration of the route)
k current load of vehicle (sum of the demands that are served up to the moment)
s total time needed to serve corresponding customer
j total time needed to serve corresponding customer in the route of corresponding vehicle
U matrix of savings (of size n) based on Clarke & Wright's heuristics
Calculate **U**
$j = d(\mathrm{x}[0], \mathrm{x}[1])$
$k = g[1]$
If $j<e[\mathrm{x}[1]]$, then $j=e[\mathrm{x}[1]]$
EndIf
$j=j+o[\mathrm{x}[1]]$
For $i=1$ to n -1
$j=j+d(\mathrm{x}[i],\mathrm{x}[i+1])$
$k=k+q[i+1]$
If $j \geq e[\mathrm{x}[i+1]] \wedge j+o[\mathrm{x}[i+1]] \leq l[\mathrm{x}[i+1]] \wedge k \leq g \wedge u[\mathrm{x}[i],\mathrm{x}[i+1]] \neq 0 \wedge j+o[\mathrm{x}[i+1]]+d(\mathrm{x}[i+1],\mathrm{x}[0]) \leq l[0]$, then $j = j + o[\mathrm{x}[i+1]]$
ElseIf $j < e[\mathrm{x}[i+1]] \wedge k \leq g \wedge u[\mathrm{x}[i],\mathrm{x}[i+1]] \neq 0$, then $v = j$, $j = e[\mathrm{x}[i+1]]+o[\mathrm{x}[i+1]]$
If j + d(x[i+1], x[0]) > l[0], then j = v - d(x[i], x[i+1])+ d(x[i], x[0]), s=s +j+p, j= d(x[0], x[i+1]), k = q[i+1]

If $j < e[x[i+1]]$, then $j = e[x[i+1]] + o[x[i+1]]$
EndIf
EndIf

ElseIf $k > g \vee u[x[i],x[i+1]] = 0 \wedge j+o[x[i+1]] + d(x[i+1], x[0]) \leq l[0]$, then $k = q[i+1]$, $j = j - d(x[i], x[i+1])$, $v = j$, $j = j+d(x[i], x[0]) + o[0] + d(x[0], x[i+1])$
If $j < e[x[i+1]]$,then $j = e[x[i+1]] + o[x[i+1]]$
EndIf
If $j + d(x[i+1], x[0]) > l[0]$, then $j = v + d(x[i], x[0])$, $s=s+j+p$, $j= d(x[0], x[i+1])$
If $j < e[x[i+1]]$, then $j = e[x[i+1]] + o[x[i+1]]$
EndIf
Endif
ElseIf $j + o[x[i+1]] > l[x[i+1]] \vee j + d(x[i+1],x[0])+o[x[i+1]] > l[0]$, then
$j = j-d(x[i],x[i+1])+d(x[i],x[0])$, $s=s+j+p$, $j= d(x[0], x[i+1]$, $k = q[i+1]$
If $j < e[x[i+1]]$, then $j = e[x[i+1]] + o[x[i+1]]$
EndIf
EndIf
EndFor
s=s+ $d(x[n], x[0])$
Fitness = s
EndProcedure

Our solution by SOMA is based on same principles as was mentioned above with the setting of control parameters (PathLength, Step and PRT) with PopSize = 3000 and Migrations = 5000.

The best result obtained from 10 realized simulations allows the use of 3 vehicles with the total duration 465.76 min.

Route A Customer sequence: 0; 20; 5; 4; 13; 7; 28; 11; 21; 26; 29; 0 (Banská Bystrica; Staré Hory; Riečka; Tajov; Králiky; Špania Dolina; Podkonice; Slovenská Ľupča; Lučatín; Medzibrod; Brusno; Banská Bystrica), total capacity 74, return due to upper time limit of a window, total time of route 154.65 min, objective value 154.65 min ⇒ the next route will be realized by another vehicle.

Route B Customer sequence: 0; 23; 14; 0 (Banská Bystrica; Dolný Harmanec; Harmanec; Banská Bystrica), total capacity 29, return due to savings = 0, total time of route 38.30 min, objective value 192.95 min ⇒ the route C will be realized with the same vehicle, service time at the centre is 30 min.

Route C Customer sequence: 0; 1; 2; 6; 19; 15; 22; 27; 8; 0 (Banská Bystrica; Kynceľová; Nemce; Selce; Priechod; Môlča; Dolná Mičiná; Čerín; Horná Mičiná; Banská Bystrica), total capacity 54, return due to upper time limit, total time of route C 94.22 min, objective value 317.17 min ⇒ the next route will be realized by another vehicle.

Table 6 Route A

Index	20	5	4	13	7	28	11	21	26	29
	Staré Hory	Riečka	Tajov	Králiky	Š. Dolina	Podkonice	S. Ľupča	Lučatín	Medzibrod	Brusno
Demand q_i	9	4	10	3	4	7	15	7	4	11
Saving t_i	5	3	5	3	3	3	8	3	3	5
Earliest limit l_i	0	0	0	60	60	0	0	30	120	0
Latest limit u_i	180	120	180	180	120	180	180	150	180	180
Arrival	12.4	35.27	43.45	51.45	78.55	104.25	113.25	124.85	129.85	134.05
Start of service	12.4	35.27	43.45	60	78.55	104.25	113.25	124.85	129.85	134.05
End of service	17.4	38.27	48.45	63	81.55	107.25	121.25	127.85	132.85	139.05
Capacity	9	13	23	26	30	37	52	59	63	74

Table 7 Route B

Index	23	14
	D. Harmanec	Harmanec
Demand q_i	12	17
Saving t_i	5	8
Earliest limit l_i	0	0
Latest limit u_i	180	180
Arrival	12.65	20.38
Start of service	12.65	20.38
End of service	17.65	28.38
Capacity	12	29

Table 8 Route C

Index	1	2	6	19	15	22	27	8
	Kynceľová	Nemce	Selce	Priechod	Môlča	D. Mičiná	Čerín	H. Mičiná
Demand q_i	5	4	12	4	4	10	5	10
Saving t_i	3	3	5	3	3	5	3	5
Earliest limit l_i	0	60	30	60	120	0	60	0
Latest limit u_i	120	180	180	150	180	180	180	180
Arrival	72.96	76.71	83.46	92.96	113.71	132.61	139.79	149.34
Start of service	72.96	76.71	83.46	92.96	120	132.61	139.79	149.34
End of service	75.96	79.71	88.46	95.96	123	137.61	142.79	154.34
Capacity	5	9	21	25	29	39	44	54

Route D Customer sequence: 0; 3; 17; 12; 16; 10; 9; 24; 25; 18; 0 (Banská Bystrica; Malachov; Hronsek; Badín; Kordíky; Horné Pršany; Vlkanová; Sliač; Kováčová; Sielnica; Banská Bystrica), total capacity 77, total time of route 148.59 min, objective value 465.76 min

The distribution was previously made on the basis of the solution that was derived with heuristic Clarke & Wright's savings algorithm with time windows[3] with the use of four vehicles and the total duration was 555.7 min. Individual routes were: Route 1: 0–8–22–27–15–0, duration 133.4; Route 2: 0–3–10–12–9–17–24–25–18–0, duration 134.3; Route 3: 0–1–6–19–2–11–28–21–26–29–0, duration 144.8; Route 4: 0–5–4–13–16–7–14–23–20–0, duration 143.2 min.

[3]http://www.ise.ncsu.edu/kay/matlog/ (1.2.2012).

Table 9 Route D

Index	3	17	12	16	10	9	24	25	18
	Malachov	Hronsek	Badín	Kordíky	Horné Pršany	Vlkanová	Sliač	Kováčová	Sielnica
Demand q_i	6	5	11	6	4	6	18	16	5
Saving t_i	3	3	5	3	3	3	8	8	3
Earliest limit l_i	0	30	0	60	30	60	0	0	120
Latest limit u_i	120	90	180	180	180	120	180	180	180
Arrival	6	26.31	38.25	63.48	85.99	102.45	112.75	123.45	134.27
Start of service	6	30	38.25	63.48	85.99	102.45	112.75	123.45	134.27
End of service	9	33	43.25	66.48	88.99	105.45	120.75	131.45	137.27
Capacity	6	11	22	28	32	38	56	72	77

Based on those results it can be stated that presented approach allows decreasing the number of vehicles simultaneously with the savings in the total time (the saving is 16.2 %).

The routes are detailed described in Tables 6, 7, 8 and 9.

6 Conclusion

Delivery is decisive for business operations of many companies. Logistic costs constitute a significant share of the total costs of every organization. This amount varies from 10 to 25 % of the total costs depending on the given industry and country; that's why many managers start to pay attention to optimization techniques that involves the reduction of logistic cost. Many variants of routing problems that can be very rewarding are known in the field of logistics. This chapter deals with some basic routing problems (traveling salesman problem, vehicle routing problem) and its variants, which take into account time restrictions, as well as the real-data vehicle routing problem with time windows is presented. The vehicle routing problem with time windows belongs to NP-hard problems, so no algorithm has been known to solve it in the polynomial time, even though with the development of information technology the number of problems that can be solved by exact algorithms has been increased. The alternative is, except for classical heuristics, the use of evolutionary algorithm, which can give after finite number of iteration an "effective" solution.

Nowadays, we follow the increased interest in methods, which are inspired by different biological evolutionary processes in nature. This technology is covered by the common name of "evolutionary algorithms". But their application to constrained problems requires some additional modifications of theirs basic versions. The chapter was focused on application of self-organizing migrating algorithm (SOMA) to routing problems. The special factors that involve the use of that algorithm were presented and the efficiency of calculations has been validated on the basis of publicly available instances. The presented approach was also used to solve real-life vehicle routing problem with time windows in Slovakia. The result was also compared with the known solution based on heuristic Clarke & Wright's savings algorithm with time windows. Based on these results the following can be stated: the number of vehicles was decreased (from 4 to 3) and the total time of distribution was improved by 16.2 %.

References

1. Ai, T.J., Kachitvichyanukul, V.: A particle swarm optimization for vehicle routing problem with time windows. Int. J. Oper. Res. **6**, 519–537 (2009)
2. Baker, E.K.: An exact algorithm for the time-constrained traveling salesman problem. Oper. Res. **31**, 938–945 (1983)

3. Bellman, R.: Combinatorial precesses and dynamic programming. Combinatorial analysis. American Mathematical Society, Providence (1960)
4. Brezina, I., Čičková, Z., Pekár, J.: Application of evolutionary approach to solving vehicle routing problem with time windows. Ekonomické rozhľady **38**, 529–539 (2009)
5. Brezina, I., Čičková, Z., Pekár, J.: Evolutionary approach as an alternative method for solving the vehicle routing problem. Ekonomické rozhľady **41**, 137–147 (2012)
6. Campbell, A., Savelsbergh, M.: Decision support for consumer direct grocery initiatives. Transp. Sci. **39**, 313–327 (2005)
7. Carlton, W.B., Barnes, J.W.: Solving the traveling salesman problem with time windows using tabu search. Inst. Ind. Eng. Trans. **28**, 617–629 (1996)
8. Chajdiak, J.: Štatistika jednoducho. Statis, Bratislava (2003)
9. Chang, Y., Chen, L.: Solve the vehicle routing problem with the time windows via a genetic algorithm. Discrete and continuous dynamical systems supplement, pp. 240–249 (2007)
10. Christofides, N.: Worst-case analysis of a new heuristic for the traveling salesman problem. Tech. Report 388, GSIA, Carnegie Mellon University (1976)
11. Christofides, N., Mingozzi, A., Toth, P.: Space state relaxation procedures for the computation of bounds to routing problems. Networks **11**, 145–164 (1981)
12. Čičková, Z., Reiff, M.: Energetická funkcia Hopfieldovej siete pri riešení úlohy o obchodnom cestujúcom. In: 1. medzinárodný seminár doktorandov Katedry operačného výskumu a ekonometrie FHI EU v Bratislave a Katedry ekonometrie FIS VŠE v Prahe, pp. 21–24. Praha (2004)
13. Čičková, Z., Brezina, I., Pekár, J.: A memetic algorithm for solving the vehicle routing problem. In: 29th International Conference on Mathematical Methods in Economics 2011, pp. 125–128. Praha (2011)
14. Čičková, Z., Brezina, I., Pekár, J.: SOMA for solving the vehicle routing problem with time windows. In: Zbornik radova: SYM-OP-IS 2008, pp. 305–308. Beograd (2008b)
15. Čičková, Z., Brezina, I.: An evolutionary approach for solving vehicle routing problem. In: Quantitative Methods in Economics Multiple Criteria Decision Making XIV, pp. 40–44. Bratislava (2008)
16. Čičková, Z., Brezina, I.: SOMA application to the travelling salesman problem. In: 24th International Conference on Mathematical Methods in Economics 2006, pp. 117–121. Plzeň (2006)
17. Čičková, Z., Brezina, I., Pekár, J.: Alternative method for solving traveling salesman problem by evolutionary algorithm. Manag. Inf. Syst. **1**, 17–22 (2008)
18. Čičková, Z., Brezina, I., Pekár, J.: Solving the real-life vehicle routing problem with time windows using self organizing migrating algorithm. Ekonomický časopis **61**, 497–513 (2013)
19. Cook, W.J.: In Pursuit of the Traveling Salesman: Mathematics at the Limits of Computation. Princeton University Press, Princeton (2012)
20. Cordeau, J.F., Desaulniers, G., Desrosiers, J., Solomon, M.M., Soumis, F.: The VRP with time windows. In: Toth, P., Vigo, D. (eds.) The Vehicle Routing Problem. SIAM Monographs on Discrete Mathematics and Applications, vol. 9, pp. 157–193 (2002)
21. Craenen, B.G.W., Eiben, A.E., Hemert, J.I.: Comparing evolutionary algorithms on binary constraint satisfaction problems. IEEE Trans. Evol. Comput. **7**, 424–444 (2003)
22. Croes, G.A.: A method for solving traveling salesman problems. Oper. Res. **6**, 791–812 (1958)
23. Dantzig, G., Fulkerson, R., Johnson, S.: Solution of a large—scale traveling-salesman problem. J. Oper. Res. Soc. Am. **2**, 393–410 (1954)
24. Desrosiers, J.Y., Dumas, Y., Solomon, M.M., Sournis, E.: Time constrained routing and scheduling. Network routing. Handb. Oper. Res. Manag. Sci. **8**, 35–139 (1995)
25. Gendreau, M., Hertz, A., Laporte, G., Stan, M.: A generalized insertion heuristic for the traveling salesman problem with time windows. Oper. Res. **46**, 330–335 (1998)
26. Ghoseiri, K., Ghannadpour, S.F.: Multi-objective vehicle routing problem with time windows using goal programming and genetic algorithm. Appl. Soft Comput. **10**, 1096–1107 (2010)
27. Grötschel, M., Holland, O.: Solution of large-scale symmetric travelling salesman problems. Math. Program. **51**, 141–202 (1991)

28. Hamida, S.B., Petrowski, A.: The need for improving the exploration operators for constrained optimization problems. In: Proceedings of the IEEE Congress on Evolutionary Computation, CEC-2000, pp. 1176–1183 (2000)
29. Helsgaun, K.: An effective implementation of the Lin-Kernighan traveling salesman heuristic. Eur. J. Oper. Res. **126**, 106–130 (2000)
30. Jr Brezina, I., Čičková, Z.: Solving the travelling salesman problem using the ant colony optimization. Manage. Inf. Syst. **6**, 10–14 (2011)
31. Lin, S., Kernighan, B.: An efficient heuristic for the traveling salesman problem. Oper. Res. **21**, 498–516 (1973)
32. Little, J.D.C., Murty, K., Sweeney, D.W., Karel, C.: An algorithm for the traveling salesman problem. Oper. Res. **11**, 972–989 (1963)
33. Mařík, V., Štěpánková, O., Lažanský, J.: Umělá Intelligence, vol. 3. Academia, Praha (2001)
34. Michalewicz, Z., Attia, N.: Evolutionary optimization of constrained problems. In: Proceeding of the 3rd Annual Conference on Evolutionary Programming, pp. 98–108 (1994)
35. Miller, C.E., Tucker, A.W., Zemlin, R.A.: Integer programming formulation of traveling salesman problems. J. ACM (JACM) **7**, 326–329 (1960)
36. Ombuki, B., Ross, B.J., Hanshar, F.: Multi-objective genetic algorithms for vehicle routing problem with time windows. Appl. Intell. **24**, 17–30 (2006)
37. Onwubolu, G.C., Babu, B.V.: New Optimization Techniques in Engineering. Springer, Berlin (2004)
38. Padberg, M., Rinaldi, G.: A branch-and-cut algorithm for the resolution of large-scale symmetric traveling salesman problems. SIAM Rev. **33**, 60–100 (1991)
39. Palúch, S., Peško, Š.: Kvantitatívne metódy v logistike. EDIS, Žilina (2006)
40. Peško, Š.: Pyramídová metóda pre úlohu obchodného cestujúceho. Komunikácie **4**, 29–34 (2000)
41. Peško, Š.: Differential evolution for small TSPs with constraints. In: Proceedings of the Fourth International Scientific Conference: Challenges in Transport and Communications, Part III, pp. 989–994. Pardubice (2006)
42. Poot, A., Kant, G., Wagelmans, A.P.M.: A savings based method for real-life vehicle routing problems. J. Oper. Res. Soc. **53**, 57–68 (2002)
43. Potvin, J.Y., Bengio, S.: The vehicle routing problem with time windows, Part II. INFORMS J. Comput. **8**, 165–172 (1996)
44. Potvin, J.Y., Rousseau, J.M.: A parallel route building algorithm for the vehicle routing and scheduling problem with time windows. Eur. J. Oper. Res. **66**, 331–340 (1993)
45. Robacker, J.T.: Some experiments on the traveling-salesman problem. RAND Research Memorandum RM-1521 (1955)
46. Rochat, Y., Taillard, D.: Probabilistic diversification and intensification in local search for vehicle routing. J. Heuristics **1**, 147–167 (1995)
47. Russell, R.A., Urban, T.L.: Vehicle routing problem with soft time windows and Erlang travel times. J. Oper. Res. Soc. **59**, 1220–1228 (2008)
48. Savelsbergh, M.W.P.: Local search in routing problems with time windows. Ann. Oper. Res. **4**, 285–305 (1985)
49. Smith, A., Tate, D.: Genetic optimization using a penalty functions. In: Proceeding of the 5th International Conference on Genetic Algorithms, pp. 499–503 (1993)
50. Solomon, M.M.: Algorithms for the vehicle routing and scheduling problems with time window constraints. Oper. Res. **35**, 254–265 (1987)
51. Thangiah, S.R.: Hybrid genetic algorithm, simulated annealing and tabu search heuristic for vehicle routing problems with time windows. In: Practical Handbook of Genetic Algorithms: Complex Coding Systems, vol III. CRC Press, Florida (1999)
52. Tung, D.V., Pinnoi, A.: Vehicle routing-scheduling for waste collection in Hanoi. Eur. J. Oper. Res. **125**, 449–468 (2000)
53. Venkatraman, S., Yen, G.: A generic framework for constrained optimization using genetic algorithms. IEEE Trans. Evol. Comput. **9**, 424–435 (2005)

54. Weigel, D., Cao, B.: Applying GIS and OR techniques to solve sears technician dispatching and home delivery problems. Interfaces **30**, 112–130 (1999)
55. Zelinka, I.: Evolutionary identification of predictive models. Nostradamus **99**, 114–122 (1999)
56. Zelinka, I.: Umělá intelligence v problémech globální optimalizace. BEN-technická literatúra, Praha (2002)
57. Zhang, Y., Tang, L.: Solving prize-collecting traveling salesman problem with time windows by chaotic neural network. Lect. Notes Comput. Sci. **4492**, 63–71 (2007)

SOMA in Financial Modeling

Juraj Pekár, Zuzana Čičková and Ivan Brezina

Abstract The basic problem in portfolio theory (based on Markowitz theory) is the selection of an appropriate mix of assets in a portfolio in order to maximize portfolio expected return and subsequently to minimize portfolio risk. Another approach takes into account portfolio performance expressed by various measurement techniques e.g. Sharpe ratio, Treynor ratio, Jensen's alpha, Information ratio, Sortino ratio, Omega function and the Sharpe Omega ratio that are focused to determine the allocation of the available resources in the selected group of assets. This chapter presents an alternative approach to the computation of weights of assets in portfolio based on the nonlinear measure techniques: Sortino ratio and Omega function. The proposed alternative includes principle of self-organizing migrating algorithm (SOMA). The experiments are set up on assets included in Dow Jones Industrial Index. Presented original approach lends itself also to other evolutionary techniques in the area of portfolio selection based on different measurement techniques.

1 Introduction

Strategic planning surely includes the process of deciding how to commit resources across lines of business. The financial side of strategic planning allocates a particular resource, capital. Finance theory has made major advances in understanding

---

J. Pekár (✉) · Z. Čičková · I. Brezina
Department of Operations Research and Econometrics, University of Economics in Bratislava, Dolnozemská cesta 1/b, 852 35 Bratislava, Slovakia
e-mail: juraj.pekar@euba.sk

Z. Čičková
e-mail: cickova@euba.sk

I. Brezina
e-mail: brezina@euba.sk

D. Davendra and I. Zelinka (eds.), *Self-Organizing Migrating Algorithm*, Studies in Computational Intelligence 626, DOI 10.1007/978-3-319-28161-2_11

how capital markets work and how risky financial assets are valued and many tools derived from finance theory are widely used in practice.

The important part of finance theory is a process of portfolio selection. The basic model in portfolio theory is known as Markowitz model focused on selection of an appropriate mix of assets in a portfolio in order to maximize portfolio expected return and subsequently to minimize investment risk [16]. Decision process on the selection of the portfolio assets can be also supported by various mathematical models e.g. [1, 10, 28]. Other group of approaches is based on maximizing the portfolio performance by using various measurement techniques [20] e.g. Sharpe ratio [24], Treynor ratio [27], Jensen's alpha [11], Information ratio [9], Sortino ratio [25], or Omega function [13]. This chapter presents the possibility of portfolio selection based on computing the Sortino ratio and Omega function. In general their computability can be difficult due to substandard structures of performance level (the nature of objective function) and therefore the use of standard techniques seems to be relatively complicated. Alternative procedures may include the principle of evolutionary algorithm that can be generally considered to be effective tool used for maximizing different performance measures in financial modeling [21].

The chapter is divided into the following interrelated parts. The brief view on finance theory as well as the motivation for the use of alternative computational techniques for determining the portfolio selection based on performance measurement is presented in the introduction. In the second part the authors present the basic lines in finance theory, namely Markowitz model used to determine effective portfolio selection and also the well-known performance measurement techniques aimed to identify the most valued portfolio. This part is mainly devoted to the wide used non-linear measurement techniques known as Sortino ratio and Omega function. Core part of the chapter is devoted to the modification of self-organizing migrating algorithm (SOMA) to solve the portfolio selection problem based on before mentioned measures. The empirical analysis is provided on the Dow Jones Industrial Index base, one of the major market indexes, as well as one of the most popular indicators of the U.S. market. The historical data published in period from January 2nd 2013 to March 31st 2014 on weekly basis were used. The experiments were divided in two parts. At first, the simulations were generated on the basis of exogenously given levels of parameters and some statistical methods were applied in order to provide analysis to determine the impact of control parameters of algorithm. At second, the best identified values of parameters are used for further experiments. Verification of presented considerations is performed based on calculation using professional system for mathematical modeling.

2 Portfolio Theory

As the cornerstone of portfolio theory may be mentioned Markowitz model created in 1952 [16], which is considered the beginning of modern portfolio theory. Markowitz contributed to the theory of financial markets, namely to portfolio selection under conditions of uncertainty. His approach has shown how it is possible to reduce multidimensional problem of investment to a large number of assets (each of different characteristics) under conditions of uncertainty to the problem of the relationship between only two elements: the expected portfolio return and its variance. The model allows determining the portfolio diversification that should maximize the expected return subsequently with minimization of its variance.

2.1 *Markowitz Model*

The classical Markowitz model has been widely recognized. Its general formulation allows determining of portfolio selection based on expected portfolio return and its variance based on the quadratic programming problem formulation. The problem can be briefly described as follows: Consider a portfolio of n assets. Let w_i, $i = 1, 2, \ldots n$ be the variables indicating weights or units of funds that an investor allocates to the ith asset in the portfolio, so that $\sum_{i=1}^{n} w_i = 1$. Let E_i, $i = 1, 2, \ldots n$ be the expected return of the ith asset in the portfolio and let parameters σ_{ij}, $i, j = 1, 2, \ldots n$, represents the covariance between the ith and jth assets, if $i \neq j$ (respectively variance when $i = j$). It is also required that no asset return is perfectly correlated with the return of the portfolio constructed from the remaining assets and that none of the assets nor portfolios are risk-free. Then the variance of portfolio return σ_P^2 can be defined as $\sigma_P^2 = \sum_{i=1}^{n} \sum_{j=1}^{n} w_i \sigma_{ij} w_j$ and the expected return of portfolio E_p can be characterized by formula $E_p = \sum_{i=1}^{n} E_i w_i$.

Markowitz formulation can be written as a multi-criteria programming problem, where the objective is to achieve maximal expected return with the minimal risk:

$$\min \sigma_P^2 = \sum_{i=1}^{n} \sum_{j=1}^{n} w_i \sigma_{ij} w_j \quad (1)$$

$$\max E_p = \sum_{i=1}^{n} E_i w_i \quad (2)$$

$$\sum_{i=1}^{n} w_i = 1 \quad (3)$$

The use of presented model allows generating the set of effective portfolios from the set of feasible portfolios that are called efficient portfolios in the space of expected return and its variance.

To obtain efficient portfolios it is necessary to solve sub-problems that can be formulated in the terms of quadratic programming (quadratic programming problems [8] can be addressed effectively by using appropriate software for optimization calculations). Each of them can be formulated as follows: finding the portfolio with the minimal risk successively for different values of the expected return of the portfolio E_p, formally:

$$\min \sigma_P^2 = \sum_{i=1}^{n} \sum_{j=1}^{n} w_i \sigma_{ij} w_j \tag{4}$$

$$\sum_{i=1}^{n} E_i w_i = E_p \tag{5}$$

$$\sum_{i=1}^{n} w_i = 1 \tag{6}$$

Equation (5) defines the search reaches the specified portfolio return, the second equation require investing just full amount of investment. The objective (4) is to minimize the risk of the portfolio.

Markowitz theory from 1952 was criticized, mainly the use of the concept of risk through investment variance and covariance between investments due to the fact that the variance is measured as dispersion of expected asset returns and deemed income below and above the expected return for the same [16]. Markowitz admitted the limitations of the model and in 1959 he has proposed a new rate lower partial risk, which measures the risk under the expected return [17]. He called it a semi-variance. His work was followed up many authors who deal with finding the appropriate level of risk, e.g. [14, 22].

2.2 *Portfolio Performance Measurement Techniques*

Investment performance measurement is the quantification of the results obtained by the chosen strategy. It is basically statistical recapitulation of rate of return and the estimation of the risk incurred. Portfolio performance measurement techniques can be considered to be a tool capable of capturing various risk characteristics unlike Markowitz's theories.

The relevant literature aimed on measuring portfolio performance goes back to the beginning of the Capital Asset Pricing Model (CAPM) developed by Sharpe in 1964 [23]. Its basis is one of the first performance techniques known as the Sharpe Ratio. Since the first measurement techniques, also called classical performance techniques (Jensen, Treynor, Sharpe), a number of other portfolio performance techniques have been proposed. One of the main disadvantages of those techniques is that they are not able to capture all characteristics of the time series of returns. Alternative performance techniques addressing this lack are essentially modifications of classical performance techniques. One of the first modifications is a Sortino ratio, which is an alternative to the Sharpe ratio [26]. Omega function is included among the latest and most progressive performance techniques. This level of performance technique is considered to be a new tool for financial analysis utilizing full information that time series of returns provides [13].

Sharpe ratio

Consider investment with known expected return and known amount of risk (measured e.g. by standard deviation). Sharpe ratio (*SP*) is determined as a proportion of the asset return less the risk-free rate and standard deviation of the returns [24]:

$$SP = \frac{E_p - R_F}{\sigma_p} \tag{7}$$

where

E_p denotes portfolio expected return,
R_F denotes risk-free rate,
σ_p denotes standard deviation of portfolio return.

Sharpe ratio is a measurement technique that takes into account the risk profile of the investment, namely the average performance of the asset above the risk-free asset return. The higher the value of the indicator returned is, the higher performance of the portfolio is (relative to its risk).

Treynor ratio

Treynor ratio (*TP*) is essentially similar to the Sharpe ratio. The relative performance of the portfolio is in contrast to the Sharpe ratio measured as the share of the additional expected return and risk (β_p) [27]:

$$TP = \frac{E_P - R_F}{\beta_P} \tag{8}$$

where

R_F denotes return of risk-free asset,
E_p denotes expected return of portfolio,
β_p is the riskiness of the return measured by systematic risk [24].

Treynor ratio is thus achieved rate of return over the risk-free investment with respect to the unit of risk. The higher the value is, the better investment is.

Sortino ratio

The Sortino ratio [25] is based on the known Sharpe ratio, but though both ratios measure an investment's risk-adjusted returns, they do so in significantly different ways. The formula for the Sharpe ratio based on standard deviation (simply the square root of variance) allows to measure a non-directionally-biased measurement of volatility to adjust the risk and therefore penalizes both upside and downside volatility equally. This concept has been criticized and many investors prefer its modification known as Sortino ratio. Instead of using standard deviation, the Sortino ratio uses downside semi-variance and penalizes only those returns falling below a user-specified rate. This is a measurement of return deviation below a minimal acceptable rate. By utilizing this value, the Sortino ratio only penalizes for “harmful” volatility.

Sortino ratio calculation assumes the existence of historical data on portfolio return for T periods, so that for each period t ($t = 1, 2, \ldots T$) the portfolio return R_t is known. The principle of the calculation is similar to the variance calculation except the fact that only returns below the expected return are taken into account. Lower semi-variance is calculated as follows [26]:

$$\frac{1}{T}\sum_{t=1}^{T} \max\left((E_p - R_t, 0)\right)^2 \tag{9}$$

where

R_t denotes the portfolio return in corresponding sub-period t ($t = 1, 2, \ldots T$),
E_p denotes expected return of portfolio,
T denotes number of sub-periods.

Sortino ratio uses specified value of minimal expected return (MAR) that is exogenously set by investor. Thus, Sortino ratio (SR) is calculated by the formula [25]:

$$SR(MAR) = \frac{E_p - MAR}{\sqrt{\frac{1}{T}\sum_{t=1}^{T}\left(\max(MAR - R_t, 0)\right)^2}} \tag{10}$$

where

T represents the number of periods,
MAR denotes the minimum acceptable return,
R_t denotes the return of portfolio in tth period, where $t = 1, 2, \ldots T$,
E_p denotes the expected return of portfolio.

Omega function

Omega function is measurement which incorporates all the distributional characteristics of a return series. The measurement is a function of the returns leveled and requires non parametric assumption on the distribution. Precisely, it considers the returns below and above a specific minimum acceptable return and provides a ratio of total probability weighted losses and gains that fully describe the risk reward properties of the distribution [13]:

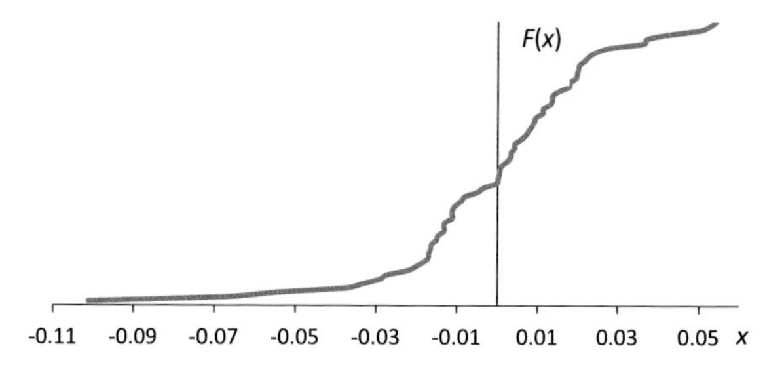

Fig. 1 Example of cumulative distribution function based on IBM asset

$$\Omega(MAR) = \frac{\int_{MAR}^{\infty} (1 - F(x))dx}{\int_{-\infty}^{MAR} F(x)dx} \tag{11}$$

where

MAR denotes the minimum acceptable return,
x variable represents asset returns,
F(*x*) represents the cumulative distribution function of asset returns.

Example of cumulative distribution function of IBM asset is given in Fig. 1, specifically its historical data published from January 2nd 2013 to March 31st 2014 on weekly basis were used.

Before mentioned formula involves also the use of discrete data (historical time series) [12]:

$$\Omega(MAR) = \frac{\sum_{t=1}^{T} \max(R_t - MAR, 0)}{\sum_{t=1}^{T} \max(MAR - R_t, 0)} \tag{12}$$

where

T represents the number of periods,
MAR denotes the minimum acceptable return,
R_t denotes the return of asset in *t*th period, where $t = 1, 2, \ldots T$.

2.2.1 Portfolio Selection by Maximizing Sortino Ratio

The formulation of the portfolio selection problem based on Sortino ratio includes the maximization of performance measurement function (10). The model deals with the variables w_i, $i = 1, 2, \ldots n$ (where *n* represents the number of assets) that represent the weights of each asset in the portfolio. Corresponding problem can be formulated as follows [26]:

$$\max SR(\mathbf{w}) = \frac{\sum_{i=1}^{n} E_i w_i - MAR}{\sqrt{\frac{1}{T}\sum_{t=1}^{T}\left(\max\left(MAR - \sum_{i=1}^{n} r_{ti} w_i, 0\right)\right)^2}} \tag{13}$$

$$\sum_{i=1}^{n} w_i = 1 \tag{14}$$

$$w_i \geq 0, \quad i = 1, 2, \ldots n \tag{15}$$

where

T represents the number of periods,
MAR denotes the minimum acceptable return,
r_{ti} denotes the return of ith portfolio asset in tth period, where t = 1, 2, … T, i = 1, 2, … n,
E_i denotes the expected returns of ith portfolio asset, i = 1, 2, … n.

2.2.2 Portfolio Selection by Maximizing Omega Function

As it was mentioned, the Omega function involves consideration of all the information contained in the time series of returns. The aim of portfolio selection problem is to maximize the level of Omega performance measurement, where the variables w_i, i = 1, 2, … n (where n represents the number of assets) represent the weights of each asset in the portfolio. Corresponding problem can be formulated as follows [3]:

$$\max \Omega(\mathbf{w}) = \frac{\sum_{t=1}^{T} \max\left(\sum_{i=1}^{n} r_{ti} w_i - MAR, 0\right)}{\sum_{t=1}^{T} \max\left(MAR - \sum_{i=1}^{n} r_{ti} w_i, 0\right)} \tag{16}$$

$$\sum_{i=1}^{n} w_i = 1 \tag{17}$$

$$w_i \geq 0, \quad i = 1, 2, \ldots n \tag{18}$$

where

T represents the number of periods,
MAR denotes the minimum acceptable return,
r_{ti} denotes the return of ith portfolio asset in tth period, where t = 1, 2, … T, i = 1, 2, … n.

3 SOMA for Maximizing Performance Measurement

The computational complexity of the problems based on presented performance measurement arises from its non-linear structure. Therefore, evolutionary algorithms seem to be a suitable alternative to standard techniques, due to their ability to achieve the suboptimal solutions in relatively short time.

Nowadays evolutionary algorithms are considered to be effective tools which can be used to search for solutions of a wide variety of optimization problems (e.g. [2, 4, 7, 18, 19, 29]). The big advantage over traditional methods is that they are designed to find global extremes (with built-in stochastic component) and that their use does not require a priori knowledge of optimized function (convexity, differential etc.), and in that way they work well to solve continuous non-linear problems, where is hard to use traditional mathematic methods.

In order to apply SOMA for solving problems of the portfolio selection based on performance measurement, it is necessary to consider the following factors: *selection of an appropriate representation of individual, transformation of unfeasible solutions, setting of the control parameters.*

Self-organizing migrating algorithm (SOMA) involves a search on population of individuals (PopSize—number of individuals), where each individual represents one candidate solution for the given problem. Particular candidate solution is represented by parameters of individual (Dim—number of parameters of individual). Also the fitness representing the relevant value of objective function (f_{cost}) is associated with each individual. Every step of the algorithm involves a competitive selection that carried out poor solutions. The steps of the algorithm are described in e.g. [18, 30].

Selection of an appropriate representation of individual. Considering problem of portfolio selection it could be appropriate to use following representation: let ml be the index corresponded with number of migrating loops, so that $ml = 0, 1, \ldots$ Migrations, where the parameter Migrations represents the maximum number of iterations. Each population is represented by matrix $\mathbf{W}^{(ml)}$, which consist of PopSize individuals (the parameter PopSize represents the number of individuals in the population) represented by vectors $\mathbf{w}_i^{(ml)}$, $i = 1, 2, \ldots \text{PopSize}$. Each parameter of individual represents the corresponding weight of the asset in the portfolio $w_{i,j}^{(ml)}$, $i = 1, 2, \ldots \text{PopSize}, j = 1, 2, \ldots \text{Dim}$ (the parameter Dim represents the number of assets). The use of mentioned representation involves the easy calculation of fitness ($f_{cost}(\mathrm{w}_i^{(ml)})$, $i = 1, 2, \ldots$ PopSize), on the base of Sortino ratio (13)–(15) as well as Omega function (16)–(18).

The population $\mathbf{W}^{(0)}$ can be randomly initialized at the beginning of evolutionary process according to the rule:

$$w_{ij}^{begin} = rnd\langle 0, 1\rangle, \quad i = 1, 2, \ldots \text{Popsize}, \; j = 1, 2, \ldots \text{Dim},$$

$$w_{ij}^{(0)} = \frac{w_{ij}^{begin}}{\sum_{l=1}^{\text{Dim}} w_{il}^{begin}}, \quad i = 1, 2, \ldots \text{PopSize}, \quad j = 1, 2, \ldots \text{Dim},$$

that ensure that the total weights of portfolio (for each individual) is equal to one. Each individual is then evaluated with the $f_{cost}(\mathbf{w}_i^{(0)})$, i = 1, 2, … PopSize.

Transformation of unfeasible solutions. The following rule can be used to ensure the feasibility of solution: Let $w_{ij}^{test(ml)}$ represent the *j*th parameter of *i*th individual after one jump in the process of moving in *ml*th migration loop, i = 1, 2, … PopSize, j = 1, 2, … Dim, ml = 1, 2, … Migrations, $test = \langle 0$, by *step* to, $mass\rangle$. Then, the following rule can be applied: if $w_{ij}^{test(ml)} < 0$, then $w_{ij}^{test(ml)} = rnd\langle 0, 1\rangle$ and then $w_{ij}^{(ml)} = \frac{w_{ij}^{test(ml)}}{\sum_{l=1}^{n} w_{il}^{test(ml)}}$.

Setting of the control parameters. Recommended values for the parameters are usually derived empirically from experiments [15, 18, 30], or one can apply some statistical methods [6].

4 Empirical Results

The portfolio analysis was based on Dow Jones Industrial Average (DJIA), which is one of the major market indexes, as well as one of the most popular indicators of the U.S. market. It is a stock market index, and one of several indices created by Wall Street Journal editor and Dow Jones & Company co-founder Charles Dow. It was founded on May 26, 1896. It is an index that shows how 30 large publicly owned companies based in the United States have traded during a standard trading session in the stock market (so called Large-Cap companies—companies with market capitalization above 10 billion USD). Data[1] are processed weekly for the period January 2nd 2013 to March 31st 2014. Time series consist of 65 data set (for each company in Table 1).

The input parameter of *MAR* (the target of required rate of return) was set to 0.005.

The algorithms were implemented in MatLab 8.3. All the experiments were run on PC with Intel® Core™ i7-3770 CPU with a frequency of 3.40 GHz and 8 GB of RAM under MS Windows 8.

[1]http://finance.yahoo.com/ (2014).

Table 1 Company overview DJIA

No.	Company name	Symbol	No.	Company name	Symbol	No.	Company name	Symbol
1.	3M	MMM	11.	Goldman Sachs	GS	21.	Pfizer	PFE
2.	American Express	AXP	12.	Chevron	CVX	22.	Procter & Gamble	PG
3.	AT&T	T	13.	IBM	IBM	23.	The Home Depot	HD
4.	Boeing	BA	14.	Intel	INTC	24.	Travelers	TRV
5.	Caterpillar	CAT	15.	Johnson & Johnson	JNJ	25.	United Technologies	UTX
6.	Cisco Systems	CSCO	16.	JPMorgan Chase	JPM	26.	UnitedHealth Group	UNH
7.	Coca-Cola	KO	17.	McDonald's	MCD	27.	Verizon	VZ
8.	DuPont	DD	18.	Merck	MRK	28.	Visa	V
9.	ExxonMobil	XOM	19.	Microsoft	MSFT	29.	Wal-Mart	WMT
10.	General Electric	GE	20.	Nike	NKE	30.	Walt Disney	DIS

Empirical results—Portfolio Selection by Maximizing Sortino ratio and Omega Function using SOMA

A disadvantage of SOMA, as well as of other evolutionary approaches, is that it has a dependence on the control parameter setting. Due to this fact, the tests were done on above mentioned data in effort to determine effective settings of the parameters. The tested values of parameters PRT and Step were set as sequence of levels 0.1, 0.2, 0.3, 0.4, 0.5, 0.6, 0.7, 0.8, 0.9 (PopSize = 250, Migrations = 100, PathLength = 3). The interval limits were not considered during testing (purely deterministic and purely stochastic nature of the algorithm). Twenty experiments were conducted for each combination of pairs. The average values of Sortino ratio *E*

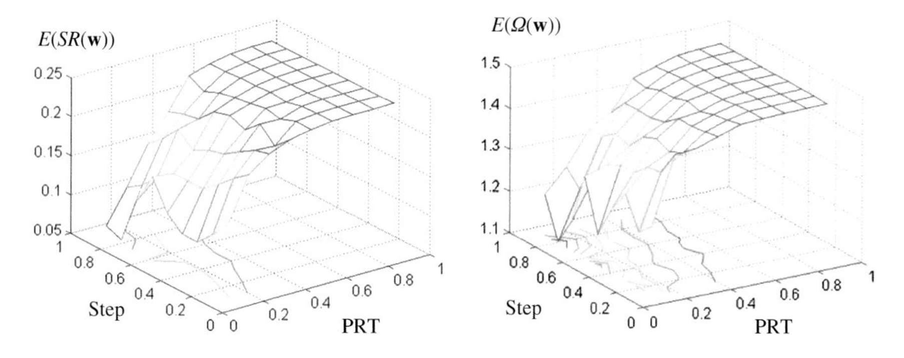

Fig. 2 Average value of $SR(\mathbf{w})$ and $\Omega(\mathbf{w})$ depending on different levels of parameters PRT and step

Table 2 Summary statistics for $SR(\mathbf{w})$ for parameter step

Step	Count	Average	Standard deviation	Minimum	Maximum	Stnd. skewness	Stnd. kurtosis
0.1	180	0.221941	0.0456411	−0.0149	0.2413	−20.875	40.3173
0.2	180	0.215258	0.0536906	−0.0147	0.2412	−16.29	23.4428
0.3	180	0.210797	0.0603353	−0.009	0.2412	−14.1648	15.8944
0.4	180	0.205605	0.0613185	−0.0246	0.2411	−12.4814	12.0825
0.5	180	0.207647	0.059782	−0.0159	0.2411	−13.3403	13.961
0.6	180	0.209214	0.05022	−0.0185	0.241	−12.4675	14.0676
0.7	180	0.199938	0.0667196	−0.2211	0.241	−14.7371	26.826
0.8	180	0.189333	0.0734058	−0.0299	0.2412	−8.46233	3.20589
0.9	180	0.187365	0.0768724	−0.0172	0.241	−8.10041	2.3545
Total	1620	0.205233	0.0624281	−0.2211	0.2413	−38.0527	41.073

($SR(\mathbf{w})$) and Omega function $E(\Omega(\mathbf{w}))$ depended on the combination of parameters PRT and Step are illustrated in Fig. 2.

The aim of the simulation was to determine the influence of parameters PRT and Step on the value of functions $SR(\mathbf{w})$ and $\Omega(\mathbf{w})$. This simulation study is set up in accordance with work [6], which describes the possibility of identify the effective parameters setting applying some statistical methods e.g. Kruskal-Wallis test, Bartlett's test, Cochran-Hartley's test [5].

All statistical tests (at the 95.0 % confidence level) indicate statistically significant differences between groups, so the levels of function $SR(\mathbf{w})$ are statistically different depended on levels of parameters PRT and Step. The descriptive statistics (Table 2) demonstrated the effectiveness of the Step on the level 0.1 (the groups with Step = 0.1, 0.2, 0.3 and 0.6 are considered to be homogeneous groups). The effective setting of parameter PRT is on 0.9 level (Table 3), while the homogeneous groups are groups with PRT = 0.6, 0.7, 0.8, 0.9.

Table 3 Summary statistics for $SR(\mathbf{w})$ for parameter PRT

PRT	Count	Average	Standard deviation	Minimum	Maximum	Stnd. skewness	Stnd. kurtosis
0.1	180	0.106288	0.0898629	−0.2211	0.2299	−2.00703	−2.10431
0.2	180	0.152956	0.0719439	0.0001	0.2333	−4.39469	−2.06396
0.3	180	0.18937	0.0490171	0.0167	0.2403	−8.74919	5.5263
0.4	180	0.211805	0.0345809	0.0397	0.2404	−14.6493	22.1466
0.5	180	0.228938	0.0139682	0.1375	0.241	−13.9828	28.1789
0.6	180	0.236603	0.00549319	0.1981	0.2412	−18.4169	44.6871
0.7	180	0.239544	0.00158376	0.2308	0.2412	−11.4224	17.9189
0.8	180	0.240688	0.000444162	0.2381	0.2412	−12.3945	24.0851
0.9	180	0.240906	0.000201437	0.2404	0.2413	−2.27972	−0.54334
Total	1620	0.205233	0.0624281	−0.2211	0.2413	−38.0527	41.073

Table 4 Summary statistics for $\Omega(\mathbf{w})$ for parameter step

Step	Count	Average	Standard deviation	Minimum	Maximum	Stnd. skewness	Stnd. kurtosis
0.1	180	1.43933	0.0832021	0.9629	1.4767	−21.7065	46.5103
0.2	180	1.42408	0.105525	0.9604	1.4767	−16.7791	24.8935
0.3	180	1.42857	0.0804901	0.9797	1.4766	−16.2926	28.3195
0.4	180	1.41827	0.0987719	0.9913	1.4764	−14.7539	19.683
0.5	180	1.40544	0.122302	0.9691	1.476	−13.0857	13.0265
0.6	180	1.41021	0.0938492	1.0006	1.4763	−11.4204	11.7889
0.7	180	1.39128	0.126901	0.9405	1.4759	−10.2244	7.17148
0.8	180	1.37422	0.142082	0.9559	1.4759	−8.49313	3.52962
0.9	180	1.38442	0.128613	0.9352	1.4762	−9.78758	6.60925
Total	1620	1.40842	0.112596	0.9352	1.4767	−38.513	40.723

Table 5 Summary statistics for $\Omega(\mathbf{w})$ for parameter PRT

PRT	Count	Average	Standard deviation	Minimum	Maximum	Stnd. skewness	Stnd. kurtosis
0.1	180	1.22368	0.164366	0.9352	1.4525	−1.58435	−3.97309
0.2	180	1.31918	0.123977	0.9818	1.4647	−5.43922	−0.440709
0.3	180	1.3776	0.0859898	1.0638	1.467	−9.81702	8.17111
0.4	180	1.41895	0.0541002	1.1684	1.4753	−12.198	14.9324
0.5	180	1.44959	0.0238172	1.3154	1.4762	−13.9346	25.5765
0.6	180	1.46468	0.00976054	1.4075	1.4762	−11.0862	19.4977
0.7	180	1.47174	0.00346256	1.4581	1.4766	−7.7651	7.92555
0.8	180	1.47482	0.00159083	1.4608	1.4767	−23.4348	90.8334
0.9	180	1.47559	0.000639299	1.4714	1.4767	−9.9662	25.1603
Total	1620	1.40842	0.112596	0.9352	1.4767	−38.513	40.723

The statistical tests also indicate statistically significant differences of the levels of function $\Omega(\mathbf{w})$ between the groups given by levels of parameters PRT and Step. The descriptive statistics (Table 4) demonstrated the effectiveness of the Step also on the level 0.1 (the groups with Step = 0.1, 0.2, 0.3 and 0.4 are considered to be homogeneous groups). The effective level of parameter PRT can be set to 0.9 (Table 5), while the homogeneous groups are groups with PRT = 0.6, 0.7, 0.8, 0.9.

Further simulation study was provided with the following settings of control parameters: Step = 0.1, PRT = 0.9, PathLength = 3, PopSize = 3000 and Migrations = 2000 in both of cases. Based on the testing parameters the problems were ten times re-solved. The results are shown in Table 6. The best obtained value of Sortino ratio was 0.24128962 and of Omega functions was 1.47687412 (convergence of the solutions can be seen in Fig. 3).

Table 6 Solutions values

$f_{cost}(\mathbf{w})$	$SR(\mathbf{w})$	$\Omega(\mathbf{w})$
Solution 1	0.24128374	1.47683268
Solution 2	0.24128374	1.47682676
Solution 3	0.24128041	1.47679080
Solution 4	0.24128319	1.47685126
Solution 5	0.24128368	1.47679642
Solution 6	0.24128374	1.47686187
Solution 7	0.24128041	1.47684886
Solution 8	0.24128319	1.47685277
Solution 9	0.24128368	1.47685322
Solution 10	0.24128374	1.47681222
MAX	0.24128374	1.47686187
MIN	0.24128041	1.47679080
Average	0.24128295	1.47683269

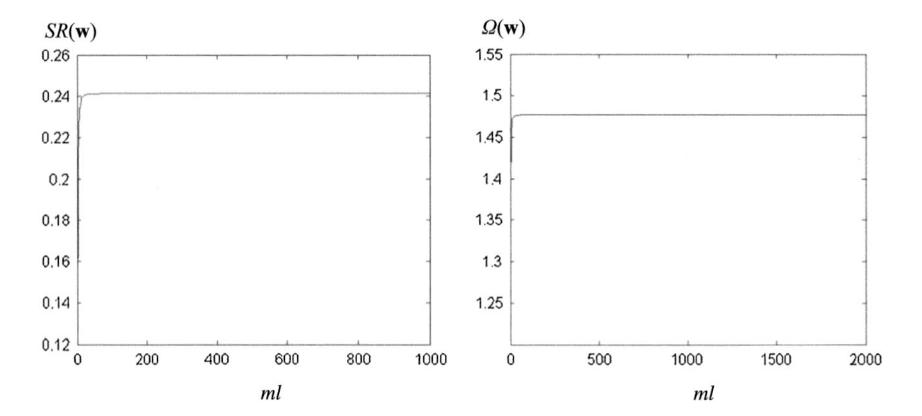

Fig. 3 Convergence of $SR(\mathbf{w})$ and $\Omega(\mathbf{w})$

The rapid convergence to the final solution at the beginning of evolutionary process is evidently seen from the Fig. 3.

The problem on the base its mathematical programming formulations (13)–(15) and (16)–(18) were also solved by software GAMS (solver CONOPT 3, version 3.14U). The best solution (from Table 6) using self-organizing migrating algorithm (SOMA) is compared to the result obtained by GAMS. Both of solutions are presented in Table 7 in detail (value of weight of corresponding asset).

According to presented results it is recommended to invest in assets of American Express (AXP), Boeing (BA), Walt Disney (DIS), Microsoft (MSFT), UnitedHealth Group (UNH) at rates presented in Table 7. The values of weights of others companies are equal to 0 %. The value of Sortino ratio and Omega function of that portfolio diversification is provided in the last row of Table 7.

Table 7 Final portfolio diversification

	Sortino ratio $SR(\mathbf{w})$		Omega function $\Omega(\mathbf{w})$	
Symbol of co.	GAMS/CONOPT	SOMA	GAMS/CONOPT	SOMA
AXP	0.0499	0.0498	0.2304	0.2295
BA	0.4029	0.4029	0.3840	0.3754
CAT	0	0	0	0
CSCO	0	0	0	0
CVX	0	0	0	0
DD	0	0	0	0
DIS	0.0165	0.0165	0.0168	0.0218
GE	0	0	0	0
GS	0	0	0	0
HD	0	0	0	0
IBM	0	0	0	0
INTC	0	0	0	0
JNJ	0	0	0	0
JPM	0	0	0	0
KO	0	0	0	0
MCD	0	0	0	0
MMM	0	0	0	0
MRK	0	0	0	0
MSFT	0.2327	0.2327	0.1167	0.1150
NKE	0	0	0	0
PFE	0	0	0	0
PG	0	0	0	0
T	0	0	0	0
TRV	0	0	0	0
UNH	0.2981	0.2980	0.2521	0.2584
UTX	0	0	0	0
V	0	0	0	0
VZ	0	0	0	0
WMT	0	0	0	0
XOM	0	0	0	0
Function values	0.24128962	0.24128374	1.47669447	1.47686187

5 Conclusions

Various sets of performance measurement tools can be used to assist the investor to allocate capital over the number of assets. The chapter is mainly focused on techniques known as Sortino ratio and Omega function with difficult computability due to non-linear structure. The chapter presents original approach enables the use of self-organizing migrating algorithm (SOMA) to maximize mentioned performance measures.

Except the presented overview of the literature devoted to the finance theory and portfolio performance measurement in the first part of article, the core part is focused on SOMA. The basic version of SOMA should provide good performance in the case of a wide variety of functions (e.g. [18, 19]), but on the other hand, if one wants to apply it for solving problem of portfolio selection, it is necessary to consider some additional factors i.e. selection of an appropriate representation of individual, setting of the control parameters, and transformation of unfeasible solutions, therefore the core of article is focused on an approach that enables solving proposed models of portfolio selection. Moreover, the algorithm depends on setting of its control parameters. Their specification could differ depended on solved problem. To discover their effective setting some statistical methods are employed.

The empirical analysis is provided on the base of the known Dow Jones Industrial index, which is one of the major market indexes (its historical data published from January 2nd 2013 to March 31st 2014 on weekly basis were used). The experiments have shown that proposed heuristics gives very good solutions in a reasonable computational time and in that way can help investor to manage the portfolio diversification. Verification of presented considerations is conducted on the base of calculation using professional modeling system GAMS (solver CONOPT 3). Due to authors meaning, presented approach offers effective tool to solve mentioned problems and can be also applied to a wide variety of portfolio selection problems based on different performance measurement.

References

1. Ágoston, K.Cs.: CVaR minimization by the SRA algorithm. Central Eur. J. Oper. Res. **20**, 623–632 (2012)
2. Ardia, D., Boudt, K., Carl, P., Mullen, K.M., Peterson, B.G.: Differential evolution with DEoptim. R J. **3**, 27–34 (2011)
3. Avouyi-Dovi, S., Morin, A., Neto, D.: Optimal asset allocation with omega function. Technical Report, Banque de France (2004)
4. Brezina, I., Čičková, Z., Pekár, J.: Application of evolutionary approach to solving vehicle routing problem with time windows. Econ. Rev. **38**, 529–539 (2009)
5. Chajdiak, J.: Štatistika jednoducho. Statis, Bratislava (2003)
6. Čičková, Z., Brezina, I., Pekár, J.: A memetic algorithm for solving the vehicle routing problem. In: 29th International Conference on Mathematical Methods in Economics 2011, Praha. 2011, pp. 125–128 (2011)
7. Čičková, Z., Brezina, I., Pekár, J.: Solving the real-life vehicle routing problem with time windows using self organizing migrating algorithm. Ekonomicky Casopis **61**(5), 497–513 (2013)
8. Fendek, M.: Nelineárne optimalizačné modely a metódy. EKONÓM, Bratislava (1998)
9. Goodwin, T.H.: The information ratio. In: Investment Performance Measurement: Evaluation and Presenting Results. Wiley, New York (2009)

10. Hasuike, T., Ishii, H.: Probability maximization models for portfolio selection under ambiguity. Central Eur. J. Oper. Res. **17**, 159–180 (2009)
11. Jensen, M.C.: The performance of mutual funds in the period 1945–1964. J. Finance **23**, 389–416 (1968)
12. Kazemi, H., Schneeweis, T., Gupta, B.: Omega as a performance measure. J. Perform. Meas. **8** (3), 16–25 (2004)
13. Keating, C., Shadwick, W.F.: A universal performance measure. J. Perform. Meas. **6**, 59–84 (2002)
14. Konno, H., Yamazaki, H.: Mean-absolute deviation portfolio optimization model and its applications to Tokyo stock market. Manage. Sci. **37**, 519–531 (1991)
15. Mařík, V., Štěpánková, O., Lažanský, J.: Umělá inteligence, vol. 4. Academia, Praha (2003)
16. Markowitz, H.M.: Portfolio selection. J. Finance **7**, 77–91 (1952)
17. Markowitz, H.M.: Portfolio Selection: Efficient Diversification of Investment. Wiley, New York (1959)
18. Onwubolu, G.C., Babu, B.V.: New Optimization Techniques in Engineering. Springer, Berlin-Heidelberg (2004)
19. Onwubolu, G.C., Davendra, D.: Differential Evolution: A Handbook for Global Permutation-Based Combinatorial Optimization. Springer, Berlin-Heidelberg (2009)
20. Pedersen, C.S., Ruddholm-Alfin, T.: Selecting risk-adjusted shareholder performance measure. J. Asset Manage. **4**, 152–172 (2003)
21. Pekár, J., Brezina, I., Čičková, Z., Reiff, M.: Portfolio selection by maximizing omega function using differential evolution. Technol. Investment **4**, 73–77 (2013)
22. Rockafellar, R.T., Uryasev, S.: Optimization of conditional value-at-risk. J. Risk **2**, 21–41 (2000)
23. Sharpe, W.F.: Capital asset prices: a theory of market equilibrium under conditions of risk. J. Finance **19**, 425–442 (1964)
24. Sharpe, W.F.: The sharpe ratio. J. Portfolio Manage. **21**, 49–58 (1994)
25. Sortino, F.A., Meer, R.: Downside risk. J. Portfolio Manage. **17**, 27–31 (1991)
26. Sortino, F.A., Price, L.N.: Performance measurement in a downside risk framework. J. Investing **3**, 59–64 (1994)
27. Treynor, J.L.: How to rate management of investment funds. Harvard Bus. Rev. **43**, 63–75 (1965)
28. Václavík, M., Jablonský, J.: Revisions of modern portfolio theory optimization model. Central Eur. J. Oper. Res. **20**, 473–483 (2012)
29. Zelinka, I.: Evolutionary identification of predictive models. Nostradamus **99**, 114–122 (1999)
30. Zelinka, I.: Umělá inteligence v problémech globální optimalizace. BEN-technická literature, Praha (2002)

Setting of Control Parameters of SOMA on the Base of Statistics

Zuzana Čičková and Martin Lukáčik

Abstract Evolutionary techniques are generally considered to be effective tool for solving a wide range of optimization problems. However, those algorithms are controlled by a special set of parameters according to their type. Control parameters of self-organizing migrating algorithm (SOMA) can be divided into several groups: the stopping parameters, parameters which depended on the type of problem to be solved and finally, parameters that are responsible for the quality of the results. The values of some parameters are directly evident from the nature of the algorithm, but the values of some may vary based on the problem and their efficient settings may significantly affect the quality of the calculation. This chapter focuses on the possibility of using some statistical methods to determine the effective values of some parameters of SOMA. The use of statistical methods is elucidated by an illustrative example.

1 Introduction

Evolutionary algorithms are successfully used for solving optimization problems of different types. Their limitation is caused by the fact that they are controlled by special set of parameters. Some of these parameters can be successfully set exogenously based on the philosophy of the algorithm, however, there is a no deeper theoretical base to adjust certain parameters (e.g. parameters determining the rate stochastics), whilst (im)proper setting can radically affect the quality of obtained results.

Z. Čičková (✉) · M. Lukáčik
Department of Operations Research and Econometrics, University of Economics in Bratislava, Dolnozemská Cesta 1/B, 852 35 Bratislava, Slovakia
e-mail: cickova@euba.sk

M. Lukáčik
e-mail: martin.lukacik@euba.sk

D. Davendra and I. Zelinka (eds.), *Self-Organizing Migrating Algorithm*,
Studies in Computational Intelligence 626, DOI 10.1007/978-3-319-28161-2_12

Based on the various tests one can conclude that SOMA is even more sensitive to the parameters setting than other algorithms [3]. The control parameters are usually set on the basis of experimental results [3, 5]. Some of the control parameters are given directly by the nature of the problem and can be changed only by its reformulation. An example of such a parameter is the dimensionality (Dim). Setting other parameter can be derived from simple geometric interpretation of SOMA. Such parameter is the parameter PathLength, where its recommended setting is 3–5. Parameters PopSize and Migrations determine “the size and length” of simulation and their settings can use philosophy “more is better” (however, increasing these parameters affect the time needed to calculation and thus is dependent on the user’s hardware). The parameter MinDiv can be set to e.g. negative value if it is desired to reach all iterations, or to positive number if one want to watch the convergence of the calculation. Parameters Step and PRT are also responsible for the quality of the results. This chapter is devoted to some statistical methods that may be helpful in clarifying their settings. To adjust the control parameters it can be suitable before final calculating to carry out several simulations with e.g. smaller population size and lower number of iterations (which are not time consuming) with different values of the other control parameters. Further on, except basic descriptive statistics (e.g. average, mode, median), which allow to acquire the initial idea of the parameters settings, also various statistical methods can be used, e.g. single and multiple-factor analysis of variance.

The chapter is divided as follows. The first part is devoted to the theoretical description of some statistical methods. The second part gives an illustrative example of setting of control parameters.

2 Single and Multiple-Factor Analysis of Variance—Theory

Analysis of variance (ANOVA) is a technique, which enables to identify if there is any difference between groups on some variable (so called factor). When two or more groups are being compared, the characteristic that distinguishes the group from one another is called the factor under investigation. Consider the evolutionary techniques; an experiment might be carried out to compare different values of control parameters of algorithm from the perspective of obtained value of the fitness of the best individual (usually value of objective function).

Further on, the following notation will be used: a *population* is the set of all observations of interest and a sample is any subset of observations selected from the *population*. Let N be the total number of observation in the data set. Consider k levels of factor under investigation and a sample for each factor level, so that the sample size by jth factor level, $j = 1, 2, \ldots k$ is designate as n_j, $\sum_{j=1}^{k} n_j = N$. Then, the ith observation for each jth factor level can be designated as x_{ij}, $j = 1, 2, \ldots k$, $i = 1, 2, \ldots n_j$. Whether the null hypothesis of a single-factor analysis of variance should be rejected depends on how substantially the samples from the different *populations* differ from one another. Let μ_j, $j = 1, 2, \ldots k$ be a mean of *population* group on corresponding factor level.

A single-factor analysis of variance problem involves a comparison of all k group means. The objective is to test the null hypothesis (H_0):

$$H_0\colon \mu_1 = \mu_2 = \cdots = \mu_k \tag{1}$$

against alternative hypothesis (H_a):

$$H_\alpha\colon \text{at least two of the } \mu_j\text{'s}, \quad j = 1, 2, \ldots k, \text{ are different} \tag{2}$$

A measure of disparity among the sample means is the between-group sum of squares, denoted by SSB and given by

$$SSB = \sum_{j=1}^{k} n_j \left(\bar{x}_j - \bar{\bar{x}}\right)^2 \tag{3}$$

where $\bar{x}_j$ is the sample mean of jth group and $\bar{\bar{x}}$ is the overall mean (ratio of sum of all observations to the total number of observations in the data set). SSB has an associated degree of freedom ($df_1 = k - 1$).

A measure of variation within the k samples, called error sum of squares and denoted by SSE, is given by

$$SSE = \sum_{j=1}^{k} \left(n_j - 1\right) s_j^2 \tag{4}$$

where s_j^2 is the sample variance of jth group. SSE has an associated degree of freedom ($df_2 = N - k$).

Total sum of squares, denoted by SST, is given by

$$SST = \sum_{j=1}^{k} \sum_{i=1}^{n_j} \left(x_{ij} - \bar{\bar{x}}\right)^2 \tag{5}$$

with associated degree of freedom ($df = N - 1$).

The relationship between those three sums of squares is called the fundamental identity and for single-factor analysis of variance is $SST = SSB + SSE$.

A mean square is a sum of squares divided by its degree of freedom. In particular:

between-group mean square: $MSB = \dfrac{SSB}{k - 1}$

within-group mean square: $MSE = \dfrac{SSE}{N - k}$.

The test statistic (F) of the single-factor analysis of variance has a Fisher distribution and it is given by the formula: $F = \frac{MSB}{MSE}$.

The validity of the analysis of variance test requires some assumptions. Peck et al. [4] present these ones:

1. Each of the k group or *population* distributions is normal.
2. The k normal distributions have identical standard deviations.
3. The observations in the sample from any particular one of the k groups or *populations* are independent of one another.
4. When comparing group or *population* means, k random samples are selected independently of one another.

The statistical significance of the F ratio is most easily judged by its P-value. If the P-value is less than 0.05, the null hypothesis of equal means is rejected at the 5 % significance level. This does not imply that every mean is significantly different from every other mean. It only implies that the means are not all the same.

All the sums of squares, degrees of freedom, mean squares and F ratio with its P-value are entered in a general format of an analysis of variance table (Table 1).

Peck, Olsen and Devore also claim that in practice, the test based on these assumptions works well as long as the assumptions are not too badly violated. If the sample sizes are reasonably large, normal probability plots or boxplots of the data in each sample are helpful in checking the assumption of normality. Often, however, sample sizes are so small, that they suggest that the F test can safely be used if the largest of the sample standard deviations is at most twice the smallest one.

When null hypothesis is rejected by the F test, it can be stated that there are differences among the k group or *population* means. Several procedures called multiple-comparison procedures exist to determine which sample means are significantly different from others. Dowdy et al. [2] discuss five different approaches: Fisher's least significant difference, Duncan's new multiple-range test, Student–Newman–Keuls' procedure, Tukey's honestly significant difference and Scheffé's method.

Next, Fisher's least significant difference (*LSD*) procedure will be performed. Fisher's *LSD* procedure could be based on the t test statistic used for the *two-population* case. It could be easier to determine how large the difference between the sample means must be to reject null hypothesis.

In this case the test statistic by Anderson et al. [1] is the difference $\bar{x}_j - \bar{x}_l$, where $j, l = 1, 2, \ldots k$, so that $j \neq l$ and the objective is to test the null hypothesis (H_0):

Table 1 General format for an analysis of variance table

Source of variation	Sum of squares	Degree of freedom	Mean square	F-ratio	P-value
Between groups	SSB	$k - 1$	MSB	F	P
Within groups	SSE	$N - k$	MSE		
Total	SST	$N - 1$			

$$H_0\colon \mu_j = \mu_l; \quad j, l = 1, 2, \ldots k; j \neq l \tag{6}$$

against alternative hypothesis (H_a):

$$H_0\colon \mu_j \neq \mu_l; \quad j, l = 1, 2, \ldots k; \quad j \neq l \tag{7}$$

The null hypothesis should be rejected if $\left|\bar{x}_j - \bar{x}_l\right| \geq LSD$, where least significant difference is given by

$$LSD = t_{\alpha/2}\sqrt{MSE\left(\frac{1}{n_j} + \frac{1}{n_l}\right)}; \quad j, l = 1, 2, \ldots k; \quad j \neq l \tag{8}$$

where α denotes the significance level and t denotes critical value of Student's distribution.

Dowdy, Weardon and Chilko recall, that Fisher's test has a drawback; it requires that the null hypothesis be rejected in the analysis of variance procedure by the F test. These authors also discuss presented assumptions. At first, the normality of the treatment groups can be roughly checked by constructing histograms of the sample from each group. The analysis of variance by them leads to valid conclusions in some cases where there are departures from normality. For small sample sizes the treatment groups should be symmetric and unimodal. For large samples, more radical departures are acceptable due to the central limit theorem. Dowdy, Weardon and Chilko assume that conditions on independence are usually satisfied if the experimental units are randomly chosen and randomly assigned to the treatments. If the treatment groups already exist the experimenter by them does not have the opportunity to assign the subjects at random to the treatments. In such cases he uses random samples from each treatment group.

The last one of the assumptions underlying the analysis of variance is that the variances of the *populations* from which the samples come are the same. Dowdy, Weardon and Chilko state that the F tests are robust with respect to departures from homogeneity; that is, moderate departures from equality of variances do not greatly affect the F statistic. If the experimenter fears a large departure from homogeneity, several procedures are available to test equality of variances.

The F_{max} test was developed by Hartley. Hartley's test may be used when all treatment groups are the same size n and involves comparing the largest sample variance with the smallest sample variance. The null hypothesis (H_0) of test (where σ_j^2 is the *population* variance of jth group, j = 1, 2, …k) is:

$$H_0\colon \sigma_1^2 = \sigma_2^2 = \cdots = \sigma_k^2 \tag{9}$$

against alternative hypothesis (H_a):

$$H_\alpha\text{: at least two of the } \sigma_j\text{'s}, \quad j = 1, 2, \ldots k, \text{ are different} \tag{10}$$

when each of the k *populations* is normal and there is a random sample of size n from each *population*. Then sample variances $s_j^2, j = 1,2\ldots k$ can be computed and it is possible to calculate

$$F_{\max} = \frac{\max\left\{s_j^2, j = 1, 2, \ldots k\right\}}{\min\left\{s_j^2, j = 1, 2, \ldots k\right\}} \tag{11}$$

Statistics F_{max} is significant if it exceeds the value given in the Fisher's table with degrees of freedom $df_1 = k$ and $df_2 = n - 1$. Dowdy, Weardon and Chilko state, that because of the sensitivity of Hartley's test to departures from normality, if statistics F_{max} is significant, it indicates either unequal variances or a lack of normality.

Two other commonly used tests of homogeneity of variances are those of Cochran and Bartlett. In most situations, Cochran's test is equivalent to Hartley's. Cochran's test compares the maximum within-sample variance to the average within-sample variance. After computing of sample variances s_j^2, $j = 1,2\ldots k$, it is calculated

$$C = \frac{\max\left\{s_j^2, j = 1, 2, \ldots k\right\}}{\sum_{j=1}^{k} s_j^2} \tag{12}$$

and statistics C is significant if value

$$A = (k - 1)\frac{C}{1 - C} \tag{13}$$

exceeds the Fisher's table value with degrees of freedom $df_1 = \frac{n}{k} - 1$ and $df_2 = \left(\frac{n}{k} - 1\right)(k - 1)$.

Bartlett's test has a more complicated test statistic but has two advantages over the other two: It can be applied to groups of unequal sample sizes, and it is more powerful. Bartlett's test: compares a weighted average of the within-sample variances to their geometric mean. The test statistics is

$$B = \frac{1}{D}\left[\left(\sum_{j=1}^{k}(n_j - 1)\right)\ln\left(\frac{1}{\sum_{j=1}^{k}(n_j - 1)}\sum_{j=1}^{k}(n_j - 1)s_j^2\right) - \sum_{j=1}^{k}(n_j - 1)\ln\left(s_j^2\right)\right] \tag{14}$$

where

$$D = 1 + \frac{1}{3(k-1)}\left[\sum_{j=1}^{k}\left(\frac{1}{n_j - 1}\right) - \frac{1}{\sum_{j=1}^{k}(n_j - 1)}\right] \tag{15}$$

Statistics B is significant if it exceeds the value given in the chi-squared distribution with $(k - 1)$ degrees of freedom.

The last presented statistics of homogeneity of variances is Levene's test. This test performs a one-way analysis of variance on the variables $z_{ij} = |x_{ij} - \bar{x}_j|$, $j = 1,2\ldots k$, where $\bar{x}_j$ is either a mean of jth group or a median of jth group. At first variables z_{ij} are computed and the F statistic of the single-factor analysis of variance for these variables is obtained. Levene's statistics is significant if it exceeds the Fisher's table value with degrees of freedom $df_1 = k - 1$ and $df_2 = N - \mathrm{k}$.

Some statistical software also presents the results of a set of two-sample F tests that compare the standard deviations for each pair of levels. This makes sense only if the initial overall test shows significant differences amongst the variances (and standard deviations). Any pair with a small P value would be a pair whose standard deviations were significantly different.

An alternative to the standard analysis of variance that compares level medians instead of means is the Kruskal-Wallis test. This test is much less sensitive to the presence of outliers than a standard one-way analysis of variance and should be used whenever the assumption of normality within levels is not reasonable. Dowdy, Weardon and Chilko indicate this procedure.

First, it is necessary to rank the data from 1 (the smallest observation) to N (the largest observation), irrespective of the group in which they are found. If two or more observations are tied for the same numerical value, the average rank for which they are tied is assigned. Then for every group, when all treatment groups are the same size n, the average rank of group denoted by $\bar{r}_j$ is computed. Finally, the test statistic is:

$$H = n\frac{\left[\sum_{j=1}^{k}\left(\bar{r}_j - \frac{N+1}{2}\right)^2\right]}{\frac{N(N+1)}{12}} \tag{16}$$

The null hypothesis (H_0) of test is

$$H_0\colon E(\bar{r}_j) = \frac{N+1}{2} \quad \text{for all } j \tag{17}$$

against alternative hypothesis (H_α):

$$H_0\colon E(\bar{r}_j) \neq \frac{N+1}{2} \quad \text{for some } j \tag{18}$$

where $E(\bar{r}_j)$ denotes the expected value by $\bar{r}_j$.

Statistics H is significant if it exceeds the value given in the chi-squared distribution with $(k - 1)$ degrees of freedom and it indicates that there are significant differences amongst the level medians.

However, in some experiments it is desirable to draw conclusions about more than one variable or factor. The term factorial is used because the experimental conditions include all possible combinations of the factors. For example, for a levels of factor A and b levels of factor B, the experiment will involve collecting data on ab combinations. In experimental design terminology, the sample size of r for each group combination indicates that there are r replications, so abr observations are needed. Additional replications ($2r$, $3r$) and larger sample size put statistical conclusions in more precise terms.

This situation brings new effect—interaction effect. If the interaction effect has a significant impact, it can be concluded that the effect of the type of factor A depends on the factor B. There are three sets of hypothesis with the two-way ANOVA.

At first, the objective is to test the null hypothesis of comparison of a group means μ_{Ai}, $i = 1, 2, \ldots a$ by different values of factor A (H_0):

$$H_0 \colon \mu_{A1} = \mu_{A2} = \cdots = \mu_{Aa} \tag{19}$$

against alternative hypothesis (H_a):

$$H_\alpha \colon \text{at least two of the } \mu_{Aj}\text{'s}, \quad j = 1, 2, \ldots a, \text{ are different} \tag{20}$$

and also comparison of b group means μ_{Bi}, $j = 1, 2, \ldots b$ by different values of factor B:

$$H_0 \colon \mu_{B1} = \mu_{B2} = \cdots = \mu_{Bb} \tag{21}$$

against alternative hypothesis (H_a):

$$H_\alpha \colon \text{at least two of the } \mu_{Bj}\text{'s}, \quad j = 1, 2, \ldots b, \text{ are different} \tag{22}$$

Second objective is comparison of ab group means μ_{AiBj}, $i = 1, 2, \ldots a$, $j = 1, 2, \ldots b$ by different values of A and B:

$$H_0 \colon \mu_{A1B1} = \mu_{A2B2} = \cdots = \mu_{AaBb} \tag{23}$$

against the alternative (H_a):

$$H_\alpha \colon \text{There is no interaction between the factors } A \text{ and } B \tag{24}$$

The analysis of variance procedure for the two-factor factorial experiment requires us to partition the total sum of squares into sum of squares for factor A, sum of squares for factor B, sum of squares for interaction and sum of squares due to error.

Table 2 General format for an analysis of variance table

Source of variation	Sum of squares	Degree of freedom	Mean square	F-ratio	P-value
Factor A	SSA	$a - 1$	MSA	F_A	P_A
Factor B	SSB	$b - 1$	MSB	F_B	P_B
Interaction	SSAB	$(a - 1)(b - 1)$	$MSAB$	F_{AB}	P_{AB}
Error	SSE	$ab(r - 1)$	MSE		
Total	SST	$N - 1$			

Sum of squares for factor A is denoted by SSA and given by

$$SSA = br\sum_{i=1}^{a}(\bar{x}_i - \bar{\bar{x}})^2 \tag{25}$$

where b is number of levels of factor B, a is number of levels of factor A, $\bar{\bar{x}}$ is the overall mean and r is the number of replications.

Sum of squares for factor B is denoted by SSB and given by

$$SSB = ar\sum_{i=1}^{b}(\bar{x}_i - \bar{\bar{x}})^2 \tag{26}$$

Sum of squares for interaction is denoted by $SSAB$ and given by

$$SSAB = r\sum_{i=1}^{a}\sum_{j-1}^{b}\left(\bar{x}_{ij} - \bar{x}_i - \bar{x}_j + \bar{\bar{x}}\right)^2 \tag{27}$$

where x_{ij} is the sample mean for the observations corresponding to the combination of group i (factor A) and group j (factor B).

Error sum of squares (SSE) and total sum of squares (SST) are given by the same relations as in the case of single-factor analysis of variance (4) and (5). All the sums of squares, degrees of freedom, mean squares and F ratios with their P-values are presented in the analysis of variance table (Table 2).

3 Parameters Setting of SOMA

Next, the setting of control parameters of SOMA will be presented based on an illustrative example of solving traveling salesman problem. Consider the matrix of shortest distances between eight cities (Table 3).

The traveling salesman needs to find the shortest route between all the cities so that each city is visited exactly once. The solving was provided with the use of natural representation (the city was represented directly with its index in an

Table 3 Matrix of shortest distances between eight cities

	City 1	City 2	City 3	City 4	City 5	City 6	City 7	City 8
City 1	–	202	197	115	191	123	161	86
City 2	202	–	389	85	393	121	48	195
City 3	197	389	–	304	35	317	351	241
City 4	115	85	304	–	308	84	46	138
City 5	191	393	35	308	–	297	355	221
City 6	123	121	317	84	297	–	73	74
City 7	161	48	351	46	355	73	–	147
City 8	86	195	241	138	221	74	147	–

individual). A simple penalty approach was used if the unfeasible solutions appeared. Some of the control and termination parameters were set as follows: Parameter PopSize was set to 80 and parameter Migrations was set to 300. The parameter MinDiv was set to negative value to reach all iterations (the small size of instance enables to reach all iteration in a relative short time). The parameter PathLength was set to the value 3. The settings of parameters Step and PRT was provided on the base of before mentioned statistical methods. Let the value of the shortest route (denoted as fc) be the response variable. Further on, we can specify the impact of factors' level (levels of parameters Step and PRT) on the variability of the response variable.

Parameter PRT can take values from 0 (purely stochastic behavior of algorithm) to 1 (purely deterministic behavior). At first, the levels of parameters PRT were set to 0.2, 0.4, 0.6 and 0.8. Parameter Step can take values from 0.1 to value of parameter PathLength, which equals 3. Following the previous simulations it was found that the value of the Step >1, increased probability of getting extremely "bad" outcome. So, the values 0.3, 0.5, 0.7 and 0.9 were used as the levels of parameter Step in testing.

First tested hypothesis is comparison of 4 group means $\overline{fc}_1, \overline{fc}_2, \overline{fc}_3, \overline{fc}_4$ by different values of parameter PRT (0.2, 0.4, 0.6 and 0.8) according to (1) and (2). The experiment is balancing—for each pair PRT-Step the same number of simulations was realized—eight replications. It was thus implemented the first phase of a total of 128 simulations (Table 4).

The summary of the descriptive statistics for every value of parameter PRT is given in Table 5.

There is big difference between the smallest and the largest standard deviation. This may cause problems since the analysis of variance assumes that the standard deviations at all levels are equal. The results are also presented by the box and whisker plot (Fig. 1).

There is evident some significant non-normality in the data, which violates the assumption that the data come from normal distributions. Someone may wish to transform the values of fc to remove any dependence of the standard deviation on

Table 4 Results of simulations

Step	PRT	*fc*	Step	PRT	*fc*	Step	PRT	*fc*	Step	PRT	*fc*
0.3	0.2	862	0.3	0.2	862	0.3	0.2	921	0.3	0.2	906
0.3	0.4	862	0.3	0.4	848	0.3	0.4	848	0.3	0.4	848
0.3	0.6	848	0.3	0.6	848	0.3	0.6	848	0.3	0.6	848
0.3	0.8	848	0.3	0.8	848	0.3	0.8	848	0.3	0.8	848
0.5	0.2	896	0.5	0.2	857	0.5	0.2	1072	0.5	0.2	896
0.5	0.4	906	0.5	0.4	896	0.5	0.4	842	0.5	0.4	848
0.5	0.6	848	0.5	0.6	848	0.5	0.6	848	0.5	0.6	848
0.5	0.8	848	0.5	0.8	848	0.5	0.8	848	0.5	0.8	848
0.7	0.2	848	0.7	0.2	907	0.7	0.2	979	0.7	0.2	963
0.7	0.4	848	0.7	0.4	871	0.7	0.4	848	0.7	0.4	857
0.7	0.6	848	0.7	0.6	905	0.7	0.6	848	0.7	0.6	848
0.7	0.8	857	0.7	0.8	848	0.7	0.8	848	0.7	0.8	848
0.9	0.2	980	0.9	0.2	848	0.9	0.2	857	0.9	0.2	905
0.9	0.4	848	0.9	0.4	857	0.9	0.4	848	0.9	0.4	848
0.9	0.6	848	0.9	0.6	848	0.9	0.6	848	0.9	0.6	848
0.9	0.8	848	0.9	0.8	857	0.9	0.8	857	0.9	0.8	848
0.3	0.2	919	0.3	0.2	988	0.3	0.2	871	0.3	0.2	954
0.3	0.4	848	0.3	0.4	848	0.3	0.4	857	0.3	0.4	848
0.3	0.6	848	0.3	0.6	848	0.3	0.6	848	0.3	0.6	848
0.3	0.8	848	0.3	0.8	848	0.3	0.8	848	0.3	0.8	848
0.5	0.2	905	0.5	0.2	919	0.5	0.2	857	0.5	0.2	896
0.5	0.4	848	0.5	0.4	848	0.5	0.4	857	0.5	0.4	848
0.5	0.6	848	0.5	0.6	848	0.5	0.6	848	0.5	0.6	848
0.5	0.8	848	0.5	0.8	848	0.5	0.8	848	0.5	0.8	848
0.7	0.2	907	0.7	0.2	919	0.7	0.2	848	0.7	0.2	862
0.7	0.4	848	0.7	0.4	848	0.7	0.4	862	0.7	0.4	848
0.7	0.6	848	0.7	0.6	857	0.7	0.6	857	0.7	0.6	857
0.7	0.8	862	0.7	0.8	848	0.7	0.8	848	0.7	0.8	848
0.9	0.2	862	0.9	0.2	907	0.9	0.2	919	0.9	0.2	1003
0.9	0.4	848	0.9	0.4	848	0.9	0.4	848	0.9	0.4	905
0.9	0.6	848	0.9	0.6	848	0.9	0.6	848	0.9	0.6	848
0.9	0.8	848	0.9	0.8	848	0.9	0.8	848	0.9	0.8	857

the mean. The analysis of variance decomposes the variance of *fc* into two components: a between-group component and a within-group component (Table 6).

The *F* ratio, which equals 33.0788, is a ratio of the between-group estimate to the within-group estimate. Since the *P*-value of the *F* test is less than 0.05, there is a statistically significant difference between the mean *fc* from one level of PRT to another at the 5.0 % significance level. Then, Fisher's least significant difference was used to determine which means are significantly different from which others (Table 7).

Table 5 Summary statistics—data grouped by PRT

PRT	Count	Average	Standard deviation	Coefficient of variation (%)	Minimum	Maximum	Range	Median
0.2	32	909.219	53.0324	5.83274	848.0	1072.0	224.0	905.5
0.4	32	855.625	16.3997	1.9167	842.0	906.0	64.0	848.0
0.6	32	850.625	10.2729	1.20769	848.0	905.0	57.0	848.0
0.8	32	849.563	3.77545	0.444399	848.0	862.0	14.0	848.0
Total	128	866.258	37.5064	4.3297	842.0	1072.0	230.0	848.0

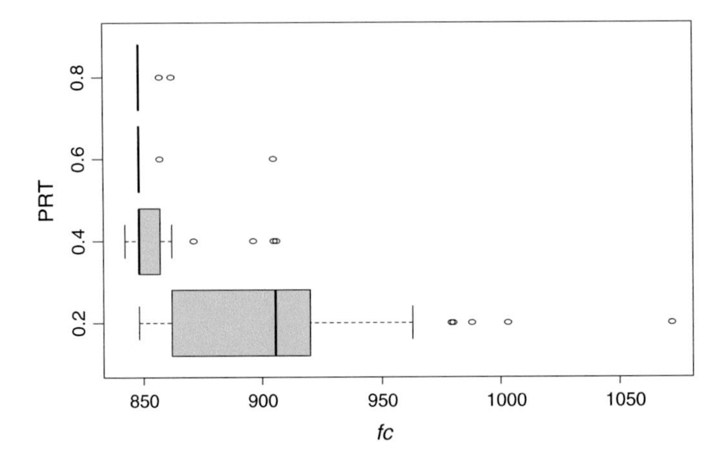

Fig. 1 Box and whisker plot—data grouped by PRT

Table 6 Analysis of variance table—data grouped by PRT

Source of variation	Sum of squares	Degree of freedom	Mean square	*F*-ratio	*P*-value
Between groups	79418.1	3	26472.7	33.08	0.0000
Within groups	99236.3	124	800.293		
Total	178654.4	127			

Table 7 Comparison procedure of Fisher's least significant difference—data grouped by PRT

Contrast	Difference of means	*LSD*	Significant differences
0.2–0.4	53.5938	13.9982	Yes
0.2–0.6	58.5938	13.9982	Yes
0.2–0.8	59.6563	13.9982	Yes
0.4–0.6	5.0	13.9982	No
0.4–0.8	6.0625	13.9982	No
0.6–0.8	1.0625	13.9982	No

Now, one can see significant difference for group of simulations where PRT equals 0.2 to other groups. It can be stated that there is a large departure from homogeneity, so all the equality of variances' tests are used (Table 8).

The statistics displayed in the Table 8 and also the *P*-values show, that there is a statistically significant difference amongst the standard deviations of groups. This violates one of the important assumptions underlying the analysis of variance.

The comparison of the standard deviations for each pair of samples is given in Table 9. All *P*-Values below 0.05 indicate statistically significant differences between standard deviations of every pair of groups.

The situation is clear; statistically different values of averages and standard deviations for groups of values *fc* by different values of parameter PRT were obtained. Due to failure of assumptions, the results of the analysis of variance cannot be taken into account. Finally, despite all the previous conclusions, the decision is to use the Kruskal-Wallis test as alternative to the standard analysis of variance to compare the medians instead of the means (Table 10).

Table 8 Tests of homogeneity of variances—data grouped by PRT

	Test	*P*-value
Levene's	23.1283	5.87053E−12
Cochran's	0.878564	0
Bartlett's	4.35742	0
Hartley's	197.308	

Table 9 Comparison of the standard deviations for each pair of groups—data grouped by PRT

Comparison	Standard deviation 1	Standard deviation 2	*F*-ratio	*P*-value
0.2/0.4	53.0324	16.3997	10.457	0.0000
0.2/0.6	53.0324	10.2729	26.65	0.0000
0.2/0.8	53.0324	3.77545	197.308	0.0000
0.4/0.6	16.3997	10.2729	2.54853	0.0111
0.4/0.8	16.3997	3.77545	18.8685	0.0000
0.6/0.8	10.2729	3.77545	7.40368	0.0000

Table 10 Kruskal-Wallis test —data grouped by PRT

PRT	Sample size	Average rank
0.2	32	104.469
0.4	32	57.4375
0.6	32	47.5
0.8	32	48.5938
Test statistic	66.7	*P*-value = 0

The null hypothesis of Kruskal-Wallis test is that the medians of *fc* within each of the four levels of PRT are the same (17). Since the *P*-value is less than 0.05, there is a statistically significant difference amongst the medians.

It is evident (from the results of tests and also from box and whisker plot) that median of group where PRT equals 0.2 is significantly different from others. It seems that values 0.6 or 0.8 for the parameter PRT are the appropriate choice. More preferred alternative is a latter value in order to eliminate possible outliers.

Second tested hypothesis is comparison of 4 group means $\overline{fc}_1, \overline{fc}_2, \overline{fc}_3, \overline{fc}_4$ by different values of parameter Step (0.3, 0.5, 0.7 and 0.9) according to (1) and (2).

The summary of the descriptive statistics by every value of Step is given in the Table 11.

In this case there is not so big difference between the smallest and the largest standard deviation as in previous case. From the box and whisker plot (Fig. 2) it is seen some significant non-normality in the data, which again violates the assumption that the data come from normal distributions. Recall that the normal distribution is symmetric with a median in the middle of the box bounded by the

Table 11 Summary statistics—data grouped by Step

Step	Count	Average	Standard dev.	Coef. of variation (%)	Minimum	Maximum	Range	Median
0.3	32	864.313	34.6656	4.01077	848.0	988.0	140.0	848.0
0.5	32	867.469	43.6984	5.03746	842.0	1072.0	230.0	848.0
0.7	32	866.813	33.9995	3.92236	848.0	979.0	131.0	848.0
0.9	32	866.438	38.6013	4.45517	848.0	1003.0	155.0	848.0
Total	128	866.258	37.5064	4.3297	842.0	1072.0	230.0	848.0

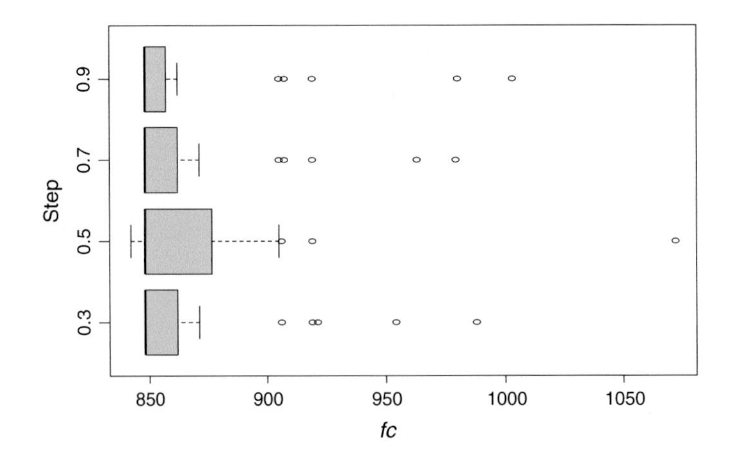

Fig. 2 Box and whisker plot—data grouped by Step

first and the third quartile. This is not the case, because the median is the smallest value in three cases and for the last one is typical outlier.

The analysis of variance decomposes the variance of *fc* once again into two components: a between-group component and a within-group component, but now data are grouped by parameter Step (Table 12).

The *F* ratio, which equals 0.0414313, is a ratio of the between-group estimate to the within-group estimate. Since the *P*-value of the *F* test is greater than 0.05, there is not a statistically significant difference amongst the mean *fc* from one level of Step to another at the 5.0 % significance level. Fisher's test requires that the null hypothesis could be rejected in the analysis of variance procedure by the *F* test, what is not the case; nevertheless its results are shown (Table 13).

Evidently, there is not a significant difference in means between groups.

The all statistics displayed in the Table 14 and also the *P*-values greater than or equal to 0.05 show, that there is not a statistically significant difference amongst the standard deviations of groups.

The comparison of the standard deviations for each pair of samples is given in Table 15. It can be stated there are no statistically significant differences between any pair of means.

Table 12 Analysis of variance table—data grouped by Step

Source of variation	Sum of squares	Degree of freedom	Mean square	*F*-ratio	*P*-value
Between groups	178.898	3	59.6328	0.04	0.9887
Within groups	178476.6	124	1439.32		
Total	178654.4	127			

Table 13 Comparison procedure of Fisher's least significant difference—data grouped by Step

Contrast	Difference of means	*LSD*	Significant differences
0.3–0.5	−3.15625	18.7727	No
0.3–0.7	−2.5	18.7727	No
0.3–0.9	−2.125	18.7727	No
0.5–0.7	0.65625	18.7727	No
0.5–0.9	1.03125	18.7727	No
0.7–0.9	0.375	18.7727	No

Table 14 Tests of homogeneity of variances—data grouped by Step

	Test	*P*-value
Levene's	0.0490732	0.985551
Cochran's C	0.331675	0.298469
Bartlett's	1.02079	0.472019
Hartley's	1.65191	

Table 15 Comparison of the standard deviations for each pair of groups—data grouped by Step

Comparison	Standard deviation 1	Standard deviation 2	F-ratio	P-value
0.3/0.5	34.6656	43.6984	0.629314	0.2029
0.3/0.7	34.6656	33.9995	1.03957	0.9147
0.3/0.9	34.6656	38.6013	0.806481	0.5529
0.5/0.7	43.6984	33.9995	1.65191	0.1679
0.5/0.9	43.6984	38.6013	1.28152	0.4939
0.7/0.9	33.9995	38.6013	0.775783	0.4838

The situation is different from the previous case; we didn't obtain statistically different values of averages and standard deviations for groups of values *fc* by different values of parameter Step. Again, we decided to use alternative to the standard analysis of variance—the Kruskal-Wallis test—to compare the medians instead of the means (Table 16).

The null hypothesis of Kruskal-Wallis test is that the medians of *fc* within each of the four levels of Step are the same (17). Since the P-value is greater than 0.05, there is not a statistically significant difference amongst the medians.

Hence, the results of tests and also from box and whisker plot show the means, medians and standard deviations of all four samples are equal. It is not a difference between arbitrary values of the parameter Step from a statistical point of view. Despite the results, it seems that values 0.7 or 0.9 for the parameter Step is the appropriate choice, since calculations are usually faster for bigger values of Step. These values generate the equivalent results with similar outliers. More preferred alternative is a latter value because of smaller interquartile range.

Last tested hypothesis is comparison of 4 group means $\overline{fc}_{A1}, \overline{fc}_{A2}, \overline{fc}_{A3}, \overline{fc}_{A4}$ by different values of parameter Step (factor A) according the test (19) and (20), where the levels of Step were set to 0.3, 0.5, 0.7 and 0.9 and also comparison of 4 group means $\overline{fc}_{B1}, \overline{fc}_{B2}, \overline{fc}_{B3}, \overline{fc}_{B4}$ by different values of parameter PRT (factor B) according the test (21) and (22), where the levels of PRT were set to 0.2, 0.4, 0.6 and 0.8, as well as the comparison of 16 group means $\overline{fc}_{A1B1}, \overline{fc}_{A2B1}, \ldots, \overline{fc}_{A4B4}$ by mentioned different values of Step and PRT (23) and (24).

The ANOVA table (Table 17) decomposes the variability of *fc* into contributions due to both factors Step and PRT. The contribution of each factor is measured having removed the effect of another factor. Since P-value of factor PRT is less than 0.05, this factor has a statistically significant effect on *fc* at the 5.0 % significance

Table 16 Kruskal-Wallis test—data grouped by Step

Step	Sample size	Average rank
0.3	32	62.0625
0.5	32	61.6875
0.7	32	69.875
0.9	32	64.375
Test statistic	1.30262	P-value = 0.728508

Table 17 Multiple analysis of variance table

Source	Sum of squares	Df	Mean square	*F*-ratio	*P*-value
Factor A: step	178.898	3	59.6328	0.07	0.9766
Factor B: prt	79418.1	3	26472.7	30.35	0.0000
Interactions AB	1359.82	9	151.091	0.17	0.9964
Error	97697.6	112	872.3		
Total	178654	127			

level. Significant interaction effects between analysed factors have not been confirmed. The results of multiple factor analysis of variance confirmed the conclusions that were obtained using a single factor analysis of variance. Different values of parameter Step didn't result to statistically different values of function *fc*. Contrary, different values of parameter PRT resulted to statistically different values of function *fc*.

It is evident (Fig. 3) that the small values of PRT (0.2 and 0.4) results to big variability of values *fc* regardless of the value of Step. The interaction plot (Fig. 4) gives the mean values of *fc* depending on combination of mentioned factors. Based on that, it seems to be an appropriate choice to set the parameter PRT to 0.6 or 0.8.

Further on, one more analysis was realized in order to choose between the two values of parameter PRT and one way factor analysis for parameter PRT was conducted. Tested hypothesis is comparison of 5 group means $\overline{fc}_1, \overline{fc}_2, \overline{fc}_3, \overline{fc}_4, \overline{fc}_5$ by different values of parameter PRT on levels 0.5, 0.6, 0.7, 0.8 and 0.9 according to (1) and (2). Parameter Step was set to value 0.9. The experiment is balancing—for each value of PRT the same number of simulations was realized—eight replications. It was thus implemented a total of 40 simulations, which results are summarized in Table 18.

The summary of the descriptive statistics by every value of parameter PRT can be seen in the Table 19.

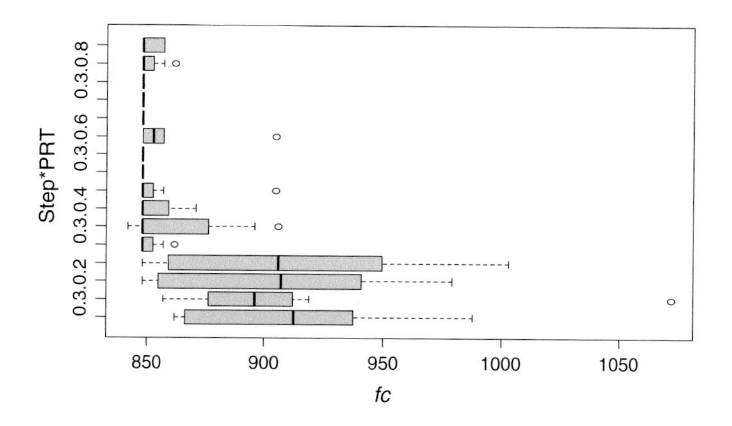

Fig. 3 Box and whisker plot—data grouped by interaction of Step and PRT

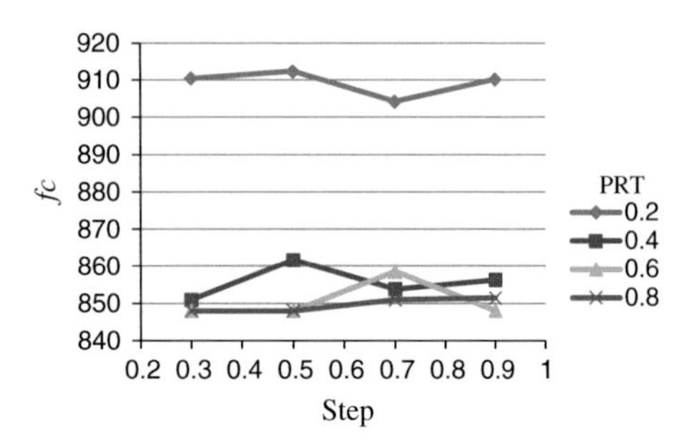

Fig. 4 Interaction plot

Table 18 Results of simulations

PRT	*fc*	PRT	*fc*	PRT	*fc*	PRT	*fc*	PRT	*fc*
0.5	848	0.6	857	0.7	848	0.8	848	0.9	848
0.5	857	0.6	848	0.7	848	0.8	848	0.9	848
0.5	862	0.6	921	0.7	857	0.8	857	0.9	848
0.5	862	0.6	848	0.7	848	0.8	848	0.9	848
0.5	959	0.6	896	0.7	857	0.8	848	0.9	848
0.5	905	0.6	871	0.7	848	0.8	857	0.9	848
0.5	1007	0.6	857	0.7	848	0.8	857	0.9	848
0.5	848	0.6	848	0.7	857	0.8	848	0.9	848

Table 19 Summary statistics—data grouped by PRT

prt	Count	Average	Standard dev.	Coef. of variation (%)	Minimum	Maximum	Range	Median
0.5	8	893.5	59.4763	6.65655	848.0	1007.0	159.0	862.5
0.6	8	868.25	26.8421	3.09152	848.0	921.0	73.0	857.0
0.7	8	851.375	4.65794	0.547108	848.0	857.0	9.0	848.0
0.8	8	851.375	4.65794	0.547108	848.0	857.0	9.0	848.0
0.9	8	848.0	0	0	848.0	848.0	0	848.0
Total	40	862.5	32.7085	3.79229	848.0	1007.0	159.0	848.0

There is again big difference between the smallest and the largest standard deviation. Remember, that this may cause problems since the analysis of variance assumes that the standard deviations at all levels are equal. It is evident also from the box and whisker plot of results (Fig. 5).

It is evident there is some significant non-normality in the data, which violates the assumption that the data come from normal distributions. The analysis of variance decomposes the variance of *fc* into two components: a between-group component and a within-group component (Table 20).

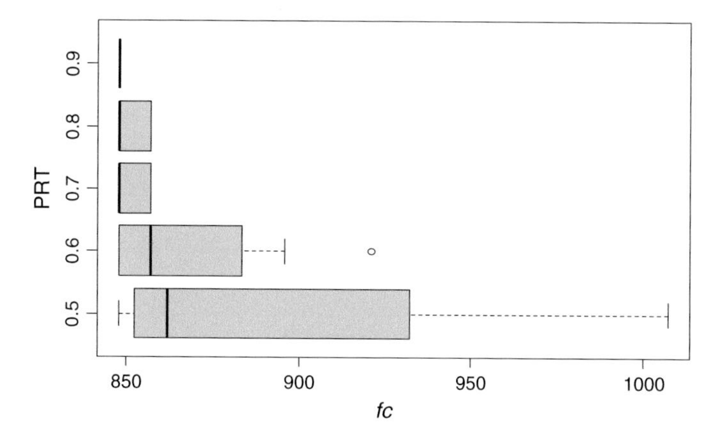

Fig. 5 Box and whisker plot—data grouped by PRT

Table 20 Analysis of variance table—data grouped by PRT

Source of variation	Sum of squares	Degree of freedom	Mean square	*F*-ratio	*P*-value
Between groups	11614.8	4	2903.69	3.3753	0.01946
Within groups	30109.3	35	860.264		
Total	41724.0	39			

The *F* ratio, which equals 3.3753, is a ratio of the between-group estimate to the within-group estimate. Since the *P*-value of the *F* test is less than 0.05, there is a statistically significant difference between the mean *fc* from one level of PRT to another at the 5.0 % significance level. Fisher's least significant difference is used to determine which means are significantly different from which others (Table 21).

It is seen significant difference for group of simulations where PRT equals 0.5 to groups where PRT equal to 0.7, 0.8 and 0.9. It is evident a large departure from homogeneity, so next all the equality of variances' tests are used (Table 22).

The statistics displayed in this table and also the *P*-values show, that there is a statistically significant difference amongst the standard deviations of groups. This violates one of the important assumptions underlying the analysis of variance.

The comparison of the standard deviations for each pair of samples is given in Table 23. *P*-Values below 0.05 indicate statistically significant differences between standard deviations of these pair of groups.

The situation is such as in the first analysis of parameter PRT; there are statistically different values of averages and standard deviations for groups of values *fc* by different values of parameter PRT. Due to failure of assumptions, the results of the analysis of variance cannot be taken into account. Finally, despite all the previous conclusions, we decided to use the Kruskal-Wallis test to compare the medians instead of the means (Table 24).

Table 21 Comparison procedure of Fisher's least significant difference—data grouped by PRT

Contrast	Difference of means	*LSD*	Significant differences
0.5–0.6	25.25	29.7719	No
0.5–0.7	42.125	29.7719	Yes
0.5–0.8	42.125	29.7719	Yes
0.5–0.9	45.5	29.7719	Yes
0.6–0.7	16.875	29.7719	No
0.6–0.8	16.875	29.7719	No
0.6–0.9	20.25	29.7719	No
0.7–0.8	0	29.7719	No
0.7–0.9	3.375	29.7719	No
0.8–0.9	3.375	29.7719	No

Table 22 Tests of homogeneity of variances—data grouped by PRT

	Test	*P*-value
Levene's	3.2163	0.023823
Cochran's	0.822405	0.00000532605
Bartlett's	5.77786	4.76942E−10
Hartley's	163.042	

Table 23 Comparison of the standard deviations for each pair of groups – data grouped by PRT

Comparison	Standard deviation 1	Standard deviation 2	*F*-ratio	*P*-value
0.5/0.6	59.4763	26.8421	4.90969	0.0523
0.5/0.7	59.4763	4.65794	163.042	0.0000
0.5/0.8	59.4763	4.65794	163.042	0.0000
0.5/0.9	59.4763	0	–	–
0.6/0.7	26.8421	4.65794	33.2082	0.0002
0.6/0.8	26.8421	4.65794	33.2082	0.0002
0.6/0.9	26.8421	0	–	–
0.7/0.8	4.65794	4.65794	1.0	1.0000
0.7/0.9	4.65794	0	–	–
0.8/0.9	4.65794	0	–	–

The null hypothesis of Kruskal-Wallis test (17) is that the medians of *fc* within each of the fiver levels of PRT are the same. Since the *P*-value is less than 0.05, there is a statistically significant difference amongst the medians. It seems that median of group where PRT equals 0.5 and 0.6 are significantly different from others. Based on mentioned above the values 0.7–0.9 for the parameter PRT are considered as the appropriate choice.

Table 24 Kruskal-Wallis test —data grouped by PRT

PRT	Sample size	Average rank
0.5	8	29.375
0.6	8	25.125
0.7	8	18.0
0.8	8	18.0
0.9	8	12.0
Test statistic	10.844	*P*-value = 0.0283745

4 Conclusions

Evolutionary algorithms are considered to be universal and effective tool for solving various optimization problems. Their effectiveness is limited by fact they are generally controlled by special set of parameters. Although some of parameters can be successfully set exogenously based on the philosophy of the algorithm or according to type of solved problem, there is a no deeper theoretical base to adjust all the parameters. This chapter focuses on the possibility of using some statistical methods that may be helpful to determine the effective values of some parameters of SOMA.

Based on the various tests one can conclude that SOMA is even more sensitive on the parameters setting than other algorithms [3, 5], thus the efficient setting may significantly affect the quality of the results. The setting of control parameter can be supported by statistical methods especially aimed at determining whether the level of some parameter brings the difference in results. A brief view to corresponding statistical methods (single factor analyze of variance, Levene's test, Cochran's test, Bartlett's test, Hartley's test, two-way analyze of variance) is given in the first half of the chapter. The second half is aimed on example of practical use based on illustrative data of traveling salesman problem.

References

1. Anderson, D.R., Sweeney, D.J., Williams, T.A.: Statistics for business and economics, 11th edn. South-Western Cengage Learning, Boston (2011)
2. Dowdy, S., Weardon, S., Chilko, D.: Statistics for research, 3rd edn. Wiley, New York (2004)
3. Onwubolu, G.C., Babu, B.V.: New Optimization Techniques in Engineering. Springer, Berlin (2004)
4. Peck, R., Olsen, C., Devore, J.: Introduction to statistics and data analysis, 4th edn. Cengage Learning, Boston (2012)
5. Zelinka, I.: Umělá inteligence v problémech globální optimalizace. BEN-technická literature, Praha (2002)

Inspired in SOMA: Perturbation Vector Embedded into the Chaotic PSO Algorithm Driven by Lozi Chaotic Map

Michal Pluhacek, Ivan Zelinka, Roman Senkerik and Donald Davendra

Abstract In this chapter a new approach for Particle Swarm Optimization (PSO) algorithm driven by chaotic pseudorandom number generator based on chaotic Lozi map is presented. This research represents the continuation of the satisfactory results obtained by means of chaos embedded (driven) swarm based algorithms, which utilize the chaotic dynamics in the place of pseudorandom number generators. The perturbation vector, which is introduced here, was inspired by the swarm based Self-organizing Migrating Algorithm (SOMA). It was embedded into the PSO algorithm to help overcome the issue of premature convergence.

Keywords Particle swarm optimization · Chaos · PSO · SOMA hybrid

1 Introduction

The Particle Swarm Optimization Algorithm (PSO) [1–4] is probably the most prominent representative of swarm inspired algorithms alongside Ant Colony Optimization [5] and SOMA [6]. The PSO is the subject of intensive research for

M. Pluhacek (✉) · R. Senkerik
Faculty of Applied Informatics, Tomas Bata University in Zlin,
T.G. Masaryka 5555, 760 01 Zlin, Czech Republic
e-mail: pluhacek@fai.utb.cz

R. Senkerik
e-mail: senkerik@fai.utb.cz

I. Zelinka · D. Davendra
Faculty of Electrical Engineering and Computer Science, VŠB-Technical University of Ostrava, 17. listopadu 15, 708 33 Ostrava, Poruba, Czech Republic
e-mail: ivan.zelinka@vsb.cz

D. Davendra
Department of Computer Science, Central Washington University,
400 E. University Way, Ellensburg, WA 98926-7520, USA
e-mail: donald.davendra@vsb.cz; DonaldD@cwu.edu

D. Davendra and I. Zelinka (eds.), *Self-Organizing Migrating Algorithm*,
Studies in Computational Intelligence 626, DOI 10.1007/978-3-319-28161-2_13

more than two decades now. As many other Evolutionary Optimization Techniques (ECTs) the PSO is being intensively studied, modified and re-designed.

Some studies indicated that using chaotic systems in the place of pseudorandom number generators (PRNG) might improve the quality of results, convergence speed or other performance indicators of various ECTs [7–14]. The chaotic approach causes the heuristic to map unique regions, since the chaotic map iterates to new regions. Several studies have already dealt with the possibilities of embedding of chaotic dynamics into the PSO algorithm and have investigated the influence to the performance of PSO [9–12]. Recently a novel approach that utilizes alternating of two different chaotic systems within one run of the algorithm was proposed [13, 14]. This paper presents a novel approach for PSO algorithm driven by Lozi chaotic map [11, 12, 15], that was developed as a response to the satisfactory results of the previous research [11–14].

Inspired by SOMA algorithm it is implemented the perturbation vector into the PSO. The perturbation vector is designed to help prevent premature convergence of the swarm—one of the biggest issues of PSO especially embedded with chaotic Lozi map.

The chapter is structured as follows: Firstly, the motivation for this research and the definition of PSO algorithm is given, followed by the basic principles of Self-organizing Migrating Algorithm (SOMA) that served as an inspiration for this novel approach. The Lozi chaotic map is described in Sect. 5. Used benchmark functions are detailed in the sixth section and experiments set up in the following section number seven. Finally the results are presented and discussed in the last two sections.

2 Motivation

In the previous research [11–14] the chaotic pseudorandom number generator (CPRNG) based on Lozi chaotic map was implemented into the PSO algorithm. It was observed that there was a significant improvement in the convergence speed of the algorithm and it was possible to obtain good solutions within less iterations of the algorithm. However the final precision of the solution was not as good because the solutions stopped improving very shortly (the issue of premature convergence).

The Self-organizing Migrating Algorithm (SOMA) [6] deals with the issue of premature convergence by means of so called *PRTVector* (or *Perturbation Vector*), that limits the movement of particles to certain dimensions only. The aim of this work is to investigate whether this tool can be successfully implemented into the concept of PSO driven by Lozi chaotic map and improve its performance.

3 Particle Swarm Optimization Algorithm

The PSO algorithm is inspired by the natural swarm behavior of animals (such as birds and fish). It was firstly introduced by Eberhart and Kennedy in 1995 [1]. The PSO became popular method for global optimization. Each particle in the population represents a possible solution of the optimization problem which is defined by the cost function (CF). In each iteration of the algorithm, a new location (combination of CF parameters) of the particle is calculated based on its previous location and velocity vector (velocity vector contains particle velocity for each dimension).

According to the method of selection of the swarm or subswarm for best solution information spreading, the PSO algorithms are noted as global PSO (GPSO) or local PSO (LPSO). Within this research the PSO algorithm with global topology (GPSO) [6] was utilized. The CPRNG is used in the main GPSO formula (1), which determines a new "velocity", thus directly affects the position of each particle in the next iteration.

$$v_{ij}^{t+1} = w \cdot v_{ij}^{t} + c_1 \cdot Rand \cdot \left(pBest_{ij} - x_{ij}^{t}\right) + c_2 \cdot Rand \cdot \left(gBest_j - x_{ij}^{t}\right) \quad (1)$$

where

v_i^{t+1}	New velocity of the ith particle in iteration $t + 1$
w	Inertia weight value
v_i^t	Current velocity of the ith particle in iteration t
c_1, c_2	Priority factors
$pBest_i$	Local (personal) best solution found by the ith particle
$gBest$	Best solution found in a population
x_{ij}^t	Current position of the ith particle (component j of the dimension D) in iteration t
$Rand$	Pseudorandom number, interval $\langle 0, 1 \rangle$. CPRNG is applied only here.

The maximum velocity was limited to 0.2 times the range as it is usual. The new position of each particle is then given by (2), where x_i^{t+1} is the new particle position:

$$x_i^{t+1} = x_i^t + v_i^{t+1} \quad (2)$$

Finally the linear decreasing inertia weight [6, 7] strategy was used in this work. The dynamic inertia weight is meant to slow the particles over time thus to improve the local search capability in the later phase of the optimization. The inertia weight has two control parameters w_{start} and w_{end}. A new w for each iteration is given by (3), where t stands for current iteration number and n stands for the total number of iterations. The values used in this study were $w_{start} = 0.9$ and $w_{end} = 0.4$.

$$w = w_{start} - \frac{((w_{start} - w_{end}) \cdot t)}{n} \quad (3)$$

4 Self-organizing Migrating Algorithm (SOMA)

SOMA [4] works with groups of individuals (population) whose behavior can be described as a competitive—cooperative strategy. The construction of a new population of individuals is not based on evolutionary principles (two parents produce offspring) but on the behavior of social group, e.g. a herd of animals looking for food. This algorithm can be classified as an algorithm of a social environment. To the same group of algorithms, PSO algorithm [1] can also be put in. In the case of SOMA, there is no velocity vector as in PSO, only the position of individuals in the search space is changed during one generation, here called 'migration loop' (*ML*).

The rules are as follows: In every migration loop the best individual is chosen, i.e. individual with the minimum cost value, which is called the *Leader*. An active individual from the population moves in the direction towards the Leader in the search space. The movement consists of jumps determined by the *Step* parameter until the individual reaches the final position given by the *PathLength* parameter. For each step, the cost function for the actual position is evaluated and the best value is saved. At the end of the crossover, the position of the individual with minimum cost value is chosen. If the cost value of the new position is better than the cost value of an individual from the old population, the new one appears in new population. Otherwise the old one remains there. The main principle is depicted in Fig. 1 and the crossover is described by Eq. (4):

$$x_{i,j}^{ML+1} = x_{i,j,START}^{ML} + \left(x_{L,j}^{ML} - x_{i,j,START}^{ML}\right) \cdot t \cdot PRTVector_j \tag{4}$$

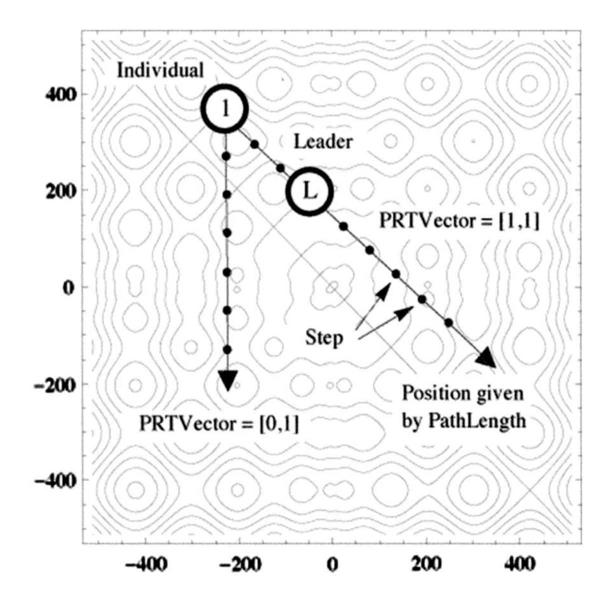

Fig. 1 The basic principle of crossover in SOMA

where

$x_{i,j}^{ML+1}$	value of *i*th individual's *j*th parameter, in step *t* in migration loop *ML* + 1,
$x_{i,j,START}^{ML}$	value of *i*th individual's *j*th parameter, Start position in actual *ML*,
$x_{L,j}^{ML}$	value of Leader's *j*th parameter in migration loop *ML*,
t	$step \in < 0$, by *step* to, $PathLength >$,
PRTVector (*Perturbation Vector*)	vector of ones and zeros dependent on *PRT* value. The *PRT* value is a predefined constant. If random number from interval $\langle 0, 1\rangle$ is less than *PRT*, then 1 is saved to *PRTVector*, otherwise it is 0. The value of *PRT* is typically set to 0.8.

5 Lozi Chaotic Map

This section contains the description of discrete chaotic Lozi map that was used as the CPRNG for PSO algorithm. Direct output iterations of variable x were transferred into the typical pseudorandom number range $\langle 0, 1\rangle$. The initial concept of embedding chaotic dynamics into evolutionary algorithms is given in [7].

The Lozi map is a simple discrete two-dimensional chaotic map. The map equations are given in (5). The parameters used in this work are: $a = 1.7$ and $b = 0.5$ with respect to [15].

The Lozi map is depicted in Fig. 2. The distribution of the output sequence transformed into the range $\langle 0, 1\rangle$ for the purposes of CPRNG (15,000 iterations) is depicted in Fig. 3.

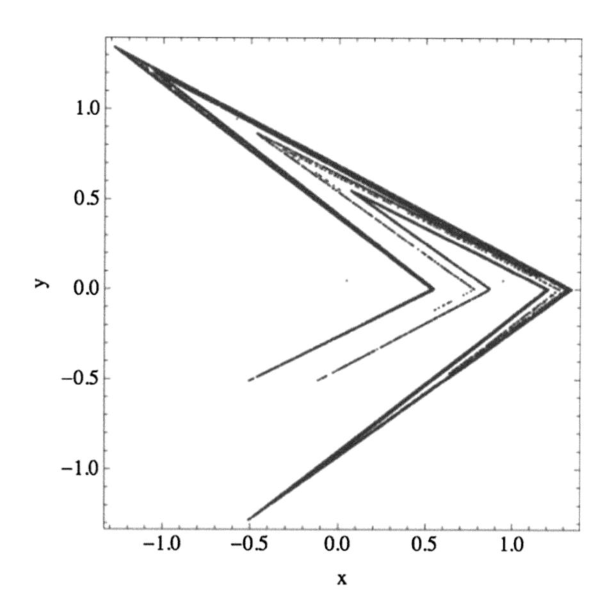

Fig. 2 *Left* x, y plot of the Lozi map

$$X_{n+1} = 1 - a|X_n| + bY_n$$
$$Y_{n+1} = X_n \tag{5}$$

6 Test Functions

Within this research following seven well-known and frequently used benchmark functions were utilized. The dimension (D) was set to 30 for all experiments. The global optimum value is 0 for the all used functions.

Sphere function.

$$f_1(x) = \sum_{i=1}^{D} x_i^2 \tag{6}$$

Search Range: $[-100, 100]^D$; Init. Range: $[-100, 50]^D$; Glob. Opt. Pos.: $[0]^D$

Schwefel's P2.22 function

$$f_2(x) = \sum_{i=1}^{D} |x_i| + \prod_{i=1}^{D} |x_i| \tag{7}$$

Search Range: $[-10, 10]^D$; Init. Range: $[-10, 5]^D$; Glob. Opt. Pos.: $[0]^D$

Rosenbrock's function.

$$f_3(x) = \sum_{i=1}^{D-1} \left[100\left(x_i^2 - x_{i+1}\right)^2 + (1 - x_i)^2\right] \tag{8}$$

Search Range: $[-10, 10]^D$; Init. Range: $[-10, 10]^D$; Glob. Opt. Pos.: $[0]^D$

Noise function.

$$f_4(x) = \sum_{i=1}^{D} x_i^4 + random[0, 1) \tag{9}$$

Search Range: $[-1.28, 1.28]^D$; Init. Range: $[-1.28, 50]^D$; Glob. Opt. Pos.: $[0]^D$

Schwefel's function.

$$f_5(x) = 418.9829 \cdot \mathrm{D} - \sum_{i=1}^{D} -x_i \sin\left(\sqrt{|x|}\right) \tag{10}$$

Search Range: $[-500, 500]^D$; Init. Range: $[-500, 500]^D$; Glob. Opt. Pos.: $[420.96]^D$

Rastrigin's function.

$$f_6(x) = \sum_{i=1}^{D} \left[x_i^2 - 10\cos(2\pi x_i) + 10\right] \quad (11)$$

Search Range: $[-5.12, 5.12]^D$; Init. Range: $[-5.12, 2]^D$; Glob. Opt. Pos.: $[0]^D$
Ackley's function.

$$f_7(x) = -20\exp\left(-0.2\sqrt{\frac{1}{D}\sum_{i=1}^{D} x_i^2}\right) - \exp\left(\frac{1}{D}\sum_{i=1}^{D} x_i^2 \cos 2\pi x_i\right) + 20 + e \quad (12)$$

Search Range: $[-32, 32]^D$; Init. Range: $[-32, 16]^D$; Glob. Opt. Pos.: $[0]^D$

7 Experiment Setup

The control parameters of PSO algorithm were set following way:

Population size	40
Iterations	5000

The *PRTVector* (*Perturbation Vector*), as described in section four, was used to multiply the velocity vector (1) in the GPSO algorithm. Therefore the movement was limited only to certain dimensions. The *PRT* value was set to 0.8 and 0.9 with respect to literature [6]. These two different setting were compared to original GPSO and GPSO driven by Lozi map without a *PRTVector* (technically the same as *PRT* = 1).

7.1 Notation

Totally four versions of GPSO algorithm were used. The notation is as follows:

- GPSO—with canonical PRNG.
- GPSO Lozi—CPRNG based on Lozi map.
- GPSO Lozi 0.8—CPRNG based on Lozi map. *PRT* set to 0.8.
- GPSO Lozi 0.9—CPRNG based on Lozi map. *PRT* set to 0.9.

As aforementioned in Sect. 2, the chaotic pseudorandom number generator was applied only in the main formula of PSO (2). For other purposes (generating of initial population etc.) default *C* language built-in pseudorandom number generator was used for all four described versions of PSO.

8 Results

In this section, the results for each test function are summarized into the simple statistical overview (Tables 1, 2, 3, 4, 5, 6 and 7). The best result (Cost Function—CF value) and the best mean result are highlighted by bold numbers. The brief results analysis follows in the next section. Furthermore the mean gBest history is depicted in Figs. 4, 5, 6, 7, 8, 9 and 10.

Table 1 Results—Sphere function

	GPSO	GPSO Lozi	GPSO Lozi 0.8	GPSO Lozi 0.9
Mean CF value	3.91E−31	1.59E−110	**3.90E−125**	2.73E−119
Std. Dev.	1.30E−30	6.77E−110	1.53E−124	9.91E−119
CF value median	8.23E−33	3.90E−113	3.58E−128	6.05E−121
Max. CF value	6.49E−30	3.67E−109	7.48E−124	5.37E−118
Min. CF value	1.02E−35	6.57E−119	**4.17E−133**	1.42E−126

Table 2 Results—Schwefel's p2.22 function

	GPSO	GPSO Lozi	GPSO Lozi 0.8	GPSO Lozi 0.9
Mean CF value	1.10E−21	1.11E−29	7.53E−30	**2.92E−31**
Std. Dev.	2.22E−21	4.40E−29	3.02E−29	1.58E−30
CF value median	3.32E−22	2.81E−40	2.41E−36	5.36E−38
Max. CF value	1.09E−20	2.13E−28	1.49E−28	8.67E−30
Min. CF value	5.38E−23	**7.71E−49**	1.38E−44	1.15E−44

Table 3 Results—Rosenbrock's function

	GPSO	GPSO Lozi	GPSO Lozi 0.8	GPSO Lozi 0.9
Mean CF value	2.78E+01	**1.40E+01**	1.73E+01	1.67E+01
Std. Dev.	2.35E+01	2.06E+01	1.98E+01	2.14E+01
CF value median	2.26E+01	8.80E+00	1.50E+01	1.18E+01
Max. CF value	7.96E+01	7.31E+01	7.39E+01	7.08E+01
Min. CF value	5.42E−02	**2.06E−07**	4.50E−03	3.72E−03

Table 4 Results—Noise function

	GPSO	GPSO Lozi	GPSO Lozi 0.8	GPSO Lozi 0.9
Mean CF value	6.96E−03	3.02E−03	**2.45E−03**	2.96E−03
Std. Dev.	2.31E−03	1.23E−03	9.22E−04	1.11E−03
CF value median	6.90E−03	2.60E−03	2.42E−03	2.87E−03
Max. CF value	1.25E−02	5.89E−03	4.92E−03	5.01E−03
Min. CF value	3.31E−03	1.45E−03	**6.52E−04**	9.06E−04

Table 5 Results—Schwefel's function

	GPSO	GPSO Lozi	GPSO Lozi 0.8	GPSO Lozi 0.9
Mean CF value	3.72E+03	4.09E+03	**3.17E+03**	3.28E+03
Std. Dev.	3.93E+02	4.00E+02	4.03E+02	3.72E+02
CF Value median	3.70E+03	4.05E+03	3.15E+03	3.37E+03
Max. CF value	4.46E+03	5.07E+03	4.11E+03	3.95E+03
Min. CF value	3.06E+03	3.02E+03	**2.27E+03**	2.33E+03

Table 6 Results—Rastrigin's function

	GPSO	GPSO Lozi	GPSO Lozi 0.8	GPSO Lozi 0.9
Mean CF value	**2.70E+01**	3.22E+01	3.62E+01	3.11E+01
Std. Dev.	7.40E+00	8.00E+00	1.04E+01	8.72E+00
CF Value median	2.69E+01	3.13E+01	3.63E+01	2.98E+01
Max. CF value	4.18E+01	5.47E+01	5.27E+01	4.68E+01
Min. CF value	**1.39E+01**	1.79E+01	1.89E+01	**1.39E+01**

Table 7 Results—Ackley's function

	GPSO	GPSO Lozi	GPSO Lozi 0.8	GPSO Lozi 0.9
Mean CF value	1.24E−14	6.84E−15	**6.37E−15**	7.08E−15
Std. Dev.	3.67E−15	1.45E−15	1.70E−15	1.23E−15
CF Value median	1.47E−14	7.55E−15	7.55E−15	7.55E−15
Max. CF value	2.18E−14	7.55E−15	7.55E−15	7.55E−15
Min. CF value	7.55E−15	**4.00E−15**	**4.00E−15**	**4.00E−15**

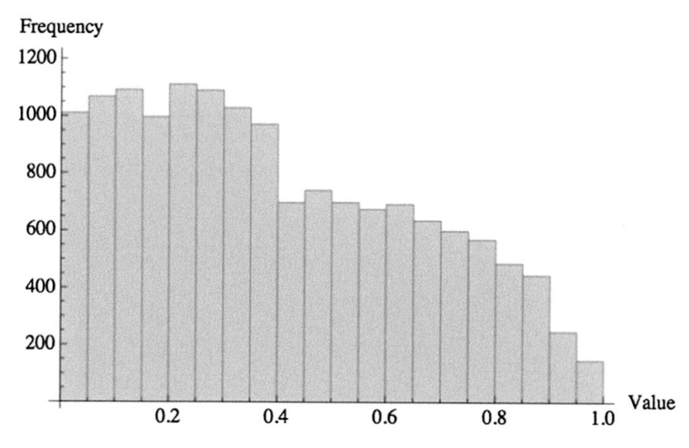

Fig. 3 CPRNG based on Lozi map—distribution histogram transferred into the range ⟨0, 1⟩ (15,000 samples)

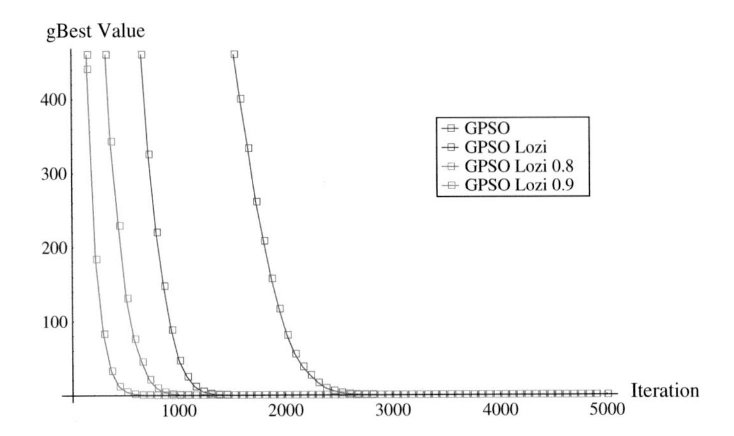

Fig. 4 History of the mean *gBest* value for Sphere function

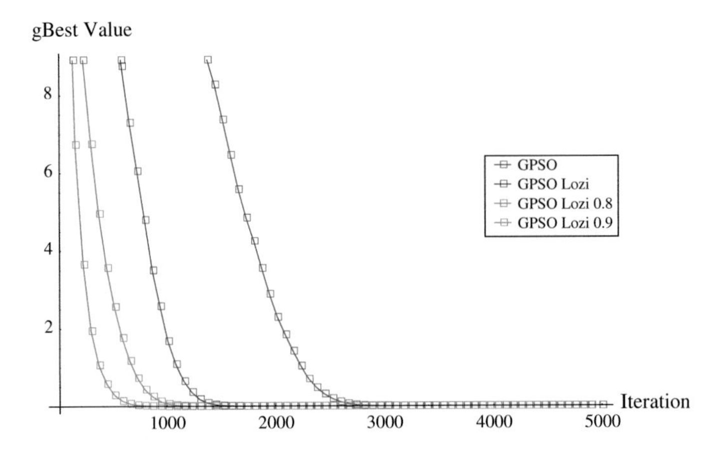

Fig. 5 History of the mean *gBest* value for Schwefel's p2.22 function

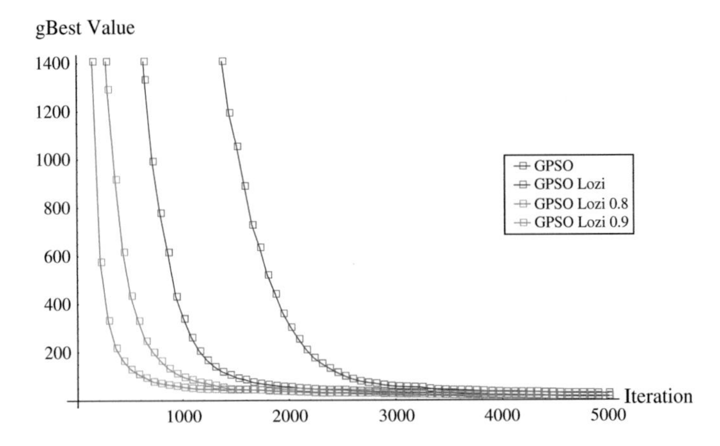

Fig. 6 History of the mean *gBest* value for Rosenbrock's function

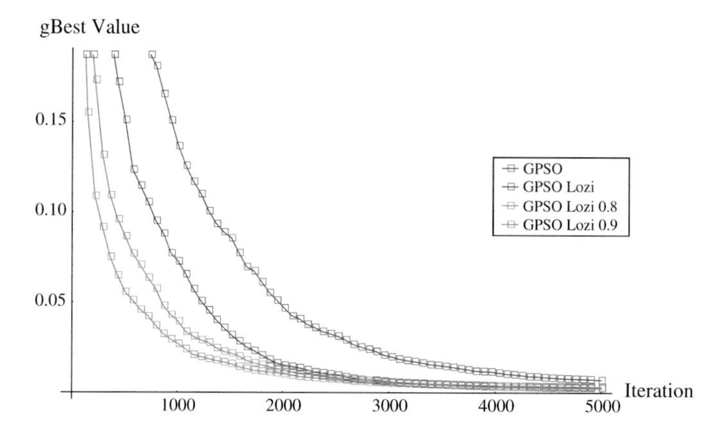

Fig. 7 History of the mean gBest value for Noise function

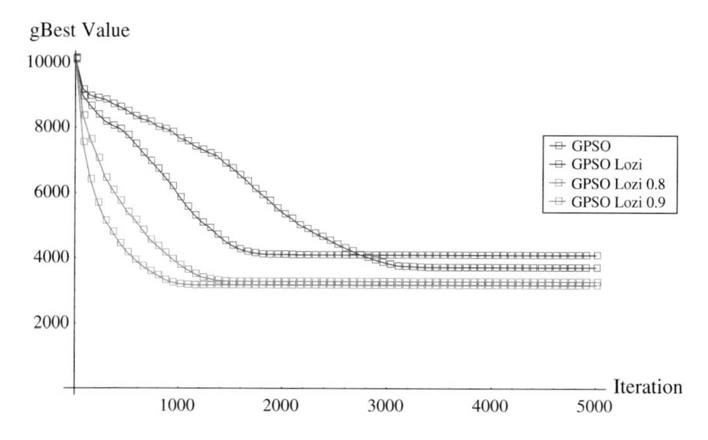

Fig. 8 History of the mean *gBest* value for Schwefel's function

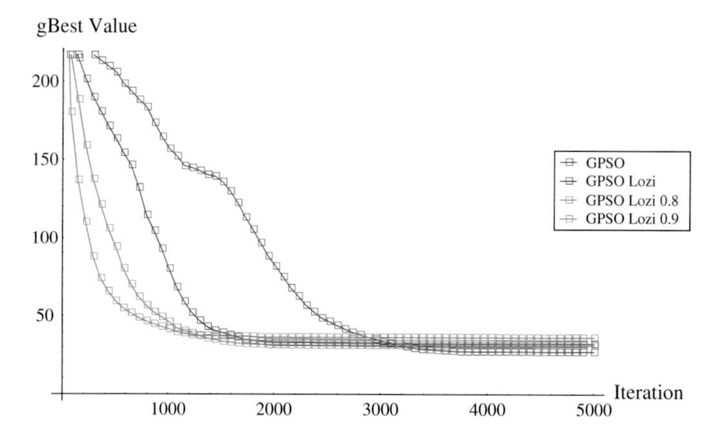

Fig. 9 History of the mean *gBest* value for Rastrigin's function

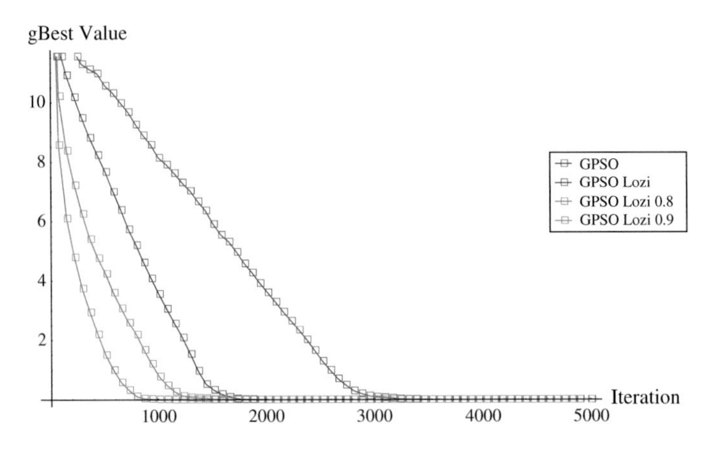

Fig. 10 History of the mean *gBest* value for Ackley's function

9 Analysis of the Results

The results presented in Tables 1, 2, 3, 4, 5, 6 and 7 indicate that it is possible to noticeably improve the performance of PSO algorithm driven by Lozi chaotic map through the implementation of the *PRTVector* mechanism from SOMA algorithm. In 5 cases (Tables 1, 2, 4, 5 and 7) the newly designed chaos embedded GPSO with *PRTVector* managed to obtain the best mean result over the 50 runs. Furthermore also in many cases the best overall result was found (min. CF value).

However as it is depicted in Fig. 4 the influence to the algorithm is very different than what was originally anticipated. It seems that the initial speed of the convergence is increased even further. Thus it is possible to find better solutions even faster than with the original design of either GPSO or Chaos GPSO driven by Lozi map.

10 Conclusion

In this research a new approach improving the performance of chaos driven PSO algorithm was investigated. The inspiration came from the SOMA algorithm The *Perturbation Vector* mechanism was implemented into the GPSO. It was observed that it is possible to noticeably improve the performance of GPSO with Lozi map based CPRNG in many cases utilizing different test problems. The Lozi map remains a very intense subject of research in the area of chaos driven ECTs thanks to its unique sequencing and other unique attributes. The future research will focus on deeper understanding of the inner dynamic of GPSO driven by Lozi map based CPRNG either with or without the *PRTVector* mechanism embedded.

Acknowledgments This work was supported by Grant Agency of the Czech Republic—GACR P103/15/06700S, further by financial support of research project NPU I No. MSMT-7778/2014 by the Ministry of Education of the Czech Republic. also by the European Regional Development Fund under the Project CEBIA-Tech No. CZ.1.05/2.1.00/03.0089, partially supported by Grant of SGS No. SP2015/142, VŠB—Technical University of Ostrava, Czech Republic and by Internal Grant Agency of Tomas Bata University under the project No. IGA/FAI/2015/057.

References

1. Kennedy, J., Eberhart, R.: Particle swarm optimization. In: Proceedings of IEEE International Conference on Neural Networks, IV, pp. 1942–1948 (1995)
2. Eberhart, R., Kennedy, J.: Swarm intelligence. The Morgan Kaufmann Series in Artificial Intelligence. Morgan Kaufmann, Burlington (2001)
3. Shi, Y.H., Eberhart R.C.: A modified particle swarm optimizer. In: IEEE International Conference on Evolutionary Computation, Anchorage, Alaska, pp. 69–73 (1998)
4. Nickabadi, A., Ebadzadeh, M.M., Safabakhsh, R.: A novel particle swarm optimization algorithm with adaptive inertia weight. Appl. Soft Comput. **11**(4), 3658–3670 (2011). ISSN 1568-4946
5. Dorigo, M.: Ant Colony Optimization and Swarm Intelligence. Springer, Berlin (2006)
6. Zelinka, I.: SOMA—self organizing migrating algorithm. In: Babu, B.V., Onwubolu, G. (eds.) New Optimization Techniques in Engineering, vol. 33. Springer, Berlin (2004). ISBN: 3-540-20167X (Chapter 7)
7. Caponetto, R., Fortuna, L., Fazzino, S., Xibilia, M.G.: Chaotic sequences to improve the performance of evolutionary algorithms. IEEE Trans. Evol. Comput. **7**(3), 289–304 (2003)
8. Davendra, D., Zelinka, I., Senkerik, R.: Chaos driven evolutionary algorithms for the task of PID control. Comput. Math. Appl. **60**(4), 1088–1104 (2010). ISSN 0898-1221
9. Araujo, E., Coelho, L.: Particle swarm approaches using Lozi map chaotic sequences to fuzzy modelling of an experimental thermal-vacuum system. Appl. Soft Comput. **8**(4), 1354–1364 (2008)
10. Alatas, B., Akin, E., Ozer, B.A.: Chaos embedded particle swarm optimization algorithms. Chaos Solitons Fractals **40**(4), 1715–1734. ISSN 0960-0779 (2009)
11. Pluhacek, M., Senkerik, R., Davendra, D., Oplatkova, Z.K, Zelinka, I.: On the behavior and performance of chaos driven PSO algorithm with inertia weight. Comput. Math. Appl. **66**, 122–134 (2013)
12. Pluhacek, M., Budikova, V., Senkerik, R., Oplatkova Z., Zelinka, I.: On the performance of enhanced PSO algorithm with Lozi Chaotic map—an initial study. In: Proceedings of the 18th International Conference on Soft Computing, MENDEL 2012, pp. 40–45 (2012). ISBN 978-80-214-4540-6
13. Pluhacek, M., Senkerik, R., Zelinka, I., Davendra, D.: Chaos PSO algorithm driven alternately by two different chaotic maps—an initial study. In: IEEE Congress on Evolutionary Computation (CEC), 2013, pp. 2444, 2449 (2013). doi:10.1109/CEC.2013.6557862, ISBN: 978-1-4799-0451-8
14. Pluhacek, M., Senkerik, R., Zelinka, I., Davendra, D.: New adaptive approach for chaos PSO algorithm driven alternately by two different chaotic maps—an initial study. Advances in Intelligent Systems and Computing, vol. 210, Nostradamus 2013: Prediction, Modeling and Analysis of Complex Systems, pp. 77–88 (2013). ISBN 978-3-319-00541-6
15. Sprott, J.C.: Chaos and Time-Series Analysis. Oxford University Press, Oxford (2003)

Printed in the United States
By Bookmasters